CRC SERIES IN PESTICIDE CHEMISTRY

Editor-in-Chief
Gunter Zweig
School of Public Health
University of California
Berkeley, California

CHLORINATED INSECTICIDES
Volume I and II
G. T. Brooks
Principal Scientific Officer
Agricultural Research Council
Unit of Invertebrate Chemistry and Physiology
The Chemical Laboratory
University of Sussex
Sussex, England

ORGANOPHOSPHORUS PESTICIDES: Organic and Biological Chemistry
Morifusa Eto
Associate Professor
Laboratory of Pesticide Chemistry
Department of Agricultural Chemistry
Kyushu University
Japan

CARBAMATE INSECTICIDES: Chemistry, Biochemistry, and Toxicology
Ronald J. Kuhr
Associate Professor of Insect Toxicology
New York State Agricultural Experiment Station
Geneva, New York

H. Wyman Dorough
Professor
Department of Entomology
University of Kentucky
Lexington, Kentucky

CONTROLLED RELEASE PESTICIDES FORMULATIONS
Nate F. Cardarelli
Associate Professor of General Technology
University of Akron
Akron, Ohio

THE PHENOXYALKANOIC HERBICIDES
Volume I - Chemistry, Analysis, and Environmental Pollution
Volume II- Biology
Shane S. Que Hee
Assistant Professor of Environmental Health
Kettering Laboratory
University of Cincinnati Medical Center
Cincinnati, Ohio

Ronald G. Sutherland
Professor of Chemistry
University of Saskatchewan
Saskatoon, Saskatchewan, Canada

INSECT SUPPRESSION WITH CONTROLLED RELEASE PHEROMONE SYSTEMS
Agis F. Kydonieus
Executive Vice President
Herculite Products, Inc.
New York, New York

M. Beroza
M. Beroza and Associates
Silver Spring, Maryland

Insect Suppression with Controlled Release Pheromone Systems

Volume I

Editors

Agis F. Kydonieus, Ph.D.
Executive Vice President
Health-Chem Corporation
New York, New York

Morton Beroza, Ph.D.
M. Beroza and Associates, Inc.
Silver Spring, Md.

Editor-in-Chief
CRC Series in Pesticide Chemistry

Gunter Zweig, Ph.D.
Visiting Scholar
School of Public Health
The University of California
Berkeley, California

CRC Press, Inc.
Boca Raton, Florida

Library of Congress Cataloging in Publication Data
Main entry under title:

Insect suppression with controlled release
pheromone systems

CRC series in pesticide chemistry)
Bibliography: v. 1, p.
Includes index.
1. Insect sex attractants — Controlled
release. I. Kydonieus, Agis F., 1938-
II. Beroza, Morton, 1917- III. Series.
SB933.4.I57 632'.7 81-38507
ISBN 0-8493-5855-8 (v. 1) AACR2
ISBN-0-8493-5856-b (v. 2)

Direct all inquiries to CRC Press, Inc., 2000 N.W. 24th Street, Boca Raton, Florida, 33431.

International Standard Book Number 0-8493-5855-8 (Volume I)
Insternational Standard Book Number 0-8493-5856-6 (Volume II)

Library of Congress Card Number 81-38507
Printed in the United States

FOREWORD

There have been many important research developments during the past two decades that offer prospects for achieving more effective and more acceptable systems of insect pest management. The most exciting and perhaps the most challenging and promising opportunity for such improvements is in the area of behavioral chemicals. The availability of synthetic replicas of natural chemicals produced by insects for purposes of aggregation and reproduction provides new tools both for detection and suppression of insect pest populations. Their availability will also make possible more effective and judicious use of insecticides that we have relied upon for more than a half century. Synthetic sex pheromones are now available for many of our most damaging and difficult to control pest species that in the aggregate cost our nation's economy several billions of dollars each year and which are directly or indirectly responsible for the use of most of the agricultural insecticides. These include the boll weevil, tobacco budworm, corn earworm, pink bollworm, codling moth, European corn borer, Japanese beetle, Gypsy moth, Douglas fir tussock moth, spruce budworm and several species of bark beetles. Research chemists throughout the world are steadily adding to the list of synthetic sex pheromones that will likely prove useful for detection and/or control of a wide range of insect pests. Synthetic pheromones also provide new tools for gaining a better understanding of the ecology, dynamics, and behavior of insects in their natural habitats, which in turn can serve as a guide for the development and application of more effective tactics and strategies for the suppression and management of insects, employing the control methods now in use.

It is not a simple matter, however, to determine how and under what circumstances these highly active and generally species-specific chemicals can be used to achiive effective and practical management of the target pests. Difficult and innovative research is necessary on suitable formulations of the pheromones and on appropriate trapping devices or dispersing systems. Equally, if not more important, consideration must be given to the dynamics and behavior of the target pest in the natural environment in devising effective methods of using pheromones to disrupt normal reproduction processes. For highly mobile species the pest population must be attacked in a large area and at strategic times and places. It also seems clear that in order to make optimum use of sex pheromone for most pests, they must be employed primarily as a preventive measure.

The defensive strategy of applying suppressive measures only where and when populations have already reached or threaten economic damage levels, as generally advocated when we rely on insecticides, will not be satisfactory in the use of pheromones for managing most pests. Not only is the degree of suppression achieved by the use of pheromones inversely proportional to the density of the target pest, the suppressive action may be too slow to produce the desired result, after a pest has already reached threatening population levels.

As noted by the various topics discussed in this publication, pheromones offer possibilities for the management of insect pests by disrupting insect reproduction through the "confusion" technique and by mass trapping. In the application of the confusion technique enough pheromone vapor is maintained in the pest ecosystem to interfere with mate finding by the aggressor sex. In mass trapping enough males (or females, depending on the responsive sex), are lured to and destroyed by the trapping device to reduce mating in a pest habitat. Pheromone vapors released from the trapping devices may also add a confusion control factor.

Much remains to be known about the action of different pheromone suppression techniques and how and when they can be employed most effectively. However, since the density and distribution of the target pest will influence the effectiveness of pheromone suppression methods, the pheromones may find their greatest use as a vital component in integrated pest management systems, that take advantage of other available control methods including the use of conventional insecticides, cultural practices, augmentation of natural biological control organisms and the growing of resistant plant varieties. The use of pheromones alone may be adequate for controlling some pests but they will likely be of most value as an adjunct to other pest population management systems.

Aside from the potential value of insect sex pheromones for insect suppression, the sex pheromones have already demonstrated their usefulness for insect detection. The lack of highly sensitive methods of detecting the presence of many of the major pests when they exist at low densities, has been a serious handicap to optimum utilization of control methods available in the past, whether the objective was management, containment, or eradication. The ability to detect where, when, and how many of a target pest are present in a pest ecosystem, even at very low densities, should in time have a major influence on insect pest population management concepts and on the most rational approach to the solution of some of our major pest problems.

The full potential of insect sex pheromones as alternatives or supplements to present insect pest management systems cannot yet be fully appraised because this is still a new area of investigation. However, the contributors to this publication and others who have pioneered research on behavioral chemicals are to be commended for the substantial progress that has already been made. It is important that available information be compiled in a volume such as this. It will lay a firm foundation for even greater advances in the future. It is a privilege for me to make this introductory statement for a publication that should be of great value to those engaged in research and practical insect pest management programs.

E. F. Knipling, Ph.D.

EDITOR-IN-CHIEF'S PREFACE

This two-volume treatise on insect suppression by the use of controlled-release pheromones, edited by two world-renowned and honored scientists A. F. Kydonieus and M. Beroza, is part of the CRC Series on Pesticides. Previous volumes have covered chlorinated insecticides, organophosphates, carbamates, and most recently, the first volume of a two-volume treatise on phenoxyalkanoic acid herbicides.

Pheromones, or insect attractants, have been known to exist for almost 200 years, but it has only been possible during the last two decades through the great advances in separation and analytical techniques to separate and identify the natural components of literally hundreds of specific insect species pheromones. Also, during the past 20 years it has become possible to develop controlled release delivery systems for pheromones which have now made it possible to consider pheromones as an alternate pest control technique. The advantages using phermones to replace or diminish the use of conventional, synthetic insecticides are their uniqueness to attract only the species to be controlled and their innate low toxicity against nontarget organisms, including man.

This most authoritative work by the best qualified experts in pheromone research and application will undoubtedly enhance increased efforts to make the use of pheromones practical and economic in the continuing battle of man against ravaging insect populations.

The editors and contributors of this monumental work ought to be congratulated for having put together all of the known, up-to-date information (1981) on insect pheromones and the associated controlled release technology.

Gunter Zweig, Ph.D.
Fall, 1981

THE EDITOR-IN-CHIEF

Gunter Zweig received his undergraduate and graduate training at the University of Maryland, where he was awarded the Ph. D. in biochemistry in 1952. Two years following his graduation, Dr. Zweig was affiliated with the late R. J. Block, pioneer in paper chromatography of amino acids. Zweig, Block, and Le Strange wrote one of the first books on paper chromatography which was published in 1952 by Academic Press and went into three editions. *Paper Chromatography* (1952) was also translated into Russian.

From 1953 till 1957, Dr. Zweig was research biochemist at the C. F. Kettering Foundation, Antioch College, Yellow Springs, Ohio, where he pursued research on the path of carbon and sulfur in plants, using the then newly developed techniques of autoradiography and paper chromatography. From 1957 till 1965, Dr. Zweig served as lecturer and chemist, University of California, Davis and worked on analytical methods for pesticide residues, mainly by chromatographic techniques. In 1965, Dr. Zweig became Director of Life Sciences, Syracuse University Research Corporation (research on environmental pollution), and in 1973 he became Chief, Environmental Fate Branch, Environmental Protection Agency in Washington, D.C. In 1980, he was appointed Senior Science Advisor in the same agency.

During his government career, Dr. Zweig continued his scientific writing and editing. Among his works are (many in collaboration with Dr. Sherma) the now 11-volume series *Analytical Methods for Pesticides and Plant Growth Regulators* (published by Academic Press); the handbook series in chromatography for CRC Press, co-editor of *Journal of Toxicology and Environmental Health*; co-author of basic review on paper and thin-layer chromatography for *Analytical Chemistry* from 1968-1980; co-author of applied chromatography review on pesticide analysis for *Analytical Chemistry*, beginning in 1981.

Among the scientific honors awarded to Dr. Zweig during his distinguished career are the Wiley Award in 1977, Rothschild Fellowship to the Weizmann Institute in 1963/64; the Bronze Medal by the EPA in 1980.

Dr. Zweig has authored or co-authored over 75 scientific papers on diverse subjects in chromatography and biochemistry, besides being the holder of three U.S. patents.

At the present time (1980/81), Dr. Zweig is Visiting Scholar in the School of Public Health, University of California, Berkeley, where he is doing research on farmworker safety as related to pesticide exposure.

PREFACE

Use of insect sex pheromones, which are innocuous chemicals secreted by insects to call a mating partner, has been proposed for suppressing highly destructive, economically important insect species, and recent reports have substantially enhanced the credibility of this approach. Success in this effort could help reduce our dependence on the highly toxic insecticides now widely used to protect our agricultural products from insect depredation.

The list of pheromones available for such exploration has now become quite extensive (see Volume II, Chapter 15 by M. Inscoe), and it continues to grow. Exhaustive coverage of the pheromone field is therefore neither possible nor desirable in this volume. Instead, we have invited leading investigators in the pheromone field to present a comprehensive account of their particular subject matter and experiences in a manner that will be valuable to the expert and still be understandable by the novice. Authors have been chosen to deal with a wide variety of insect problems and a broad spectrum of situations in order to cover the current status of the pheromone field thoroughly. Emphasis has been placed on the practical rather than on theoretical premises. The presentation of concepts rather than detail has been encouraged. In essence, a major goal of this book has been to document the latest accomplishments and technology relating to pheromone use along with the information that lead to these advances.

We believe that the foregoing approach to assembling information on the exciting new field of pheromone use in insect pest management will best serve the reader's interests and will appeal to scientists and technologists from the many disciplines that are expected to participate in this development; e.g., entomologists, biologists, chemists, biochemists, pest control specialists, insect physiologists, agronomists, agricultural engineers, animal behaviorists, ecologists, and last, but not least, industrialists seeking a new technology with a promising future.

Finally, this volume, through its references, constitutes a ready reservoir of pheromone information now scattered throughout the literature. Thus, the aforementioned listing of pheromones (and insects affected) includes references to the original reports describing each item listed, thereby bringing together in one place for rapid reference more than 500 pheromone citations. The listing provides an up-to-date background of material to help bring both the advanced and the new worker abreast of the rapidly growing pheromone field.

Agis F. Kydonieus
Morton Beroza

We are most grateful to the many authors who contributed to this volume for their willingness to share their knowledge and experiences with the reader. Like many in our society, they recognize the urgent need for improved methods of insect control. We wish also, to acknowledge the valuable editorial assistance provided by Laura Zeoli and KarenAnn Lalsingh.

THE EDITORS

Agis F. Kydonieus, Ph.D., is Executive Vice President of Hercon Division, Health-Chem Corporation, New York, New York.

Dr. Kydonieus graduated in 1959 from the University of Florida, Gainesville, with a B.S. degree in Chemical Engineering (summa cum laude) and obtained his Ph.D. degree from the same school in 1964.

Dr. Kydonieus is Vice President and a member of the Board of Governors of The Controlled Release Society as well as a member of The American Institute of Chemical Engineers, The Entomological Society of America, American Chemical Society, Pest Control Association, and The Society of Plastics Engineers.

Dr. Kydonieus is the author of over 100 patents, publications and presentations in the fields of controlled release and biomedical devices. He is the editor of *CONTROLLED RELEASE TECHNOLOGIES,* published by CRC Press in 1980.

Dr. Morton Beroza has been a consultant for the past six years and earlier was Chief of the Organic Chemical Synthesis Laboratory, Agricultural Research Service, U.S. Department of Agriculture, Beltsville, MD, where he was a research chemist since 1948.

He received his B.S. degree from George Washington University in 1943 and the M.S. and Ph.D. degrees from Georgetown University in 1946 and 1950, with majors in organic chemistry and biochemistry.

Dr. Beroza has been engaged in research on the chemical aspects of entomology. His fields of endeavor have included synthesis of insecticides, insect attractants and repellents, synergists, alkaloids, spectroscopy, isolation and determination of constituents of plants and insects, pesticide analysis, and methods of determining the structure of compounds. He has authored more than 300 publications, including 29 patents. His most recent work has been on the identification of insect sex pheromones and on developing the technology to use these chemicals effectively in pest management.

Dr. Beroza received the Burdick and Jackson International Award in Pesticide Chemistry from the Pesticide Chemistry Division of the American Chemical Society 1977, the Gold Medal Award in Environmental Chemistry of the Synthetic Organic Chemicals Manufacturer's Association 1973, the Harvey W. Wiley Award of the Association of Official Analytical Chemists 1970, the American Chemical Society Award in Chromatography and Electrophoresis 1969, and ten awards from USDA (Seven Certificates of Merit and three Unit Awards for Superior Service). He served for several years on the United Nations Food and Agricultural Organization Expert Working Party on Pesticide Residues. He has served as president of several local societies in Washington. He serves or has served on the advisory board of six journals.

CONTRIBUTORS

Edward Ashare, Ph.D.
Engineering Director
Dynatech R/D
Cambridge, Massachusetts

Alf Bakke, Dr. philos.
Chief Scientific Officer
Forest Protection
Norwegian Forest Research Institute
As-NLH, Norway

L. A. Bariola, Ph.D.
Research Entomologist
Western Cotton Research Laboratory
United States Department of Agriculture
Phoenix, Arizona

Frederick S. Betz, Ph.D.
Aquatic Biologist
Hazard Evaluation Division
Office of Pesticides and Toxic Substances
United States Environmental Protection Agency
Washington, D.C.

B. A. Bierl-Leonhardt, Ph.D.
Research Chemist
Organic Chemical Synthesis Laboratory
United States Department of Agriculture
Beltsville, Maryland

T. W. Brooks, Ph.D.
Group Technical Director
Environmental Chemicals Group
Albany International Corporation
Columbus, Ohio

George D. Butler, Jr. Ph.D.
Research Entomologist
Western Cotton Research Laboratory
United States Department of Agriculture
Phoenix, Arizona

Joseph H. Caro
Retired
Research Chemist
Soil Nitrogen and Environmental Chemistry Laboratory
United States Department of Agriculture
Beltsville, Maryland

Stuart Z. Cohen, Ph.D.
Senior Chemist
Hazard Evaluation Division
Office of Pesticides and Toxic Substances
United States Environmental Protection Agency
Washington, D.C.

Gary E. Daterman, Ph.D.
Research Entomologist
Pacific Northwest Forest and Range Experiment Station
United States Department of Agriculture, Forest Service
Corvallis, Oregon

Hollis M. Flint, Ph.D.
Research Entomologist
Western Cotton Research Laboratory
United States Department of Agriculture
Phoenix, Arizona

C. R. Gentry
Research Entomologist
Southeastern Fruit and Tree Nut Research Laboratory
United States Department of Agriculture
Byron, Georgia

Janice M. Gillespie, Ph.D.
Manager, Pheromone Programs Western States
Hercon Division
Health-Chem Corporation
New York, New York

D. D. Hardee, Ph.D.
President
Pest Management Specialists, Inc.
Starkville, Mississippi

Thomas J. Henneberry, Ph.D.
Research Entomologist
Director
Western Cotton Research Laboratory
United States Department of Agriculture
Phoenix, Arizona

May N. Inscoe, Ph.D.
Research Chemist
Organic Chemicals Synthesis Laboratory
United States Department of Agriculture
Beltsville, Maryland

Waldemar Klassen, Ph.D.
Staff Scientist
National Program Staff
United States Department of Agriculture
Beltsville, Maryland

Michael G. Klein, Ph.D.
Research Entomologist
Japanese Beetle and Horticultural Insect Pests Research Laboartory
United States Department of Agricutlure
Ohio Agricultural Research and Development Center
Wooster, Ohio

Thyril L. Ladd, Jr., Ph.D.
Research Leader
Japanese Beetle and Horticultural Insect Pests Research Laboratory
United States Department of Agriculture
Ohio Agricultural Research and Development Center
Wooster, Ohio

Peter D. Lingren, Ph.D.
Research Entomologist
Western Cotton Research Laboratory
United States Department of Agriculture
Phoenix, Arizona

Jack R. Plimmer, Ph.D.
Chief
Organic Chemical Synthesis Laboratory
United States Department of Agriculture
Beltsville, Maryland

Alberto R. Quisumbing, Ph.D.
Director, Product Development
Hercon Division
Health-Chem Corporation
New York, New York

J. R. Raulston, M.S.
Research Entomologist
Cotton Insects Research Laboratory
United States Department of Agriculture
Brownsville, Texas

R. L. Ridgway, Ph.D.
National Research Program Leader
Cotton and Tobacco Insects
United States Department of Agriculture
Beltsville, Maryland

Leif Riege, Dr.
Research Chemist
Borregaard Gud. Ltd.
Sarpsborg, Norway

G. H. L. Rothschild, Ph.D.
Senior Principal Research Scientist
Division of Entomology
Commonwealth Scientific and Industrial Research Organization
Canberra, Australia

C. J. Sanders, Ph.D.
Research Scientist
Great Lakes Forest Research Centre
Sault Ste Marie, Ontario, Canada

W. D. Seabrook, Ph.D.
Professor
Department of Biology
University of New Brunswick
Fredericton, New Brunswick, Canada

Lonne L. Sower, Ph.D.
Research Entomologist
Pacific Northwest Forest and Range Experiment Station
United States Department of Agriculture
Corvallis, Oregon

Alton N. Sparks, Ph.D.
Research Entomologist
Southern Grain Insect Research Laboratory
United States Department of Agriculture
Tifton, Georgia

David W. Swenson
Plant Manager
Controlled Release Division
Albany International, Inc.
Needham Heights, Massachusetts

Ralph E. Webb, Ph.D.
Research Entomologist
Horticultural Science Institute
United States Department of Agriculture
Beltsville, Maryland

Wayne W. Wolf
Agricultural Engineer
Southern Grain Insect Research Laboratory
United States Department of Agriculture
Tifton, Georgia

C. E. Yonce, Ph.D.
Research Entomologist
Southeastern Fruit and Nut Research Laboratory
United States Department of Agriculture
Byron, Georgia

Laura T. Zeoli, Ph.D.
Manager, Product Registration
Hercon Division
Health-Chem Corporation
New York, New York

Gunter Zweig, Ph.D.
Visiting Scholar and Lecturer
School of Public Health
University of California
Berkeley, California

TABLE OF CONTENTS

Volume I

Volume II

Section I
Fundamental Concepts and Government Procedures

Chapter 1

PHEROMONES AND THEIR USE

Agis F. Kydonieus and Morton Beroza

TABLE OF CONTENTS

I. INTRODUCTION

Mounting demands for food, fiber, lumber, and other crops by an expanding world population are intensifying the search by scientists and responsible officials for means that are safer and more efficient than the conventional insecticides now used to combat the insect hordes that regularly take their toll of our agricultural production. Insects have become resistant to insecticides. Insecticides have killed beneficial insects along with the targeted species, so nonpestiferous insects, ordinarily held in check by beneficial species, have in many instances become the dominant pest. Deleterious effects of insecticide residues on man, wildlife, fish, and other biota, including the bioaccumulation of such residues, continue to surface from time to time. Finally, there is the ever-present uncertainty about the level of conventional insecticides that can be considered truly safe for humans and other biota in a lifetime of exposure.

As one promising approach, the U.S. Department of Agriculture (USDA) has pioneered the possibility of using insect sex pheromones to combat our insect competitors. Pheromones are chemicals released by one organism to direct the behavior of other members of its species. Scientists have known since the 1800s that insects are able to call a mating partner by releasing a pheromone (sex attractant scent) into the atmosphere. Insects detecting the chemical respond by flying or moving upwind toward the source of emission and mate when contact is made (see Volume I, Chapter 4 for a detailed description of pheromone concentration and behavior in air).

Major advantages of using pheromones are that, unlike the conventional insecticides, they are not toxic chemicals, and they are effective at incredibly low concentrations.

II. EARLY USE OF PHEROMONES AND ATTRACTANTS

As early as the 1940s, the USDA used the sex attractant of the gypsy moth (as a crude extract of female abdominal tips) in traps to monitor large areas for the whereabouts of this insect.[1] The moth is a serious pest of hardwoods, which was accidentally imported from Europe into the northeastern part of the U.S. about 100 years ago, and federal and local authorities have been battling it ever since. Through the use of a USDA-developed process for stabilizing the lure,[2] the attractant could be stockpiled without the rapid deterioration previously encountered,[3] and this made the monitoring program possible. Each season, hundreds of thousands of gypsy moth pupae were laboriously collected in the field to furnish the lure needed to bait each trap with extract equivalent to 12 female insects. Responding males entered the traps and became enmired on an adhesive coating within. Where insects were caught outside the generally infested area, insecticides were applied to prevent the spread of the moth. This pheromone detection system was highly successful, and, for quite a few years, spread of the moth was indeed restricted. Unfortunately, the control aspects of this program suffered a setback when use of DDT, applied to suppress the moth, was terminated in the early 1960s owing to unacceptable side effects from use of this insecticide.

The USDA has nevertheless continued to survey successfully huge areas of this country for the gypsy moth to this day. In 1970, when the gypsy moth pheromone, called disparlure, was identified and synthesized,[4] the natural extract bait was replaced by the inexpensive synthetic lure. Likewise, when the more effective plus-disparlure (optically active version of disparlure) became available in the mid-1970s,[5] it has been used in trapping (see chapters by Webb, Plimmer, and Bierl-Leonhardt for accounts of the extensive work with the gypsy moth).

In another early, closely related program started in 1956, USDA scientists explored the use of synthetic insect attractants (nonpheromone types, e.g., food lures) as a means of detecting and controlling nonindigenous insect pests.[6] Many chemicals were screened for attractiveness to given species, and when an attractant was found (actually by chance), related compounds were synthesized and screened to try to find a better attractant. The program experienced an unexpected upturn when the highly destructive Mediterranean fruit fly (Medfly) was found in the state of Florida infesting about one million acres. Fortunately, an effective attractant was quickly found and synthesized for this operation.[7] About a year later, the Medfly was eradicated from Florida.[8]

As with the gypsy moth, the synthetic attractant, called siglure, enticed the Medflies into a trap and thereby signaled the presence of the flies in the trap vicinity. Insecticide was applied only where the Medfly was found and only as long as such applications were needed, thereby reducing pesticide pollution to a minimum. A superior lure, called trimedlure, was subsequently synthesized.[9] It is currently used in traps deployed across the southern part of the U.S. as part of an early warning system to prevent an accidental entry of the Medfly into this country. Several reintroductions have occurred[10] (one on the West coast at this writing), and the traps have been used to direct eradication measures before the insect could become established. Trimedlure is now used worldwide to guide control measures against the Medfly.

Although the Medfly attractants are not pheromones, the spectacular speed and efficiency with which the Medfly was eradicated from Florida sparked considerable interest in the possibility that the sex pheromones of insects, if they could be identified and synthesized, could be used in the same way to help in the control of many other insect pests in a safe and efficient manner.

III. MORE RECENT ACTIVITIES WITH PHEROMONES

The early experiences with the gypsy moth and Medfly triggered a considerable fallout of research activities aimed at the identification and utilization of pheromones to help suppress destructive insect populations. These efforts culminated in the late 1960s and the 1970s with a massive amount of progress, which included identification and synthesis of the pheromones of many insect species, the development of improved means of using pheromones, and the reporting of a vast amount of information on insect behavior and biology related to the use of pheromones.

Also included in this array of advances were the development of much better means of timing insecticide sprays with pheromone-baited monitor traps, better trap designs for many of our most important economic pests, improved controlled release formulations and devices, new and more effective methods for the distribution of pheromones (particularly by air) for disorienting insects seeking mates in order to reduce or prevent their reproduction.

Most important has been the experimentation on how pheromones can be used effectively in concert with other pest control techniques, an approach often referred to as Integrated Pest Management. This topic is discussed in the next chapter by two USDA scientists with extensive knowledge and background on the subject. They have included in their chapter in tabular form a report on the current status of research and development of pheromone and some nonpheromone attractants (Table 1), which workers in the field are likely to find invaluable.

Encouraged by the numerous reports from the field, industrial organizations have, within the past several years, entered the pheromone arena and made available controlled release formulations, traps, pheromones, pheromone dispensers, and pheromone delivery equipment. A number of sales organizations have concurrently begun to market pheromone products to the U.S. Government, the farmer, and even to the public.

Although not applicable to all pest control problems, pheromones have been shown to be very effective against certain highly destructive, economically important insect pests. For example, the growing of cotton consumes about 30% of the insecticide used in the U.S. It is significant that excellent results have been obtained with pheromones against two of the major cotton pests in this country, the pink bollworm (see chapter by Henneberry et al.) and the boll weevil, and at least three commercial pheromonal products are now available for use against these insects. Also noteworthy, the highly effective pheromone of the tobacco budworm (another serious pest of cotton) is now at hand and undergoing trials to determine the best way to use it.

IV. HOW SEX PHEROMONES WORK

Insects live in a hostile environment and normally encounter great difficulty in surviving. However, nature has offset the high mortality of many species by endowing them with a high reproductive potential, which in many instances rests heavily on the ability of the sexes to find one another for mating and propagation. This process is frequently mediated by a sex attractant pheromone. With some species, the males emit the sex pheromones and the females respond; with other species, the females emit the pheromone that attracts the males. With yet other species, one of the sexes emits what is called an aggregation pheromone that causes both sexes to congregate for mating.

In all of these cases, the pheromone emitted is highly specific in inducing a response only from individuals of the emitting species, and only the adults (sexually mature) respond. Furthermore, the effect of the sex pheromone on insects is generally powerful enough, in terms of infinitesimal amounts needed and intensity of response, to make them potentially valuable for the direct control of economically important species. Our interest in these chemicals is, therefore, a very practical one. We think they can be used to control insect pests more safely.

Attractants have been used in combination with toxicants, traps (mass trapping), or other insect control agents, e.g., insect pathogens. In at least one pilot experiment, in which a synthetic lure rather than a sex attractant was used, eradication was achieved.[11] Small cane-fiber squares impregnated with a fast-acting insecticide and methyleugenol, a powerful lure for the male oriental fruit fly, were dropped by aircraft onto an isolated small island in the Pacific Ocean. In a matter of months, all the males died, and the females, unable to propagate without males, also perished. With a powerful sex attractant, the technique should be similarly effective. The advantages of this approach are that only the target species is attacked, wildlife is not affected, and the amount of insecticide used is far less than that required in conventional control.

In the use of insect sex attractants, the following four methods have been given major attention:

1. Monitoring (with traps)
2. Mass trapping
3. Trap cropping
4. Mating disruption

A. Monitoring

As already noted, traps may be baited with the sex lures for detection (as well as a rough estimation) of certain species in specific areas, and for indicating when and where control measures should be applied. This procedure is useful in limiting insect spread into new areas, preventing damage from excessively high populations, and in timing of insecticide sprays. Because monitoring traps are not used for the direct control of insects, the U.S. Environmental Protection Agency does not require that they be registered (see Chapter 5 for EPA registration requirements for pheromones).

B. Mass Trapping

The possibility that pheromone traps can be used in the direct control of insects and thereby minimize dependence on insecticides has sparked considerable research. Early limited trials generally verified predictions[12,13] that large-scale trapping would be effective only against low-level insect populations. By one approach, incipient low-level insect populations found outside a generally infested area would be eliminated by mass trapping, e.g., spot infestations of gypsy moth found in Wisconsin, which are well away from the generally infested areas in the northeastern U.S. In another approach, insects can be suppressed with insecticide to a low level in the larval stage, and then further suppressed by mass trapping the adults that manage to survive. In this way, insect populations, once brought to a low level, may at least *theoretically* be maintained at a low level by a continued mass trapping effort, making additional insecticide applications unnecessary. Thus, if mass trapping is initiated early in the growing season, and should the traps remove enough insects to prevent the natural buildup of the insect population, the traps can continue as an effective suppressant for the entire season.

Major difficulties in mass trapping are the large number of traps usually required and the high cost of deploying and maintaining them. There are, nevertheless, instances in which mass trapping can be most effective and, in fact, the method of choice.

A most useful advantage of the mass trapping approach is the built-in monitoring feature; i.e., the trap captures indicate which areas continue to be infested and which areas (by lack of captures) are essentially insect-free. Thus, with proper surveillance, there need be no surprises in this type of control operation.

Whether a trap is used for monitoring or mass trapping, its efficiency is of prime importance. To maximize captures, a variety of parameters must be considered. Trap design is most important, and many different types of traps have been devised to cater to the idiosyncracies of each species. The pheromone dispenser within the trap should emit an optimum amount of chemical, an amount that will draw insects from a great distance and yet not be intense enough to inhibit entrance into the trap.

As for means of retaining responding insects, adhesive has been smeared inside some traps; in others, "one-way" mechanical baffles have been installed; a fast-acting insecticide, such as vapona, has been used to overcome entering insects; and sometimes a trap is equipped with a funnel having a hole at its apex small enough to allow the insect to enter the trap readily but not escape.

Height of a trap above ground level can affect trapping efficiency. For example, a trap for the oriental fruit moth caught almost no insects at ground level but was highly effective at a 1-m height.[14] While small traps are usually adequate for detection, high-capacity ones may be needed for mass trapping. Since pheromone vapors must get out to advertise the presence of the trap, placement of traps is important. Even the most efficient trap will catch poorly if placed in an area of restricted air movement, as in a hollow or where it is surrounded by dense foliage.

Deciding on the distance between traps can be a problem because it is difficult to determine the so-called "effective" distance of a trap. Amount of lure emitted, wind direction and velocity, sensitivity of the insect to its own pheromone (or more accurately the pheromone bait), and the amount of effort that can be expended in a trapping operation are some of the parameters that influence the decision on the optimum distance between traps for a particular application.

Field and laboratory trials, much observation, and trial-and-error type experimentation are usually required to optimize the many variables encountered in a trapping program. Acquiring this information is not difficult, but it is time-consuming because the needed data must be obtained for each species.

The U.S. Environmental Protection Agency *does* require registration of pheromone traps used to directly suppress insect populations. The Agency has registered the Hercon product containing grandlure, the sex pheromone of the boll weevil, for use on cotton, and this product is now commercially available.

C. Trap Cropping

In trap cropping, several rows of a crop are planted to emerge before the main planting, and insects attacking the plants are killed with an insecticide applied to the plant, either directly or systemically. When insects are killed in this way, the buildup of damaging insect populations during the season is markedly delayed or minimized.

Investigators have found that in many instances this procedure can be improved substantially by treating the trap crop with pheromone, thereby attracting insects to the lethal trap crop in much greater numbers and from greater distances. With the insects concentrated in a small part of the field, they are readily eliminated with a minimum of insecticide, and less insecticide is required to bring the main planting through to harvest. Also, the trap crop treatment has a minimal effect on beneficial insects since they are not drawn to the area by the pheromone.

The U.S. Environmental Protection Agency requires registration of trap cropping formulations.

D. Mating Disruption

Perhaps the most promising means of controlling insects with sex attractants is by permeating the atmosphere with the sex attractant,[6] which is usually accomplished by distributing lure-emitting particles. With attractant everywhere, the insect will have great difficulty in locating a mating partner and therefore be prevented from propagating. Undoubtedly, the searching insect becomes confused in the treated area because it cannot distinguish between pheromone from the lure-emitting particles and that from a potential sex partner.

Mating disruption trials have now been conducted successfully against quite a few insect species. These tests have demonstrated that controlled release formulations can maintain the pheromone in the atmosphere at levels sufficient to disrupt communication of the sexes for prolonged periods. In later chapters of this volume, these formulations and the results obtained with them in mating disruption are described. The most prominent of the formulations are Hercon flakes, Conrel hollow fibers, and NCR microcapsules. These materials are generally applied by aircraft with a sticker to cause the lure-bearing particulates to adhere where they land on a crop.

Mating disruption has also been accomplished by systematically hand-placing lure-emitting dispensers on trees throughout an area (see chapter by Plimmer).

As with mass trapping, mating disruptants are most effective against low populations of insects,[13] i.e., against insects that are sufficiently separated to require the lure for mate-finding. Conversely, mating disruption is not expected to be effective when insects are found in close proximity to each other, as in high populations, since the sexes are likely to see or hear one another and therefore get together. Thus, a firm prerequisite for the use of mating disruption is reduction of high insect populations to a low level, as through the use of an insecticide or other insect control agent, or by starting with a low-level population, as occurs naturally early in the growing season with many species or in newly infested areas.

Since the air-permeation technique becomes more efficient as the insect population level declines, which is a logical consequence of the increasing difficulties insects experience in searching for the diminished number of mating partners in a treated area, it should *theoretically* be possible to eradicate an insect species with its pheromone.[13] This possibility assumes that the treated area is isolated, which will avoid the influx of gravid females from nontreated areas. For essentially the same reason, i.e., to minimize the influx of gravid females, the treatment of large areas with pheromones is advocated by knowledgeable workers (see the recent book by Knipling for a comprehensive discussion of this subject[15]). By way of contrast, the application of an insecticide will eliminate a certain percentage of an insect population, e.g., 98%, regardless of whether the population is large or small. Use of an insecticide is therefore justified with a high population, and use of pheromones with a low population. The coordinated use of the two techniques holds high promise as a superior means of combatting insect pests. For more on this subject, see the chapter by Klassen et al.

Products for disrupting the mating of a number of species have now been registered by the U.S. Environmental Protection Agency and are commercially available (see chapter by Quisumbing). For applications of these materials, particularly by aircraft, equipment has been specially designed and built to distribute each type of formulation.

V. CHEMISTRY OF PHEROMONES

Because insect pheromones are highly potent, only minute amounts are generally present in insects. In early explorations, enormous numbers of insects were required to isolate even tiny amounts of lure. As a result, little progress in the identification of pheromones was registered until the late 1960s when a number of developments made it possible to undertake with confidence the difficult task of pheromone identification. Thus, new chemical instrumentation, especially mass, infrared, ultraviolet, and nuclear magnetic resonance spectrometers made possible the acquisition of chemical structural data from minute amounts of sample. Gas, thin layer, and other forms of chromatography greatly facilitated the efficient separation of the small amounts of compound; and new ancillary equipment, methods, and devices allowed microgram quantities of compound to be manipulated.

The ensuing burst of activity in the pheromone area may best be surmised by scanning the long list of pheromonal chemicals identified in the past decade (see chapter by Inscoe). This comprehensive listing of the pheromonal attractants includes pertinent references to allow the reader to rapidly acquire basic information on each compound. The listing also shows the types of chemicals insects use as sex attractants.

Most of the pheromones are multicomponent, i.e., they consist of more than one chemical. In many instances, different insect species respond to the same chemicals, but in different proportions. For example, the European corn borer in Iowa responds to a 95:5 combination of (*Z*)- and (*E*)-11-tetradecen-1-ol acetate, while the smartweed borer responds to a 50:50 combination of these chemicals.[16] Despite the existence of these insects in the same area, they are able to maintain their reproductive isolation through the use of the different ratios of chemical in their pheromones.

Most of the sex attractants of *Lepidoptera* (moths and butterflies) are unsaturated long straight-chain acetates, alcohols, aldehydes, or epoxides, while the sex lures of *Coleoptera* (beetles) are often terpene-like compounds and are more complex in nature.

In attempts to use synthetic pheromones in insect control, certain problems have arisen. A major one resulted from the fact that in some instances the complete pheromone had not been identified; e.g., if only two components of a three-component pheromone were identified, and the presence of the third component was not recognized, tests with the two-component mixture may have been inadequate because it was less active than the complete pheromone.

Another problem relates to the purity of the synthetic pheromone. In the synthesis of the chemicals, certain intermediates or side-reaction products present in a preparation may inhibit attraction of the target insect species; e.g., as little as 1% of the (*Z,Z*)-isomer in (*E,Z*)-3,13-octadecadien-1-ol acetate, the lesser peachtree borer (*Synanthe-*

don pictipes) pheromone, is enough to seriously reduce its attraction for the insect.[17] Obviously, every effort must be made to remove these inhibitory chemicals. This demand for high-purity chemicals can increase the cost of many of the synthetics. Furthermore, some pheromones are made up of exotic chemicals, at least partially, and means of preparing them must be developed. In some instances, a pheromonal chemical is optically active, and the racemic mixture available from conventional-type synthesis is much less effective. Unfortunately, excessively high cost of an exotic chemical can preclude its use in insect control.

Because many important pheromonal chemicals are unstable, their stability must be checked. Aldehydes, in particular, are unstable in air. Since pheromones must remain effective for weeks and sometimes for months, degradation of these chemicals must be prevented, especially if the products of degradation turn out to inhibit the attraction of the pheromone, in which case, attraction of the chemical is quickly negated.

VI. CONTROLLED RELEASE OF PHEROMONES

Ideally, a pheromone in a trap should be released in a controlled manner, much as a calling insect releases its pheromone to attract a mate. Releasing too high a concentration is not only wasteful, it can inhibit the entry of insects into a trap. Releasing too low a concentration lowers the trap catch, mainly by reducing the distance from which insects respond. With the inherent volatility of pheromone chemicals ranging from very high to very low, it is apparent that pheromone chemicals cannot be exposed directly; rather, their release must be regulated.

The various devices employed for controlling pheromone release are described in the chapter on controlled release technologies (see Volume I, Chapter 3). Each release system must be examined relative to each application for advantages and disadvantages in terms of convenience, release rate, protection of chemical from degradation, applicability to end use, ease of manufacture, cost, and ease and means of distribution.

Controlled release devices or systems are also used in mating disruption formulations, and it is similarly apparent that pheromone release in these products must be controlled to meter out the chemical at the most effective rate. An excessive release of pheromone can curtail the duration of effectiveness of a product as well as increase costs unnecessarily, while treatment with too low a release rate, or a rapidly declining one, will be ineffective in disrupting mating communication for the desired time interval, which usually is the entire mating season.

The level of chemical release, particularly for mating disruption, is not just a volatility parameter. Some pheromones are inherently more potent than others, so less of the more potent ones are required for a given application. Since the insects are not killed by permeating the air with the very low levels of pheromones used, they remain available for mating. Thus, the levels of chemical that are released must be maintained high enough to disrupt insect communication for the entire mating season, or multiple applications must be made to continue this disruption.

VII. TOXICITY OF PHEROMONES

Whether we like it or not, pheromones are being emitted by insects into the atmosphere during their mating season, so humans, animals, and other biota have been and are being exposed to these chemicals. Because of this situation, even before pheromones were identified and characterized, it was assumed that they were innocuous chemicals. When identities of sex attractants were finally elucidated, toxicity tests were conducted on many of them.[18] To date, to the authors' knowledge, no pheromone has been found to be truly toxic. Most appear to be similar to chemicals that make up food or food constituents. Thus, many are fatty acid derivatives, and others are similar to flavor chemicals. Furthermore, the amounts of chemical projected for use in insect control result in such low concentrations in the atmosphere that they cannot be measured except through use of greatly exaggerated rates of application and even then the chemicals must be derivatized with electron-capturing groups to allow them to be detected (see Volume I, Chapter 4). Indeed, with such low concentrations of a nontoxic natural product (the pheromone) being dispensed, one wonders why government agencies bother to regulate the use of pheromones at all.

VIII. SCOPE OF TREATISE

This chapter on pheromones and their use is intended to serve as a brief introduction to the subject matter of this treatise. It is followed by four chapters (constituting Section I) that amplify the concepts and background needed to understand pheromone technology as it relates to insect suppression.

The second section (under II) concentrates on practical aspects of pheromone technology and deals mainly with hands-on experiences with definite pheromonal formulations and products, and methods of using them for achieving optimum results in the field.

The third and fourth sections (III and IV) describe applications of pheromones for the direct control of insects and the results obtained.

The fifth and last section provides a view to the future. In one chapter, marketing and economic considerations are discussed realistically so the high promise of future expansion of the use of sex attractants in insect pest management may be properly assessed. The last chapter presents, as already noted, a listing of known pheromones by chemical name and insects affected plus pertinent references to each listing. This compilation constitutes a reservoir of knowledge for future exploration on pheromones.

Before completing this chapter, we must make reference to the recently issued book on basic principles of insect population suppression and management by Dr. Edward F. Knipling,[15] who kindly consented to prepare the Foreword of this volume. The contributions of few men in the history of entomology stand out as much as those of Dr. Knipling. His thinking on insect suppression by the use of attractants, which is outlined in Chapter 12 of his book, is therefore especially pertinent, and the reader is referred to this chapter for a scholarly and comprehensive presentation of the subject.

REFERENCES

1. **Holbrook, R. F., Beroza, M., and Burgess, E. D.,** Gypsy moth (*Porthetria dispar* (L.)) detection with the natural female sex lure, *J. Econ. Entomol.*, 53, 751, 1960.
2. **Haller, H. L., Acree, F., Jr., and Potts, S. F.,** The nature of the sex attractant of the female gypsy moth, *J. Am. Chem. Soc.*, 66, 1659, 1944.
3. **Acree, F., Jr., Beroza, M., Holbrook, R. F., and Haller, H. L.,** The stability of the hydrogenated gypsy moth sex attractant, *J. Econ. Entomol.*, 52, 82, 1959.
4. **Bierl, B. A., Beroza, M., and Collier, C. W.,** Potent sex attractant of the gypsy moth, *Porthetria dispar* (L.): its isolation, identification and synthesis, *Science*, 170, 87, 1970.
5. **Iwakï, S., Marumo, S., Saito, T., Yamada, M., and Katagiri, K.,** Synthesis and activity of optically active disparlure, *J. Am. Chem. Soc.*, 96, 7842, 1974.
6. **Beroza, M.,** Insect attractants are taking hold, *Agr. Chem.*, 15, 37, 1960.
7. **Gertler, S. I., Steiner, L. F., Mitchell, W. C., and Barthel, W. F.,** Esters of 6-methyl-3-cyclohexene-1-carboxylic acid as attractants for the Mediterranean fruit fly, *J. Agr. Food Chem.*, 6, 592, 1958.

8. **Steiner, L. F., Rohwer, G. G., Ayers, E. L., and Christenson, L. D.,** The role of attractants in the recent Mediterranean fruit fly eradication program in Florida, *J. Econ. Entomol.,* 54, 30, 1961.
9. **Beroza, M., Green, N., Gertler, S. I., Steiner, L. F., and Miyashita, D.,** New attractants for the Mediterranean fruit fly, *J. Agr. Food Chem.,* 9, 361, 1961.
10. **Reagen, E. P.,** Preventive pest-control measures, in *Scientific Aspects of Pest Control,* Publ. 1402, National Academy of Science — National Research Council, Washington, D.C., 1966, 185.
11. **Steiner, L. F., Mitchell, W. D., Harris, E. J., Kozuma, T. T., and Fujimoto, M. S.,** Oriental fruit fly eradication by male annihilation, *J. Econ. Entomol.,* 58, 961, 1965.
12. **Knipling, E. F. and McGuire, J. U., Jr.,** Population models to test theoretical effects of sex attractants used for insect control, *U.S. Department Agric. Inf. Bull.,* 308, 1966.
13. **Beroza, M. and Knipling, E. F.,** Gypsy moth control with the sex attractant pheromone, *Science,* 177, 19, 1972.
14. **Beroza, M., Gentry, C. R., Blythe, J. L., and Muschik, G. M.,** Isomer content and other factors influencing captures of oriental fruit moth by synthetic pheromone traps, *J. Econ. Entomol.,* 66, 1307, 1973.
15. **Knipling, E. F.,** The basic principles of insect population suppression and management, Agriculture Handbook 512, U.S. Department of Agriculture, Washington, D. C., 1979.
16. **Klun, J. A., Chapman, O. L., Mattes, K. C., Wojtkowski, P. W., Beroza, M., and Sonnet, P. E.,** Insect sex pheromones: minor amounts of opposite geometrical isomer critical to attraction, *Science,* 181, 661, 1973.
17. **Tumlinson, J. H.,** The need for biological information in developing strategies for applying semiochemicals, in *Chemical Ecology: Odor Communication in Animals,* Ritter, F. J., Ed., Elsevier/North-Holland, New York, 1979, 301.
18. **Beroza, M., Inscoe, M. N., Schwartz, P. H., Keplinger, M. L., and Mastri, C. W.,** Acute toxicity studies with insect attractants, *Toxicol. Appl. Pharmacol.,* 31, 421, 1975.

Chapter 2

CHEMICAL ATTRACTANTS IN INTEGRATED PEST MANAGEMENT PROGRAMS

Waldemar Klassen, Richard L. Ridgway, and May Inscoe

TABLE OF CONTENTS

I. INTRODUCTION

In the U.S. losses caused by insects to agricultural crops are estimated at 13% before harvest and 6% after harvest,[1-3] the equivalent of more than $10 billion annually. These losses occur despite the expenditure of more than $1 billion a year for the management of insect pests. Plainly, management of insect pests is essential to agricultural production and overall commerce as well as to public health. However, experience has shown that judicious pest control can have desirable effects, whereas nonjudicious pest control can have highly undesirable effects on both agriculture and forestry. Also, nonjudicious use of pest control methods is potentially harmful to natural ecosystems and to the health and welfare of people. General understanding of this latter point has led to increased emphasis on integrated pest management.[4,5]

Integrated pest management is a systems approach to designing and using control methods in a manner that will meet established economic, ecological, and sociological criteria.

The terminology of integrated pest management (IPM) has not yet become adequately stabilized. In this chapter the terms "tactics", "systems", and "strategies" are not used interchangeably as synonyms. Each of these terms has a definite meaning as defined in Secretary's Memorandum 1929.[6] Thus, an IPM tactic is a method of pest suppression, detection, or monitoring. An IPM system is the joint or sequential use of two or more tactics in a manner intended to minimize negative interactions against the ecosystem and to maximize desired interactions against the pest population. An IPM strategy is a broad plan for deploying tactics and systems to deal with a pest problem. According to USDA,[6] the major IPM strategies recognized by the U.S. Department of Agriculture are (1) uncoordinated management of local pest populations, as on a field-by-field or home-by-home basis; (2) prevention, e.g., using pest-free seed stock, prophylactic treatments of seed or soil, quarantines, etc.; (3) area-wide management of major segments or even entire pest populations; and (4) eradication from defined areas, e.g., reduction of a pest population to a nondetectable level for a prolonged period. This classification of major pest-management strategies was first formulated by Rabb.[7]

Integrated pest management includes basic research, control components research, systems research, economics research, and extension and education activities involving all pests .... insects, weeds, bacteria, fungi, viruses, nematodes, snails, slugs, birds, rodents, and other organisms.[6]

Presently, highly effective synthetic organic pesticides (insecticides, herbicides, fungicides, nematicides, etc.) constitute the predominant technology of pest control. Amounts of these chemicals used have increased during the past 3 decades from less than 50,000 lb a year to over 1.4 billion lb/year. Of these pesticides used in agriculture in 1976, about 24% of the poundage was insecticide.[8] These materials have largely displaced very large quantities of inorganic materials that were used in the U.S. as late as 1950, for example, calcium arsenate, 38,842 lb; copper sulfate, 124,573 lb; lead arsenate, 27,490 lb; and ground sulfur, 538,592 lb.[9]

The new materials that have come into common use since then are exceedingly effective, and this use for protection of agricultural commodities from insect damage has yielded very great economic benefits. However, the benefits in many cases have been attended by serious difficulties. Paramount among these difficulties for agriculture and vector control has been the development of insecticide resistance. Serious insect pests in mid- and southern temperate zones have several generations per year. They often develop insecticide resistance (or overcome crop varietal resistance) in 3 to 10 years.[2] For example, by 1979, populations of 414 insect species had developed resistance to insecticides.[10] Thus, evolution through man-imposed selection has rendered obsolete a great number of valuable insecticide use patterns with recurring severe economic hardship for the producer. Also, cross resistance may make an entire class of insecticides ineffective. Since the number of chemical classes of insecticides is few, there may be no replacements capable of controlling some very important species.

Another undesirable side effect of the use of broad spectrum insecticides is cataclysmic reduction in populations of parasites and predators. As a result, populations of the target insect may increase and outbreaks of secondary pests occur.[11]

In fact, entomologists have long recognized the need for ecologically selective methods of insect suppression, i.e., methods that destroy the pest without destroying its natural enemies;[11,12] as well as the need to slow the growth of expenditures for pest management.[13]

Further, costs of insecticide development are increasing rapidly, in part because of regulations resulting from acute concern about the effects of pesticides on the environment and human health. For example, costs of discovery and development to registration of a conventional pesticide have risen from about $1 million in 1955 to an estimated $18 million in 1980,[14] and new synthetic organic insecticides are now being registered at a rate of only about 1 per year.[15] Finally, public concern continues to build over involuntary exposure of people to insecticide residues.[16] People are unwilling to tolerate risks, no matter how small, that are imposed without prior consent.

There is a wide range of tactics or methods of pest management now emerging that is socially and ecologically desirable. In time some of these are likely to become even more feasible. These emerging methods include use of pest-resistant crop cultivars, cultural controls, parasites and predators, microbial agents, genetical methods, insect growth regulators, and pheromones and other behavior-modifying chemicals.

Pheromones are among the most ecologically selective pest suppression agents now being investigated. They are naturally occurring perfumes whose function is to mediate communication between individuals of a single species. Also, toxicological studies of a number of pheromones indicate that they are likely to have little, if any, effect on nontarget organisms. However, the practical use of pheromones and other attractants for insect pest management involves unfamiliar complexities since their properties and modes of action differ greatly from those of insecticides.

Insecticides are extremely valuable tools. Although systems of pest management that depend little or not at all on insecticides are likely to be devised against some of the most damaging insect pest species, it seems unlikely that replacements for insecticides will be found for the majority of insect pests that are sporadic in occurrence or that are annually serious on a small scale only. Therefore, integrated pest management also entails the development of systems that will prolong the usefulness of insecticides—and concomitantly of resistant crop cultivars.

II. STATUS OF RESEARCH AND DEVELOPMENT ON PHEROMONES AND ATTRACTANTS

To an extraordinary degree the lives of insects are governed by their olfactory senses.[17-20] For example, many insect species are attracted from long distances to their host plants, host animals, or other food sources; many hosts and other substrates produce odors that stimulate egg laying; certain chemicals in the insects' food stimulate the feeding response; and insects themselves emit odors known as pheromones for intraspecific (within species) communication. These pheromones may be involved in assembling the two sexes and in mating, in causing aggregation on a food or oviposition resource, in signaling alarm when an individual is attacked, in controlling the behavior of various castes of social insects, in stimulating maturation or phase transformation of migratory species, in deterring multiple ovipositions on a fruit, and in a variety of other aspects of behavior and physiology.

Some insect species produce chemicals known as allemones in order to defend themselves or to exploit other species. Some emit chemicals that are detrimental because they attract the natural enemies; these are known as kairomones.

Pheromones involved in assembling the two sexes for mating were observed more than a century ago.[20] The use of extracts of crude pheromone for detection was initiated by Collins and Potts[21] who showed that extracts of virgin gypsy moths, *Lymantria dispar*, were highly attractive to males. The first elucidation of the chemical identity of a pheromone was that of Butenandt et al.,[22] who worked with the pheromone of the female silkworm moth, *Bombyx mori*. Since that time, pheromones and other attractants have been isolated, identified, and synthesized for 674 species of arthropods (see Volume II, Chapter 15, Table 1).[23] Also, significant progress has been made towards the development and use of pheromones for survey, detection, monitoring, and suppression. However, the chemical identities of the pheromones of many important arthropods are still unknown.

A. Current Technology

The review of the status of current technology of pheromones and attractants for arthropods indicated that 475 of the 674 species for which chemical structures of pheromones or attractants have been identified are Lepidoptera (moths and butterflies), 82 are Coleoptera (beetles), 37 are Diptera (flies), 2 are Isoptera (termites), 14 are Acarina (ticks and mites), 18 are Homoptera (scales and mealybugs), 4 are Hemiptera (true bugs), 35 are Hymenoptera (wasps and sawflies), 3 are Neuroptera, and 4 are Orthoptera. (The distribution says a good deal about the level of knowledge of chemical communication in the various orders.)

The status of research and development of pheromones and nonpheromonal attractants is summarized in Table 1.

The predominant practical use of pheromones and attractants is for detection and survey. Indeed, attractant materials are used to some extent for this purpose against 81 species of Lepidoptera, 47 species of Coleoptera, 6 species of Homoptera, 1 species of Hemiptera, 12 species of Diptera, 4 species of Hymenoptera, 2 species of Orthoptera, and 6 species of Acarina. In addition, in the U.S., five pheromones have now been registered as pesticides for managing insect populations. Feeding stimulants, food attractants, and other synthetic attractants are also used extensively, especially for a number of Diptera.

B. Unidentified Pheromones

There remains a large number of important arthropod pests for which pheromones and other attractants have not been identified. Important pests for which pheromones have not been reported, that are known to cause substantial losses[24] or that would likely develop into major pests if inadvertently introduced into U.S.,[25] are included in Table 1.

Thus, there continues to be an important need for further research on chemical identification of pheromones and other behavior-modifying chemicals. Indeed, behavior-modifying chemicals are proving so useful that it would be irresponsible not to vigorously press research in this area.

III. CHARACTERISTICS OF ATTRACTANTS AND OTHER CONTROL METHODS IN RELATION TO THE DESIGN AND USE OF INTEGRATED INSECT MANAGEMENT SYSTEMS

The usefulness of attractants (including pheromones) and other methods in managing particular pest populations is determined by a number of factors. These are capac-

Table 1
STATUS OF RESEARCH, DEVELOPMENT, AND USE OF PHEROMONES AND NONPHEROMONAL ATTRACTANTS FOR ARTHROPODS

The names of the compounds can be found by noting the compound number(s) alongside the Latin names of the insect species in Table 2 of the Inscoe chapter (last in Volume II) and then referring to the compound number(s) in Inscoe's Table 1 for the identity of the compound.

The degree of progress is rated on the following scale: (5) work apparently accomplished; (4) high degree of progress; (3) considerable progress; (2) preliminary identification of useful principle; (1) no information; and (0) not considered applicable. Abbreviations are defined as follows: (a) alarm pheromone; (af) adult feeding attractant; (ag) aggregating pheromone; (agr) aggression pheromone; (ap) aphrodisiac pheromone; (b) brood pheromone; (c) contact pheromone; (ha) hair pencil aphrodisiac pheromone; (hr) hair pencil repellent pheromone; (k) kairomone; (lag) larval aggregating pheromone; (lf) larval feeding attractant; (m) maturation acceleration pheromone; (np) nonpheromonal attractant; (oa) oviposition attractant — nonpheromonal; (od) oviposition deterrent pheromone; (op) oviposition pheromone; (pt) phase transformation pheromone; (q) queen cell building pheromone; (s) sex pheromone; (sa) sex attractant — nonpheromonal; (sr) sex ratio shift pheromone; (t) trail pheromone; (te) territorial pheromone; and (w) worker control pheromone. The chemical structures of the attractants may be found in Volume II, Chapter 15, Table 1.

An attractant is listed as a pheromone only if it has been isolated from an arthropod. Arthropod species are listed alphabetically by Latin name under Orders (Lepidoptera, Coleoptera, etc.). Common names approved by the Entomological Society of America[660] are marked with an asterisk.

Species		References	Pheromone identified	Nonpheromonal behavior modifier identified	Annihilative trapping or bait demonstrated	Chemical used for predictive monitoring	Chemical used for survey and detection	Chemical used to disrupt mating	In operational use: attractant or pheromone
LEPIDOPTERA									
Achroia grisella (F.)	lesser wax moth*	497	5s	1					
Acleris emargana (F.)		70	1	5sa					
Acleris paradiseana (Walsingham)		110	1	5sa					
Acleris rhombana (Denis & Schiffermüller)		106	1	5sa					
Acleris scabrana (Denis & Schiffermüller)		514	1	5sa					
Acossus centerensis (Lintner)	aspen carpenterworm	380,381	1	5sa					
Acrobasis rufilimbalis Wileman		74	1	5sa					

Table 1 (continued)
STATUS OF RESEARCH, DEVELOPMENT, AND USE OF PHEROMONES AND NONPHEROMONAL ATTRACTANTS FOR ARTHROPODS

Species		References	Pheromone identified	Nonpheromonal behavior modifier identified	Annihilative trapping or bait demonstrated	Chemical used for predictive monitoring	Chemical used for survey and detection	Chemical used to disrupt mating	In operational use: attractant or pheromone
Acrobasis vaccinii Riley	cranberry fruitworm*	662	3s	1			3		
Acrolepiopsis alliella (Semenov & Kuznetsov)		138	1	5sa					
Acronicta grisea Walker		63		5sa				2	
Acronicta sperata Grote	dagger moth	68	1	5sa					
Adoxophyes fasciata (in Japan), see *Adoxophyes* sp.									
Adoxophyes orana (F. v. Röslerstamm)	summer fruit tortrix	314,315, 351	5s	1	1	5	5	4	Yes
Adoxophyes reticulana, see *Adoxophyes orana* (F. v. Röslerstamm)									
Adoxophyes sp. (in Japan)	smaller tea tortrix	209,302, 385	5s	1	1	1	5	1	Yes
Aethes sp.		67	1	5sa			3		
Agonopterix sp.		74	1	5sa					
Agroperina sp.	grass seed moth	662	4s	1			4		
Agrotis exclamationis (L.)		336	5s	1			5		No
Agrotis fucosa Butler	common cutworm	201,202	5s	1					
Agrotis gladiara (Morrison)	claybacked cutworm	662	1	1					
Agrotis ipsilon (Hufnagel)	black cutworm*	226	5s	1					
Agrotis orthogonia Morrison	pale western cutworm*	64		5sa				2	

Agrotis segetum (Denis & Schiffermüller)	cutworm on vegetables in Europe	203							
Agrotis venerabilis Walker	dusky cutworm	65		4sa				2	
Agrotis volubilis Harvey		68	1	5sa					
Albuna fraxini (Hy. Edwards)		162	1	5sa					
Albuna pyrimidalis (Walker)		169	1	5sa					
Alcathoe carolinensis Engelhardt		164	1	5sa					
Aletia oxygala luteopallens (Smith)	cutworm on grass	663	1s,5ha	1	structures of ha (663)				No
Allagrapha aereoides (Grote)	false looper	121	1	5sa			3		
Amathes c-nigrum (L.) (large form), see *Xestia dolosa* Franclemont									
Amathes c-nigrum (L.) (small form), see *Xestia adela* Franclemont									
Amorbia cuneana (Walsingham)	avocado leafroller	662	5s	1			4		No
Amphipoea interoceanica (Smith)		67	1	5sa					
Amphipoea velata (Walker)		67		5sa					
Amphipyra monolitha Guenée		74,107	1	5sa					
Amyelois transitella (Walker)	navel orangeworm*	530	5s	1			5	4	Yes
Anagasta kuehniella (Zeller)	Mediterranean flour moth*	386,387, 680	5s,3op	1			5	2	Yes
Anagrapha falcifera (Kirby)	celery looper*	121,189	1	4sa	1	1	3	1	Yes
Anarsia lineatella Zeller	peach twig borer*	55—57	5s	1	1	5	5	1	Yes
Andropolia contacta (Walker)		63	1	5sa					
Antheraea polyphemus (Cramer)	polyphemus moth*	424,425	5s	1			5		No
Anticarsia gemmatalis Hübner	velvetbean caterpillar*	662a	2s	1					
Apamea monoglypha (Hufnagel)	dark arch moth	663,664	5ha	1	(ha = pinocarvone)				
Aphania auricristana Walsingham, see *Apotomis auricristana* (Walsingham)									
Aphania infida (Heinrich), see *Apotomis infida* (Heinrich)									

Table 1 (continued)
STATUS OF RESEARCH, DEVELOPMENT, AND USE OF PHEROMONES AND NONPHEROMONAL ATTRACTANTS FOR ARTHROPODS

Species		References	Pheromone identified	Nonpheromonal behavior modifier identified	Annihilative trapping or bait demonstrated	Chemical used for predictive monitoring	Chemical used for survey and detection	Chemical used to disrupt mating	In operational use: attractant or pheromone
Aphelia paleana (Hübner)		57	1	5sa					
Apotomis auricristana (Walsingham)		107	1	5sa					
Apotomis corticana, see *Apotomis lineana* (Denis & Schiffermuller)									
Apotomis infida (Heinrich)		67	1	5sa					
Apotomis lineana (Denis & Schiffermüller)		242	1	5sa					
Apotomis sp.		74	1	5sa					
Archepandemis borealis (Freeman)		357	1	5sa					
Archippus, see *Archips*									
Archips argyrospilus (Walker)	fruittree leafroller*	111,114	5s	1	1	1	5	3	Yes
Archips breviplicanus Walsingham	Asiatic leafroller	303	5s	1			5		No
Archips ingentanus (Christoph)		74,110	1	5sa					
Archips longicellanus (Walsingham)		107	1	5sa					
Archips mortuanus Kearfott		210	1	5sa					
Archips peratratus Yasuda		74		5sa					
Archips piceanus (L.)		74,110	1	5sa					
Archips podanus (Scopoli)	fruit tree tortrix moth	352,354	5s	1			5	3	Yes
Archips rosanus (L.)	European leafroller	112,114	5s	1			5		Yes
Archips semiferanus (Walker)	oak leafroller*	355	5s	1	1	1	5	2	Yes
Archips tsuganus Powell		357	1	5sa					

Archips xylosteanus (L.)		57,74	4s						
Argynnis sp., see *Argynnis paphia* (L.)									
Argynnis paphia (L.)		664	2ap	1	glands in wings of male				
Argyrogramma basigera (Walker)		227	1	5sa					
Argyrogramma verruca (F.)		227,228	1	5sa					
Argyroploce aurofasciana (Haworth)		77	1	5sa					
Argyroploce lacunana (Denis & Schiffermüller)		242	1	5sa					
Argyroploce leucotreta (Meyrick), see *Pseudogalleria leucotreta* (Meyrick)									
Argyroploce schistaceana (Snellen), see *Tetramoera schistaceana* (Snellen)									
Argyrotaenia angustilineata (Walsingham), see *Epagoge angustilineata* (Walsingham)									
Argyrotaenia citrana (Fernald)	orange tortrix*	356	5s	1			5	2	Yes
Argyrotaenia dorsalana (Dyar)		63,68,357	1	5sa					
Argyrotaenia pulchellana (Hübner)		57,358	1	5sa					
Argyrotaenia quadrifasciana (Fernald)		116	1	5sa					
Argyrotaenia velutinana (Walker)	redbanded leaf-roller*	118,121, 213,367	5s	1			5	4	Yes
Artogeia melete Ménétriès		710	5ha	1					No
Artogeia napi (L.)	mustard white	664	2ap		glands on wings				
Artogeia rapae (L.)	imported cabbageworm*	664	1s,2ap	1	ap from wings				
Ascalapha odorata (L.)		664	2ap		glands on metathoracic legs				
Athetis dissimilis Hampson		74	1	5sa					
Athrips sp.		307	1	5sa			3		
Autographa ampla (Walker)		67	1	5sa					
Autographa biloba (Stephens)		67	1	5sa					
Autographa californica (Speyer)	alfalfa looper*	66,189, 229,230	5s	4sa	1	3	5	3	Yes

Table 1 (continued)
STATUS OF RESEARCH, DEVELOPMENT, AND USE OF PHEROMONES AND NONPHEROMONAL ATTRACTANTS FOR ARTHROPODS

Species		References	Pheromone identified	Nonpheromonal behavior modifier identified	Annihilative trapping or bait demonstrated	Chemical used for predictive monitoring	Chemical used for survey and detection	Chemical used to disrupt mating	In operational use: attractant or pheromone
Autographa gamma (L.)	anti-European false looper	231,663	2ap	4sa					
Autographa precationis (Guenée)		67		4sa					
Autoplusia egena (Guenée)	bean leafskeletonizer*	225	1	5sa					
Bapta temerata (Denis & Schiffermüller), see *Lomographa temerata* (Denis & Schiffermüller)									
Barbera colfaxiana (Kearfott)	Douglas-fir cone moth*	71	1	5sa					
Battaristis sp.		67	1	5sa					
Boarmia obliqua hypulina Wehrli	tea torticid in China	665	3s	1			3		
Bombyx mandarina (Moore)		155	5s	1			5		No
Bombyx mori (L.)	silkworm*	156,156a, 525,526	5s				5	2	No
Brachmia arotraea (Meyrick)		74	1	5sa					
Brachmia macroscopa (Meyrick)	sweet potato leaftier	423a	5s				5		No
Bryotropha senectella (Zeller)		204	1	5sa					
Bryotropha similis (Stainton)	gelechiid moth	322	5s	1			5		
Bryotropha sp.	moth	67		5sa					
Bryotropha terrella (Denis & Schiffermüller)		77	1	5sa					
Cacoecimorpha pronubana (Hübner)		323,361		5sa			3		

Cadra cautella (Walker)	almond moth*	263,327, 390,681	5s,2ha	1		5	4	Yes
Cadra figulilella (Gregson)	raisin moth*	665a	5s	1		5		Yes
Caligo arisbe Hübner		664	2ap		glands on 4th & 5th abdomen segment			
Caradrina morpheus (Hufnagel)		77	1	5sa				
Carmenta bassiformis (Walker)		163,164	1	5sa				
Carmenta odda Duckworth & Eichlin		434	1	5sa				
Carmenta suffusata Engelhardt		164	1	5sa				
Carmenta texana (Hy. Edwards)		164	1	5sa				
Carmenta welchelorum Duckworth & Eichlin		435		5sa				
Carposina niponensis Walsingham	peach fruit moth of China	540—542	5s	1		5		No
Celypha rurestrana (Duponchel)		242	1	5sa				
Celypha striana (Denis & Schiffer-müller)		77	1	5sa				
Cenopis acerivorana (MacKay), see *Sparganothis acerivorana* (MacKay)								
Ceramica picta (Harris), see *Melanchra picta* (Harris)								
Chilo partellus (Swinhoe)	spotted stalk borer	139	5s	1		5	3	Yes
Chilo sacchariphagus (Bojer)	sugarcane borer	666	5s				4	No
Chilo suppressalis (Walker)	Asiatic rice borer, striped rice stem borer	191,517, 518,521	5s	1		5	4	Yes
Chionodes fuscomaculella (Chambers)		67		5sa				
Chionodes psiloptera (Barnes & Busck)	grass seed pest	426	1	3sa				
Choristoneura biennis Freeman		374	1	5sa				
Choristoneura conflictana (Walker)	large aspen tortrix*	115,507	1	5sa				

Table 1 (continued)
STATUS OF RESEARCH, DEVELOPMENT, AND USE OF PHEROMONES AND NONPHEROMONAL ATTRACTANTS FOR ARTHROPODS

Species		References	Pheromone identified	Nonpheromonal behavior modifier identified	Annihilative trapping or bait demonstrated	Chemical used for predictive monitoring	Chemical used for survey and detection	Chemical used to disrupt mating	In operational use: attractant or pheromone
Choristoneura fractivittana (Clemens)		67,115,116	1	5sa			5	3	Yes
Choristoneura fumiferana (Clemens)	spruce budworm*	117,129, 130,510	5s	1		3	5	3	Yes
Choristoneura lambertiana (Busck)	sugar pine tortrix	357	1	5sa					
Choristoneura occidentalis (Freeman)	western spruce budworm*	357,515	5s	1	1	3	5	1	Yes
Choristoneura retiniana (Walsingham)	modoc spruce budworm	357,374	1	5sa					
Choristoneura rosaceana (Harris)	obliquebanded leafroller*	118,119, 214	5s	1	1	5	5	3	Yes
Choristoneura viridis, see *Choristoneura retiniana* (Walsingham)									
Chrysapidia spp., see *Plusia*									
Chrysochrista sp. (*Chrysoclista* sp.?)		74	1	5sa					
Chrysoteuchia topiaria Zeller	cranberry girdler*	141,142	5s	1		5	5	3	Yes
Cisthene fasciata (Schaus)	lichen moth	664	2ap		glands on wings and abdomen				
Clepsis fucana (Walsingham)		357	1	5sa					
Clepsis melaleucana (Walker)		67	1	5sa					
Clepsis spectrana (Treitschke)	cabbage leafroller	295,324	5s	1			5		No
Cnephasia alternella, see *Cnephasia stephensiana* (Doubleday)									

Cnephasia communana (Herrich-Schäffer)		243	1	5sa				
Cnephasia stephensiana (Doubleday)		57,86,244, 245	1	5sa				
Coleophora sibirica (?)	casebearer in U.S.S.R.	667	2s	1				
Colias croceus (Goeffroy)		664	2ap		wing scales emit ap			
Colias edusa (F.), see *Colias croceus* (Goeffroy)								
Colias eurytheme Boisduval	alfalfa caterpillar*	668	5s	1		5		No
Colias philodice Godart	clouded sulfur caterpillar	668	5s	1		5		No
Commophila umbratosana (Kearfott), see *Henricus umbrabsanus* (Kearfott)								
Conistra griescens Draudt		74	1	5sa		3		
Conistra unimacula Sugi		74	1	5sa		3		
Cosmopterix fulminella Stringer		74	1	5sa		3		
Cosmopterix spp.		74,107	1	5sa		3		
Cosmopterix victor Stringer		74	1	5sa		3		
Creatonus gangis (L.)	arctiid	663	2ap	1	glands on inflatable tubes			
Croesia askoldana (Christoph)		110	1	5sa		3		
Croesia bergmanniana (L.)		106	1	5sa		3		
Croesia conchyloides (Walsingham)		110	1	5sa		3		
Croesia holmiana (L.)		77,106	1	5sa		3		
Croesia semipurpurana (Kearfott)	oak leaftier*	115	1	5sa		3		
Crymodes devastator (Brace)	glassy cutworm*	68,143,325	5s	1		5	3	Yes
Cryptocala acadiensis Bethune		102	1	5sa		3		
Cryptophlebia leucotreta (Meyrick)	false codling moth (Africa)	239,240, 247	5s	1		5		Yes
Cryptophlebia ombrodelta (Lower)	litchi fruit moth*	74,107	1	5sa		3		
Cucullia intermedia Speyer		67	1	5sa		3		
Cucullia umbratica (L.)		326	1	5sa		3		

Table 1 (continued)
STATUS OF RESEARCH, DEVELOPMENT, AND USE OF PHEROMONES AND NONPHEROMONAL ATTRACTANTS FOR ARTHROPODS

Species		References	Pheromone identified	Nonpheromonal behavior modifier identified	Annihilative trapping or bait demonstrated	Chemical used for predictive monitoring	Chemical used for survey and detection	Chemical used to disrupt mating	In operational use: attractant or pheromone
Cydia aurana (F.)		57,106,245	1	5sa					
Cydia funebrana (Treitschke)	plum fruit moth	249,250, 669,670	5s	1	1	1	5	4	Yes
Cydia gallicana (Guenée)		242	1	5sa					
Cydia janthinana (Duponchel)	hawthorne leafroller	214a,244	1	5sa					
Cydia nigricana (F.)	pea moth*	281,671, 672	1	5sa	1	1	5		
Cydia pomonella (L.)	codling moth*	88,89,92, 292	5s	1	2	4	5	4	Yes
Cydia tenebrosana (Duponchel)		57,86,673	1	5sa	1	1	5		
Cydia youngana (Kearfott)	spruce cutworm; spruce seed moth*	71		5sa			4		
Danaus chrysippus (L.)	African monarch butterfly	674,675	5ha	1	chemical structures (675)				No
Danaus gilippus berenice (Cramer)	Florida queen butterfly	676	5ha	1	chemical structures (675)				No
Danaus plexippus (L.)	monarch butterfly*	663	5ha	1	chemical structures (675)				No
Dargida procincta Grote	olive green cutworm		1	1					
Darsia cynicia perumbrosa, see *Diarsia rubifera perumbrosa* (Dyar)									

Darsia pseudorosaria, see *Diarsia rosaria* (Grote)									
Dasychira grisefacta ella Bryk	tussock moth	543	1	5sa					
Dasychira plagiata (Walker)	pine tussock moth	544,545	1	5sa					
Dasychira vagans grisea (Barnes & McDunnough)	tussock moth	543	1	5sa		5	4		Yes
Decadarchis simulans (Butler)		436	1	5sa			4		
Decodes horarianus (Walsingham)		63		5sa		5			
Dendrolimus pini (L.)		864	4s						
Dendrolimus punctatus (Walker)	pine caterpillar of China	677	5s	1			5		
Dendrolimus sibericus Tshetverikov	U.S.S.R.	670	4a						
Dendrolimus spectabilis Butler	pine moth	84,85	5s				5		No
Diarsia deparca (Butler)		74	5sa						
Diarsia rosaria (Grote)	grass seed moth	662	4s	1			4		
Diarsia rubifera perumbrosa (Dyar)	grass seed moth	662	4s	1			4		
Diasemia litterata (Scopoli)		74		5sa					
Diatraea grandiosella Dyar	southwestern corn borer*	678	3s	1			5		Yes
Diatraea saccharalis (F.)	sugarcane borer*	679	3s	1					
Dicentria semirufescens (Walker), see *Oligocentria semirufescens* (Walker)									
Dicestra trifolii (Hufnagel)	clover cutworm*	57,154	5s	5sa			5		No
Dichomeris ligulella Hübner	palmerworm*	67		5sa					
Dichomeris sparsella (Christoph)		74	1	5sa			3		
Dichroramphа sp.		67		5sa					
Dioryctria amatella (Hulst)		865	2s						
Diparopsis castanea (Hampson)	red bollworm	215,260, 261,284	5s	1	1	1	5	4	Yes
Doratopteryx plumigera Butler		127	1	5sa					
Earias biplaga Walker		136	1	5sa					
Earias insulana (Boisduval)	spiny bollworm	190,528, 529	5s	1			5	3	Yes

Table 1 (continued)
STATUS OF RESEARCH, DEVELOPMENT, AND USE OF PHEROMONES AND NONPHEROMONAL ATTRACTANTS FOR ARTHROPODS

Species		References	Pheromone identified	Nonpheromonal behavior modifier identified	Annihilative trapping or bait demonstrated	Chemical used for predictive monitoring	Chemical used for survey and detection	Chemical used to disrupt mating	In operational use: attractant or pheromone
Ecdytolopha insiticiana Zeller	locust twig borer*	67		5sa				3	
Egira curialis (Grote)	citrus cutworm		1	1					
Elasmopalpus lignosellus (Zeller)	lesser cornstalk borer*	103,388		5sa				2	
Elkneria pudibunda (L.)	European tussock moth		1	1					
Enarmonia formosana (Scopoli)	cherrybark tortrix moth	57,262		5sa				3	
Endopiza viteana Clemens	grape berry moth	199,264, 269	5s	1	2	1	5	4	Yes
Endothenia banausopis (Meyrick)		74		5sa				3	
Endothenia carbonana Douglas, see *Endothenia ustulana* (Haworth)									
Endothenia ustulana (Haworth)		77		5sa				3	
Epagoge angustilineata (Walsingham)		110	1	5sa					
Ephestia cautella (Walker), see *Cadra cautella* (Walker)									
Ephestia elutella (Hübner)	tobacco moth*	396,664, 689	5s,2ha	1			5	2	Yes
Ephestia kuehniella (Zeller), see *Anagasta kuehniella* (Zeller)									
Ephestiodes infimella Ragonot		397,664		5sa					
Epiblema desertana (Zeller)		67		5sa					
Epiblema foenella (L.)		57,74,244		5sa					
Epiblema scudderiana (Clemens)		86		5sa					

Epiblema scutulana (Denis & Schiffermüller)		86		5sa				
Epinotia zandana, see *Epinotia* sp.								
Epinotia sp.		67		5sa				
Epiphyas postvittana (Walker)	light brown apple moth*	375	4s					
Epiplema moza Butler		107		5sa				
Epiplema plagifera Butler		107		5sa				
Episimus argutanus (Clemens)		67,264		5sa		4	3	
Erannis defoliaria (Clerck)	U.S.S.R. winter moth	670	2s	1				
Ereunetis simulans (Butler), see *Decadarchis simulans* (Butler)								
Erynnis tages (L.)	dingy skipper	664	2ap		ap emitted from wing costal margins			
Ethmia monticola (Walsingham)	pest of Borraginaceae	357	1	5sa		4		
Eucosma campoliliana (Denis & Schiffermüller)		86	1	5sa				
Eucosoma nigromaculana, see Eucosma campoliliana (Denis & Schiffermüller)								
Eucosoma sonomana Kearfott	western pine shoot borer	265,266	5s			5	5	Yes
Eumenis semele (L.), see *Hipparchia semele* (L.)								
Euplagia quadripunctaria (Poda)	Jersey tiger moth	664	2ha					
Euploea phaenarata Schaller		664	2ap					
Eupoecilia ambiguella (Hübner)	European grape moth	82,83,216, 217	5s	1		5	4	Yes
Eupoecilia angustana (Hübner)		57		5sa				
Eupoecilia quadrilinea, see *Eupsilia quadrilinea* Leech								
Euproctis chrysorrhoea (L.)	browntail moth	685,865	2s					
Euproctis karghalica Moore		685	3s					
Euproctis pseudoconspersa Strand	tea tussock moth	684	3s	1		3		

Table 1 (continued)
STATUS OF RESEARCH, DEVELOPMENT, AND USE OF PHEROMONES AND NONPHEROMONAL ATTRACTANTS FOR ARTHROPODS

Species		References	Pheromone identified	Nonpheromonal behavior modifier identified	Annihilative trapping or bait demonstrated	Chemical used for predictive monitoring	Chemical used for survey and detection	Chemical used to disrupt mating	In operational use: attractant or pheromone
Euproctis similis (Fuessly)	yellowtail moth in U.S.S.R.	685,686	3s				3	3	
Eupsilia quadrilinea Leech		74	1	5sa			3		
Eurema fimbriata (Wallengren)		664	2ap		ap from wing scales				
Eurema hecabe (L.)		664	2ap		ap from wing scales				
Eurois astricta Morrison		68		5sa			3	3	
Eurois occulta (L.)	Great Dart	102,108		5sa			3		
Eurrhypara terrealis (Treitschke)		67	1	5sa					
Eurythmia hospitella Zeller		63		5sa			3		
Eurytides protesilaus (L.)		664	2ap		glands on wings				
Euxoa acornis (Smith)	cutworm	68		5sa			3		
Euxoa auxiliaris (Grote)	army cutworm*	101	1	5sa			4	2	
Euxoa lutulenta (Smith)		102	1	5sa			4		
Euxoa messoria (Harris)	darksided cutworm*	275,409	1	5sa			4		
Euxoa ochrogaster (Guenée)	redbacked cutworm*	199a,200, 687	5s				5	3	Yes
Euxoa pleuritica (Grote)	cutworm	68,69	1	5sa			4		
Euxoa tessellata (Harris)	striped cutworm*	67	1	5sa			4		
Euxoa tristicula (Morrison)	early cutworm	218	1	5sa			4	3	
Evergestis forficalis (L.)	garden pebble moth	106,376	1	5sa			4		
Exartema appendiceum (Zeller), see *Olethreutes appendicea* (Zeller)									

Exartema spp., see *Olethreutes* spp.									
Faronta diffusa (Walker)	wheat head armyworm*	68	1	5sa					
Feltia ducens Walker, see *Feltia jaculifera* (Guenée)									
Feltia jaculifera (Guenée)	dingy cutworm*	68	5s	1			5		
Fishia derelicta Hampson, see *Fishia instruta* Smith									
Fishia instruta Smith		68	1	5sa			3		
Gaesa sparsella (Christoph), see *Dichomeris sparsella* (Christoph)									
Galleria mellonella (L.)	greater wax moth*	495,498, 664	5s,5ha	1	ap from wings		5s		Yes
Geina tenuidactyla (Fitch)		67	1	5sa			3		
Gelechiidae sp.		74	1	5sa			3		
Gnorimoschemini sp.		74	1	5sa					
Grapholita funebrana (Treitschke), see *Cydia funebrana* (Treitschke)									
Grapholita inopinata Heinrich		74	1	5sa					
Grapholita janthinana (Duponchel), see *Cydia janthinana* (Duponchel)									
Grapholita molesta (Busck)	oriental fruit moth*; eastern fruit moth in U.S.S.R.	60,78,90, 254,695	5s	3af	1	5	5	4	Yes
Grapholita packardi Zeller	cherry fruitworm*	254	1	5sa			4		
Grapholita prunivora (Walsh)	lesser appleworm*	220,254	3s	1	1	1	5	3	Yes
Grapholita tenebrosana Duponchel, see *Cydia tenebrosana* (Duponchel)									
Grapholitha spp., see *Grapholita* spp.									
Gretchena bolliana (Slingerland)	pecan bud moth*	257	1	3	1	1	2	1	No

Table 1 (continued)
STATUS OF RESEARCH, DEVELOPMENT, AND USE OF PHEROMONES AND NONPHEROMONAL ATTRACTANTS FOR ARTHROPODS

Species		References	Pheromone identified	Nonpheromonal behavior modifier identified	Annihilative trapping or bait demonstrated	Chemical used for predictive monitoring	Chemical used for survey and detection	Chemical used to disrupt mating	In operational use: attractant or pheromone
Gynaephora ginghainensis (?)	grassland moth of China	684	5s	1			5		
Hahncappsia neobliteralis (Capps)		67		5sa			3		
Hammaptera frondosata Guenée		664	2ap		glands on wings and hair pencils				
Harpipteryx xylostella (L.), see *Ypsolopha dentella* (F.)									
Hedia, see *Hedya*									
Hedya atropunctana (Zetterstedt)		106	1	5sa			4		
Hedya chionosema (Zeller)		67	1	5sa			3		
Hedya nubiferana (Haworth)	green bud-worm*	77,221	5s	1			5	3	Yes
Hedya ochroleucana (Frölich)		221	5s	1			5		No
Heliconius erato (L.)		688	2		male-contributed antiaphrodiasiac				
Heliothis armiger (Hübner)	African boll-worm	136,414, 513,522	5s	1			5	2	Yes
Heliothis punctiger (Wallengren)		414	5s	1					
Heliothis scutosa (Denis & Schiffer-müller)		864		5sa					
Heliothis subflexa (Guenée)	ground cherry noctuid	689	4s						
Heliothis virescens (F.)	tobacco bud-worm*	144,504, 505,506	5s,4hr	1	1	1	5	3	Yes
Heliothis viriplaca (F.)	alfalfa worm in U.S.S.R.	865	2s	1					
Heliothis zea (Boddie)	bollworm*; corn ear-worm*, to-mato fruit worm*	144,194, 504,690	5s	5oa	1	1	5	4	Yes

Hemileuca oliviae Cockerell	range caterpillar	691	31ag					
Henricus umbrabsanus (Kearfott)		357	1	5sa		3		
Hepialus hecta (L.)		664	2ap		glands on legs			
Hermonassa cecilia Butler		74,107	1	5sa				
Hipparchia semele (L.)	grayling butterfly	664,682, 683	2ap	1	glands on wings			
Holomelina aurantiaca (Hübner)	tiger moth	635	5s	1		5		No
Holomelina spp.		635	5s	1				
Homoeosoma electellum (Hulst)	sunflower moth*	99	5s	1		4		
Homoglaea sp.		357	1	5sa		4		
Homona coffearia (Neitner)	tea tortrix	61	5s	1		5		
Homona magnanima Diakonoff	tea tortrix moth	268	5s	1		5		
Homonopsis illotana (Kennell)		110	1	5sa				
Hoshinoa adumbratana (Walsingham)		110	1	5sa				
Hoshinoa longicellana (Walsingham)		74	1	5sa				
Hyphantria cunea (Drury)	fall webworm*	100	5s	1				
Hysterosia cartwrightana Kearfott		362	1	5sa				
Idaea biselata (Hufnagel)		290	1	5sa		3		
Illice fasciata (Schaus), see *Cisthene fasciata* (Schaus)								
Isotrias hybridada Hübner, see *Isotrias rectifasciana* (Haworth)								
Isotrias rectifasciana (Haworth)		242	1	5sa			3	
Kawabea razowskii (Kawabe)		74	1	5sa			3	
Keiferia lycopersicella (Walsingham)	tomato pinworm	692—694	4s	1	3	4	4	Yes
Lacinipolia lorea (Guenée)	cutworm	67		5sa		3	2	
Lacnobia atlantica (Grote)	noctuid caterpillar	63	1	5sa		4	2	
Lacnobia grandis (Boisduval)	noctuid caterpillar	67	1	5sa		4		

Table 1 (continued)
STATUS OF RESEARCH, DEVELOPMENT, AND USE OF PHEROMONES AND NONPHEROMONAL ATTRACTANTS FOR ARTHROPODS

Species		References	Pheromone identified	Nonpheromonal behavior modifier identified	Annihilative trapping or bait demonstrated	Chemical used for predictive monitoring	Chemical used for survey and detection	Chemical used to disrupt mating	In operational use: attractant or pheromone
Lacnobia oleracea (L.)		149	1	5sa			4		
Lacnobia tacoma (Strecker)	noctuid caterpillar	68	1	5sa					
Lamprothripa lactaria Graeser		74		5sa			3		
Laspeyresia molesta (Busck), see *Grapholita molesta* (Busck)									
Laspeyresia nigricana (Stephens), see *Cydia nigricana* (F.)									
Laspeyresia pomonella (L.), see *Cydia pomonella* (L.)									
Laspeyresia youngana (Kearfott), see *Cydia youngana* (Kearfott)									
Lethe rohria (F.)		664	2ap						
Leucania commoides Guenée	cutworm	104,108, 275		5sa			5	2	
Leucania conigera (Denis & Schiffermüller)	noctuid cutworm	696	5ha		structure in (663)				No
Leucania impura (Hübner)	smoky wainscot moth	696	5ha		emits benzaldehyde				No
Leucania linda Franclemont		329	1	5sa			3		
Leucania loreyi (Duponchel)		74,147	5s	4sa			5		No
Leucania pallens luteopallens, see *Aletia oxygala luteopallens* (Smith)									
Leucania phragmitidicola Guenée		67		5sa					

Leucania separata (Walker), see *Pseudaletia separata* (Walker)									
Limnaecia phragmitella Stainton		362	1	5sa			3		
Lithocolletis blancardella (F.), see *Phyllonorycter blancardella* (F.)									
Lithocolletis corylifoliella (Hübner), see *Phyllonorycter corylifoliella* (Hübner)									
Lithocolletis orientalis (Kumata), see *Phyllonorycter orientalis* (Kumata)									
Lithocolletis pygmaea (Kumata), see *Phyllonorycter pygmaea* (Kumata)									
Lithocolletis watanabei (Kumata), see *Phyllonorycter watanabei* (Kumata)									
Lithomoia solidaginis (Hübner)		68	1	5sa			4		
Lixodessa ochrofasciella (Toll)		204	1	5sa			4		
Lobesia aelopa Meyrick		74,107	1	5sa			4		
Lobesia botrana (Denis & Schiffermüller)	European grapevine moth	286—289	5s	1			5	4	Yes
Lobesia cunninghamiacola (Liu & Pai)	olethreutid tree defoliator	684,700	2	1					
Lobesia sp.		74		5sa			4		
Lomographa temerata (Denis & Schiffermüller)	geometrid, clouded silver moth	663,664	5ha	1	ha = *n*-butyric acid				No
Loxagrotis albicosta (Smith)	western bean cutworm		1	1					
Loxostege chortalis (Grote), see *Sitochroa chortalis* (Grote)									
Loxostege neobliteralis (Capps), see *Hahncappsia neobliteralis* (Capps)									
Loxostege sticticalis (L.)	beet webworm*	362,664	3s	5sa	1	1	4	2	No
Lycaena sp.			2ap		glands on wings				
Lycorea ceres ceres (Cramer), see *Lycorea cleobaea* (Godart)									

Table 1 (continued)
STATUS OF RESEARCH, DEVELOPMENT, AND USE OF PHEROMONES AND NONPHEROMONAL ATTRACTANTS FOR ARTHROPODS

Species		References	Pheromone identified	Nonpheromonal behavior modifier identified	Annihilative trapping or bait demonstrated	Chemical used for predictive monitoring	Chemical used for survey and detection	Chemical used to disrupt mating	In operational use: attractant or pheromone
Lycorea cleobaea (Godart)		664	5ha	1					No
Lymantria dispar L.	gypsy moth*	576,589, 697	5s,21ag	1			5	5	Yes
Lymantria dispar japonica (Motschulsky)		582		5sa			3		
Lymantria fumida Butler		582		5sa			3		
Lymantria monacha (L.)	nun moth	583,584, 595,596	5s	1			5	2	Yes
Lymantria obfuscata Walker		585,586	1	5sa			3		
Malacosoma americanum (F.)	eastern tent caterpillar	698,699	31ag						
Malacosoma neustria (L.)	European tent caterpillar		1	1					
Mamestra brassicae (L.)	cabbage armyworm	148,149, 417,700	5s	1			5		Yes
Mamestra configurata Walker	bertha armyworm*	330,331, 418,701	5s	1			5		Yes
Mamestra illoba (Butler) - status uncertain		74	1	5sa			4		
Mamestra oleracea (L.), see *Lacnobia oleracea* (L.)									
Mamestra persicariae (L.), see *Melanchra persicariae* (L.)									
Mamestra suasa (Denis & Schiffermüller)		864	3s						

Manduca sexta (L.)	tobacco hornworm*, sphinx moth	527	4s	1			4		No
Melanchra persicariae (L.)		663	5s,5ha	1	ha(663)		5		No
Melanchra picta (Harris)	zebra caterpillar*	67,68	1	5sa					
Melissopus latiferreanus (Walsingham)	filbertworm moth	702	1	5sa			5		Yes
Micardia argentata (Butler)		74	5sa			3			
Morrisonia confusa (Hübner)		67		5sa		3			
Mutuuraia mysippusalis (Walker)		63	1	5sa			3	2	
Mycalesia suaveolens (Wood-Mason & Niceville)		664	2ap						
Naranga aenescens (Moore)	rice green caterpillar	332,332a	5s	1			5		Yes
Nedra ramosula (Guenée), see *Nedra variatellus* (Clemens)									
Nedra variatellus (Clemens)		121		5sa			3		
Nemapogon picisignatellus (Dietz)	tineid	67		5sa					
Nephelodes emmedonia (Cramer), see *Nephelodes minians* Guenée									
Nephelodes minians Guenée	bronzed cutworm*	68,143		5sa			5		
Nephelodes sp.	cutworm	68		5sa					
Nippoptilia issikii Yano		74,107	1	5sa					
Nippoptilia vitis (Sasaki)		74	1	5sa					
Nudaurelia cytherea (F.)	pine emperor moth	464	5s				4		
Ochropleura plecta (L.)		106	1	5sa			3		
Ochropleura triangularis Moore		74	1	5sa					
Odites leucostola (Meyrick)		74	1	5sa			3		
Oidaematophorus guttatus Walsingham		357	1	5sa			4		
Oidaematophorus homodactylus (Walker)		357,499	5s	1			5		yes
Oidaematophorus mathewianus (Zeller)		357		5sa					

Table 1 (continued)
STATUS OF RESEARCH, DEVELOPMENT, AND USE OF PHEROMONES AND NONPHEROMONAL ATTRACTANTS FOR ARTHROPODS

Species		References	Pheromone identified	Nonpheromonal behavior modifier identified	Annihilative trapping or bait demonstrated	Chemical used for predictive monitoring	Chemical used for survey and detection	Chemical used to disrupt mating	In operational use: attractant or pheromone
Olethreutes appendicea (Zeller)	leafroller	63	1	5sa			4		
Olethreutes spp.		70,116	1	5sa			3		
Olethreutidae sp., see Tortricidae, Olethreutinae									
Oligocentria semirufescens (Walker)		67		5sa					
Oncocnemis piffardi (Walker)		68	1	5sa			4		
Oncoptera sp.		664	2ap						
Operophtera brumata (L.)	winter moth	670	4s	1			4		Yes
Opsiphanes invirae isagoras (Fruhstorfer)		664	2ap		glands on abdomen				
Orgyia antiqua (L.)	rusty tussock moth*	546,547	5s	1			4		
Orgyia gonostigma (L.)		670,703	2s	1					
Orgyia leucostigma (J. E. Smith)	whitemarked tussock moth*	544,546	1	5sa			3	3	
Orgyia pseudotsugata (Mc-Dunnough)	Douglas-fir tussock moth*	543,548, 549,550	5s	1	1	4	5	4	Yes
Orthodes crenulata (Butler)		67	1	5sa			3		
Orthogonia sera Felder		74,107	1	5sa			4		
Orthosia angustipennis Matsumura		74	1	5sa					
Orthosia ella (Butler)		74	1	5sa					
Orthosia odiosa (Butler)		74	1	5sa					
Orthosia paromoea (Hampson)		74	1	5sa					
Ostrinia furnacalis (Guenée)	Asian corn borer	304	5s	1			5		
Ostrinia nubilalis (Hübner)	European corn borer*	67,305, 363—369	5s	5oa	1	2	5	3	Yes

Ostrinia obumbratalis (Lederer)	smartweed borer*	366	1	5sa			
Otosema odorata L., see *Ascalapha odorata* (L.)							
Pammene albuginana (Guenée)		86,242	1	5sa			
Pammene argyrana (Hübner)		86,242,244	1	5sa			
Pammene aurana Hübner, see *Pammene rhediella* (Clerck)							
Pammene aurantiana (Staudinger)		242	1	5sa			
Pammene fasciana (L.)		86,242,245	1	5sa			
Pammene gallicana (Guenée), see *Cydia gallicana* (Guenée)							
Pammene gallicolana, see *Pammene albuginana* (Guenée)							
Pammene inquiliana Fletcher		86,242	1	5sa			
Pammene insulana (Guenée)		242	1	5sa			
Pammene nemorosa V. Kuznetsov		74,107	1	5sa			
Pammene querceti Gozmany		242	1	5sa			
Pammene regiana (Zeller)		242	1	5sa			
Pammene rhediella (Clerck)	fruitlet mining tortrix	96,242	1	5sa			
Pammene spiniana (Duponchel)		242	1	5sa			
Pammene splendidulana (Guenée)		242	1	5sa			
Pammene suspectana (Zeller)		57	1	5sa			
Pandemis canadana Kearfott		63	1	5sa			
Pandemis dumetana (Treitschke)		242	1	5sa			
Pandemis heparana (Denis & Schiffermüller)		106,122, 323	5s	1	4	3	No
Pandemis limitata (Robinson)	threelined leafroller*	122,329	5s	1	3	3	No
Pandemis pyrusana Kearfott		329,333	5s	1	4	3	No
Panlymnas chrysippus (L.), see *Danaus chrysippus* (L.)							
Panolis flammea (Denis & Schiffermüller)	pine beauty moth	334,335	1	4sa	4	3	No
Pantherodes paradalaria (Hübner)		664	2ha				
Papilio aristolociae F.		664	2ap				

Table 1 (continued)
STATUS OF RESEARCH, DEVELOPMENT, AND USE OF PHEROMONES AND NONPHEROMONAL ATTRACTANTS FOR ARTHROPODS

Species	References	Pheromone identified	Nonpheromonal behavior modifier identified	Annihilative trapping or bait demonstrated	Chemical used for predictive monitoring	Chemical used for survey and detection	Chemical used to disrupt mating	In operational use: attractant or pheromone
Paralobesia viteana (Clemens), see *Endopiza viteana* Clemens								
Paramesia gnomana (Clerck)	242	1	5sa			4		
Paramyelosis transitella (Walker), see *Amyelois transitella* (Walker)								
Paranthrene asilipennis (Boisduval)	164	1	5sa			4		
Paranthrene dollii (Neumoegen)	164,165	1	5sa			4		
Paranthrene pellucida (= *perlucida* (Busck), see *Paranthrene robiniae* (Hy. Edwards)								
Paranthrene robiniae (Hy. Edwards)	437	1	5sa			4		
Paranthrene simulans (Grote)	164,437, 438	1	5sa					
Paranthrene simulans palmii, see *Paranthrene simulans* (Grote)								
Paranthrene tabaniformis (Rottemburg)	162	1	5sa					
Parapandemis borealis (Freeman), see *Archepandemis borealis* (Freeman)								
Pareromene exsectella (Christoph)	74	1	5sa					
Pechipogon barbalis (Clemens), see *Polypogon strigilata* (L.)								
Pectinophora endema Common	427		5sa					
Pectinophora gossypiella (Saunders) pink bollworm*	62,410, 428, 704—707	5s	5	4	2	5	5	Yes

Pectinophora scutigera (Holdaway)	pink-spotted bollworm	427		5sa			4		
Peridroma saucia (Hübner)	variegated cutworm*	109,708	5ha	5sa			5		Yes
Perizoma grandis Hulst		357	1	5sa			3		
Petrova luculentana (Heinrich)		72	1	5sa					
Petrova metallica (Busck)	ponderosa pine caterpillar	73	1	5sa					
Phalaenoides glycinae Lewin	vine moth	709	5ha	1					No
Phalonia sp., see *Aethes* sp.									
Phalonidia minimana (Caradja)		74	1	5sa			3		
Phaneta latens (Heinrich)	an olethreutid moth	867		5sa			5		No
Phassus schamyl (Christoph)		664	2ap		glands on legs				
Phlogophora meticulosa (L.)	noctuid, angleshade moth	663	5ha	1	structure (663)				No
Phlyctaenia terrealis (Treitschke), see *Eurrhypara terrealis* (Treitschke)									
Phragmatobia fuliginosa (L.)	ruby tiger moth	664	2ha						
Phthorimaea operculella (Zeller)	potato tuberworm*	297—300	5s	1	1	3	5	1	Yes
Phyllonorycter blancardella (F.)	spotted tentiform leafminer	282,283	1	5sa			4		
Phyllonorycter corylifoliella (Hübner)	leafminer	283	1	5sa			4		
Phyllonorycter orientalis (Kumata)	leafminer	283	1	5sa			4		
Phyllonorycter orientalis (Kumata)	leafminer	74	1	5sa			4		
Phyllonorycter pygmaea (Kumata)	leafminer	74	1	5sa			4		
Phyllonorycter watanabei (Kumata)	leafminer	74	1	5sa		4			
Pieris melete Ménétriès, see *Artogeia melete* Ménétriès									
Pieris napi (L.), see *Artogeia napi* (L.)									
Pieris rapae (L.), see *Artogeia rapae* (L.)									

Table 1 (continued)
STATUS OF RESEARCH, DEVELOPMENT, AND USE OF PHEROMONES AND NONPHEROMONAL ATTRACTANTS FOR ARTHROPODS

Species		References	Pheromone identified	Nonpheromonal behavior modifier identified	Annihilative trapping or bait demonstrated	Chemical used for predictive monitoring	Chemical used for survey and detection	Chemical used to disrupt mating	In operational use: attractant or pheromone
Plathypena scabra (F.)	green cloverworm*	1	1						
Platynota flavedana Clemens	variegated leafroller	123	5s	1			5	2	Yes
Platynota idaeusalis (Walker)	tufted apple budmoth	67,123, 124,377	5s	1			4	2	
Platynota stultana Walsingham	omnivorous leafroller	125,126	5s	1	1	5	5	3	yes
Platyptilia carduidactyla (Riley)	artichoke plume moth*	524	5s	1			5		Yes
Plodia interpunctella (Hübner)	Indianmeal moth*	135,321, 390,391, 664	5s,2ap		ap from wings		5	4	Yes
Plusia aereoides (Grote), see *Allagrapha aereoides* (Grote)									
Plusia gamma (L.), see *Autographa gamma* (L.)									
Plutella xylostella (L.)	diamondback moth*	325,419, 420,422	5s				5	3	Yes
Podosesia aureocincta Purrington & Nielsen	banded ash clearwing*	162,164, 166	1	5sa			5		
Podosesia syringae (Harris)	lilac borer*; ash borer*	162,164, 166	1	5sa			5		
Polia atlantica (Grote), see *Lacnobia atlantica* (Grote)									
Polia grandis (Boisduval), see *Lacnobia grandis* (Boisduval)									

Polia hepatica (Clerck)	noctuid caterpillar	700	5ha	1					
Polia nebulosa (Hufnagel)	noctuid	663	5ha		structures in (663)				No
Polia tacoma (Strecker), see *Lacnobia tacoma* (Strecker)									
Polia tincta, see *Polia hepatica* (Clerck)									
Polychrosis cuninghamicola, see *Lobesia cunninghamiacola* (Liu & Pai)									
Polypogon strigilata (L.)			2ap		glands on leg				
Porthetria dispar (L.), see *Lymantria dispar* (L.)									
Prays citri (Milliere)	citrus flower moth	188,500	5s	1	5		5		Yes
Prays oleae (Bernard)	olive moth	501,502	5s	1			5	2	Yes
Prinooxystus robiniae (Peck)	carpenterworm*	378,379	4s	1	2	1	5	1	Yes
Prodenia eridania (Cramer), *Spodoptera eridania* (Cramer)									
Protobathra leucostola (Meyrick), see *Odites leucostola* (Meyrick)									
Pseudaletia separata (Walker)		145,701, 711,868	5s,5ha	1	structure in (663)		5		No
Pseudaletia unipuncta (Haworth)	armyworm*	151,663, 712	5s,ha,5hr	(ha = benzyl alc.)			4		No
Pseudexentera maracana Kearfott		67		5sa			3		
Pseudogalleria leucotreta (Meyrick)	false codling moth	239,240, 247	5s	1			5		Yes
Pseudoplusia includens (Walker)	soybean looper*	194,230, 232	5s	3	1	1	5	3	Yes
Pseudorthodes crenulata (Butler), see *Orthodes crenulata* (Butler)									
Pseudorthodes vecors (Guenée)		67	1	5sa			3		
Pterophorus tenuidactylus (Fitch), see *Geina tenuidactyla* (Fitch)									
Ptycholoma lecheana circumclusana (Christoph)		74,110	1	5sa			3	2	

Table 1 (continued)
STATUS OF RESEARCH, DEVELOPMENT, AND USE OF PHEROMONES AND NONPHEROMONAL ATTRACTANTS FOR ARTHROPODS

Species		References	Pheromone identified	Nonpheromonal behavior modifier identified	Annihilative trapping or bait demonstrated	Chemical used for predictive monitoring	Chemical used for survey and detection	Chemical used to disrupt mating	In operational use: attractant or pheromone
Pyrausta aurata (Scopoli)		369	5s	1			5		No
Pyrausta ochosalis Dyar, see *Pyrausta orphisalis* Walker									
Pyrausta orphisalis Walker		67	1	5sa			3		
Pyrausta purpuralis L.		361	1	5sa			3		
Pyreferra citromba Franclemont		67	1	5sa					
Pyrgus malvae (L.)		664	2ap		glands on wings				
Pyroderces sp.		107	1	5sa			3		
Raphia frater Grote		67,70	1	5sa					
Rhopobota naevana naevana (Hübner)	blackheaded fireworm* (olethreutid)	74	1	5sa					
Rhyaciona frustrana bushnelli (Busck), see *Rhyacionia bushnelli* (Busck)									
Rhyacionia buoliana (Denis & Schiffermüller)	European pine shoot moth*	224,233, 270,276	5s	1	1	1	5	3	Yes
Rhyacionia bushnelli (Busck)	western pine tip moth		1	1					
Rhyacionia frustrana (Comstock)	Nantucket pine tip moth*	276,293	1	5sa				2	
Rhyacionia neomexicana (Dyar)	southwestern pine tip moth*	234	2s	5sa			5		No
Rhyacionia rigidana (Fernald)	pitch pine tip moth	277	5s	1			5		Yes
Rhyacionia subtropica Miller	subtropical pine tip moth*	80	5s	1			5	2	No

Rhyacionia zozana (Kearfott)		265	1	5sa			4		
Rhynchaglaea fuscipennis Sugi		74	1	5sa			4		
Rhynchaglaea scitula (Butler)		74	1	5sa			4		
Rhynchopacha sp., see *Athrips* sp.									
Rusidrina depravata (Butler)		74,107	1	5sa			3		
Rycitinae sp.		74	1	5sa			3		
Sabulodes aegrotata (Guenée)	omnivorous looper	662	4s	1			4		No
Sannina uroceriformis Walker	persimmon borer*	164,165	1	5sa			3		
Sathrobrota sp., see *Pyroderces* sp.									
Schinia bina (Guenée)		68	1	5sa			3		
Scotia exclamationis (L.), see *Agrotis exclamationis* (L.)									
Scotogramma trifolii (Hufnagel), see *Dicestra trifolii* (Hufnagel)									
Scrobipalpa atriplicella (F. v. Röslerstamm)		67	1	5sa			3		
Sesamia cretica Lederer	stemborer	105	1	5sa			3		
Sesamia inferens (Walker)	purple stem-borer moth	423	5s	1			5		No
Sesia tibialis (Harris)	cottonwood crown borer	167	5s	1			3		
Sideridis rosea (Harvey)		102,109	1	5sa			3		
Sitochroa chortalis (Grote)		63	1	5sa			4		
Sitotroga cerealella (Olivier)	Angoumois grain moth*	412,432, 433	5s,51ag		1	1	5	3	Yes
Sparganothis acerivorana (MacKay)	maple leaf-roller	115	1	5sa					
Sparganothis albicaudana Busck	tortricid	67	1	5sa			3		
Sparganothis directana (Walker)	chokecherry leafroller	714	5s	1			4		
Sparganothis niveana (Walsingham)		67,121	1	5sa			3		
Sparganothis pilleriana (Denis & Schiffermüller)		81	1	5sa			3		
Sparganothis sp.		357	1	5sa			3		
Sparganothis sulfureana (Clemens)		67	2s	4sa					
Spatalistis bifasciana (Hübner)		110	1	5sa					

Table 1 (continued)
STATUS OF RESEARCH, DEVELOPMENT, AND USE OF PHEROMONES AND NONPHEROMONAL ATTRACTANTS FOR ARTHROPODS

Species		References	Pheromone identified	Nonpheromonal behavior modifier identified	Annihilative trapping or bait demonstrated	Chemical used for predictive monitoring	Chemical used for survey and detection	Chemical used to disrupt mating	In operational use: attractant or pheromone
Sphinx ligustri L.		664	2ap		glands on abdomen				
Spilonota ocellana (F.)	eye-spotted bud moth*	77,313	5s	1			4		
Spilosoma lubricipeda (L.)	ermine tiger moth	664	2ha						
Spodoptera cilium Guenée	grassland armyworm	337	1	5sa			3		
Spodoptera dolichos (F.)		235	1	5sa			3	2	
Spodoptera eridania (Cramer)	southern armyworm*	712, 715—717	5s	1			5		Yes
Spodoptera exempta (Walker)	African armyworm	338,339	5s	1	1	3	5	1	No
Spodoptera exigua (Hübner)	beet armyworm*	271,337, 399	5s	5sa	1	1	4	3	No
Spodoptera frugiperda (J. E. Smith)	fall armyworm*	195,236, 272,273	5s	1	1	1	5	3	Yes
Spodoptera littoralis (Boisduval)	Egyptian cotton leafworm	215,342, 382	5s	1			5	4	Yes
Spodoptera litura (F.)	tobacco cutworm	343,344, 383,401	5s	1			5	4	Yes
Spodoptera mauritia (Boisduval)		347—349	5s	1			3		
Spodoptera ornithogalli (Guenée)	yellowstriped armyworm*	230		1					
Spodoptera praefica (Grote)	western yellowstriped armyworm	662	4s	1			3		

Spodoptera triturata (Walker), see *Spodoptera mauritia* (Boisduval)									
Staphylinochrous whytei Butler		127	1	5sa			3		
Sterrha biselata (Hufnagel), see *Idaea biselata* (Hufnagel)									
Sterrha imbecilla Inoue		74	1	5sa			3		
Stichophthalma camadeva (Westwood)		866	2ap						
Strophedra nitidana (F.)		74	1	5sa			3		
Synanthedon acerni (Clemens)	maple callus borer*	162	1	5sa			4		
Synanthedon alleri (Engelhardt)		164	1	5sa			3		
Synanthedon arkansasensis Duckworth & Eichlin		164	1	5sa			3		
Synanthedon bibionipennis (Boisduval)	strawberry crown moth*	168	1	5sa			3		
Synanthedon decipiens (Hy. Edwards)		164	1	5sa			3		
Synanthedon exitiosa (Say)	peachtree borer*	163,439, 470,718	5s,2ha	1	2	2	5	4	Yes
Synanthedon fatifera Hodges		162,164, 166	1	5sa			3		
Synanthedon fulvipes (Harris)		162	1	5sa			3		
Synanthedon geliformis (Walker)		164	1	5sa			3		
Synanthedon hector (Butler)	cherrytree borer	441—443	1	5sa			3		
Synanthedon kathyae Duckworth & Eichlin		434	1	5sa			3		
Synanthedon myopaeformis (Borkhausen)	apple borer in U.S.S.R.	444	5s		4			4	
Synanthedon pictipes (Grote & Robinson)	lesser peachtree borer*	439,440, 445,718	5s	1	2	2	5	4	
Synanthedon quercus (Matsumura)		446	1	5sa			3	2	
Synanthedon rhododendri (Beutenmüller)	rhododendron borer*	165	1	5sa			3		
Synanthedon rileyana (Hy. Edwards)	sequoia pitch moth	165	1	5sa			3		

Table 1 (continued)
STATUS OF RESEARCH, DEVELOPMENT, AND USE OF PHEROMONES AND NONPHEROMONAL ATTRACTANTS FOR ARTHROPODS

Species		References	Pheromone identified	Nonpheromonal behavior modifier identified	Annihilative trapping or bait demonstrated	Chemical used for predictive monitoring	Chemical used for survey and detection	Chemical used to disrupt mating	In operational use: attractant or pheromone
Synanthedon rubrofascia (Hy. Edwards)		438	1	5sa			3		
Synanthedon scitula (Harris)	dogwood borer*	438,445	1	5sa			3	2	
Synanthedon sequoiae (Hy. Edwards)	sequoia pitch moth*	168	1	5sa			3		
Synanthedon tenuis Butler		446	1	5sa			3	2	
Syncopacma cinctella (Clerck)		204	1	5sa			3		
Syncopacma vorticella Scopoli, see *Syncopacma cinctella* (Clerck)									
Syntomis phegia (L.)		587	1	5sa			3		
Syrichtus malvae (L.), see *Pyrgus malvae* (L.)									
Tahamia bonifatella, see *Tehama bonifatella* (Hulst)									
Tehama bonifatella (Hulst)	lawn moth	662	5s	1			5		Yes
Telorta divergens (Butler)		74,107	1	5sa			3		
Telorta edentata (Leech)		74,107							
Teratoglaea pacifica Sugi		74	1	5sa			3		
Terias hecabe fimbriata (Wallengren), see *Eurema fimbriata* (Wallengren) and *Eurema hecabe* (L.)									
Terricula violetana Kawabe (?)		110	1	5sa			3		
Tetramoera schistaceana (Snellen)		259		5s					
Thamnosphecia alleri (Engelhardt), see *Synanthedon alleri* (Engelhardt)									
Thyris maculata Harris		67	1	5sa			3		

Thyris usitata Butler		74	1	5sa			3		
Tineola bisselliella (Hummel)	webbing clothes moth*	664	1s,2ap	1					
Tortricidae, Olethreutinae		74	1	5sa			4		
Tortrix sinapina Butler		110	1	5sa			3		
Tortrix viridana L.	tortricid in U.S.S.R.	57,106,335	3s	1					
Trichoplusia ni (Hübner)	cabbage looper*	76,237, 271,663	5s,5hr	1	1	1	5	4	Yes
Trichoplusia oxygramma (Geyer)		228,230		5sa			4	2	
Tryporyza incertulas (Walker)	rice paddy borer	665	2	1					
Udea rubigalis (Guenée)	celery leaftier*		1	1					
Uthethesia spp.	tiger moth	721	5ha						No
Valeroides viridimacula (Graeser)		74	1	5sa			3		
Vitacea scepsiformis (Hy. Edwards)		164	1	5sa					
Xanthorhoe ferrugata (Clerck)	geometrid	663	2ap	1	ap from coremata				
Xanthorhoe fluctuata (L.)	garden carpet moth	663	2ap	1	ap from coremata				
Xenomicta cupreifera (Butler)		74		5sa			3		
Xestia adela Franclemont (small form)	spotted cutworm	67,74,311	5s	1					
Xestia dolosa Franclemont (large form)	spotted cutworm	67,309, 310	5s	1			5		
Xylena formosa Butler		74		5sa			3		
Xylomyges curialis (Grote), see *Egira curialis* (Grote)									
Xylophasia monoglypha (Hufnagel), see *Apamea monoglypha* (Hufnagel)									
Yponomeuta cagnagella (Hübner), see *Yponomeuta padella* (L.)									
Yponomeuta evonymella (L.)		404		5sa			3		
Yponomeuta padella (L.)		308	1	5sa			4	2	
Yponomeuta padellus-malinellus (Clerck), see *Yponomeuta padella* (L.)		77		5sa			3		

Table 1 (continued)
STATUS OF RESEARCH, DEVELOPMENT, AND USE OF PHEROMONES AND NONPHEROMONAL ATTRACTANTS FOR ARTHROPODS

Species		References	Pheromone identified	Nonpheromonal behavior modifier identified	Annihilative trapping or bait demonstrated	Chemical used for predictive monitoring	Chemical used for survey and detection	Chemical used to disrupt mating	In operational use: attractant or pheromone
Yponomeuta plumbella (Denis & Schiffermüller)		404	1	5sa					
Yponomeuta rorella (Hübner), see *Yponomeuta padella* (L.)									
Yponomeuta yanagawanus Matsumura (?)		74	1	5sa					
Ypsolopha dentella (F.)	European honeysuckle leafroller*	67	1	5sa			4		
Zanclognatha lunalis (Scopoli)		588	5s						
Zeiraphera diniana (Guenée)	larch bud moth, larch form	128,278, 312	1	5sa			3	2	
Zeiraphera diniana (Guenée)	larch bud moth, cembran pine form	278,279		5sa			3		
Zeiraphera griseana (Hübner)	a larch bud moth		1	1					
Zeiraphera improbana (Walker)	a larch bud moth		1	1					
Zenodoxus tineiformis (Esper)		165	1	5sa			3		
Zygaena ephialtes (L.)		57	1	5sa			3		
Zygaena transalpina Esper		128	1	5sa			3		

COLEOPTERA

Acalymma vittatum (F.)	striped cucumber beetle*		1s	1					
Acanthoscelides obtectus (Say)	bean weevil*	470,471	5s				5		Yes
Agrilus viridus (L.)	a two lined beetle		1s	1					
Agriotes gurgistanus (Faldermann)	a wireworm	722	2s	1					
Agriotes lineatus (L.)	a lined click beetle	723	2s	1					
Agriotes litigiosus Rossi	Kuban click beetle	448, 724—728	3s	1			4		Yes
Agriotes obscurus (L.)	a wireworm		1s	1					
Agriotes reitteri Schwarz	an elaterid beetle	729	4s						
Agriotes sputator (L.)	a wireworm		1s	1					
Ahasverus advena (Waltl)	foreign grain beetle*; a stored product beetle	730		2af			4		Yes
Amphimallon majalis (Razoumowsky)	European chafer*	465,483	1s	5af			5		Yes
Anomala errans (F.)	a scarabid beetle in U.S.S.R.	731	3s						
Anthonomous grandis Boheman	boll weevil*	174—177, 732	5s	3	5	5	5	2	Yes
Anthrenus flavipes LeConte	furniture carpet beetle*	467,733	5s				5		
Attagenus elongatulus Casey		469,488	5s	2af			5		Yes (af)
Attagenus megatoma (F.)	black carpet beetle*	489	5s	2af			5		Yes (s,af)
Blastophagus piniperda (L.), see *Tomicus piniperda* (L.)									
Byctiscus betulae (L.)			1	1					
Carabus problematicus Herbst		491		5					

Table 1 (continued)
STATUS OF RESEARCH, DEVELOPMENT, AND USE OF PHEROMONES AND NONPHEROMONAL ATTRACTANTS FOR ARTHROPODS

Species		References	Pheromone identified	Nonpheromonal behavior modifier identified	Annihilative trapping or bait demonstrated	Chemical used for predictive monitoring	Chemical used for survey and detection	Chemical used to disrupt mating	In operational use: attractant or pheromone
Cerambyx cerdo L.	a borer in broad leaf trees		1	1					
Cerambyx scopolii Füssely	a borer of shrubs		1	1					
Chaetocnema sp.	flea beetles		1	1					
Collops vittatus (Say)	beneficial predator	737	1	5sa			5		No
Costelytra zealandica (White)	N.Z. grass grub beetle	615,617		5sa			5		Yes
Cotinis nitida (L.)	green June beetle*	738		2af					Yes
Cryptolestes sp.	grain beetles	730		2af			4		Yes
Curculio caryae (Horn)	pecan weevil*	180,181	3ag				5		No
Cyclocephala insulicola Arrow		616		5sa					
Cylas formicarius elegantulus (Summers)	sweetpotato weevil*	739	3s	1	1	1	1	1	No
Cylindrocopturus adspersus (LeConte)	sunflower stem weevil		1	1					
Dectes texanus texanus LeConte		178	1	5sa					
Dendroctonus adjunctus Blandford	roundheaded pine beetle*	740—744	5s,ag				5		No
Dendroctonus brevicomis LeConte	western pine beetle*	596,659, 745,746	5s,ag		5	1	5	2	Yes
Dendroctonus frontalis Zimmermann	southern pine beetle*	570,600, 602	5s,ag		5	1	5	3	Yes

Dendroctonus jeffreyi Hopkins	Jeffrey pine beetle*	19	5s,ag				5		No
Dendroctonus ponderosae Hopkins	mountain pine beetle*	183,185	5s,ag		5		5		Yes
Dendroctonus pseudotsugae Hopkins	Douglas-fir beetle*	563—566, 603,656	5s,ag		5		5	3	Yes
Dendroctonus rufipennis (Kirby)	spruce beetle*	171,567, 568,605	5s,5ag		4	1	5	2	Yes
Dendroctonus spp.		533	5ag			5			
Dendroctonus vitei Wood		186	5s,5ag				5		
Dermestes ater DeGeer	black larder beetle*	468	1	5sa					
Dermestes lardarius (L.)	larder beetle*	468	1	5sa					
Dermestes maculatus DeGeer	hide beetle*	468	5s	2			5		Yes(af)
Diabrotica longicornis (Say)	northern corn rootworm*	689	3s				4		No
Diabrotica undecimpunctata howardi Barber	southern corn rootworm*; spotted cucumber beetle*	689	4s				4		No
Diabrotica virgifera LeConte	western corn rootworm*	689	5s				5		No
Diaprepes abbreviatus (L.)	sugarcane rootstalk borer weevil	689	2s						
Enocleris lecontei Wolcott	blackbellied clerid*	35		5k					
Epilachna varivestis Mulsant	Mexican bean beetle*		1	1					
Gibbium psylloides (Czenpinski)		730		2af			4		Yes
Gnathotrichus retusus (LeConte)	an ambrosia beetle	29,30	5s				5		No
Gnathotrichus sulcatus LeConte	ambrosia beetle	6,12	5s		4		5		Yes
Hylobius abietis (L.)	pine weevil	658		5sa					
Hylobius pales (Herbst)	pales weevil*	747,748	5s				5		Yes
Hylotrupes bajulus (L.)	old house borer*	571	5s,ag				5		

Table 1 (continued)
STATUS OF RESEARCH, DEVELOPMENT, AND USE OF PHEROMONES AND NONPHEROMONAL ATTRACTANTS FOR ARTHROPODS

Species		References	Pheromone identified	Nonpheromonal behavior modifier identified	Annihilative trapping or bait demonstrated	Chemical used for predictive monitoring	Chemical used for survey and detection	Chemical used to disrupt mating	In operational use: attractant or pheromone
Ips acuminatus (Gyllenhal)	a bark beetle	14	5s,ag				5	2	No
Ips amitinus (Eichhoff)	a bark beetle	889	5s,ag				5		No
Ips avulsus (Eichhoff)	small southern pine engraver*	43	5s	5	1	1	5	1	Yes
Ips calligraphus (Germar)	six-spined engraver beetle	43,47	5s,ag				5		Yes
Ips cembrae (Heer)	larch bark beetle	13	5s,ag				5		Yes
Ips confusus (LeConte)	California five-spined ips*; pine bark beetle	33,36,37	5s	5	1	1	5	1	Yes
Ips duplicatus (Sahlberg)	a bark beetle	48	5s,ag				5		No
Ips grandicollis (Eichhoff)	five-spined engraver beetle	38,44	5s,ag				5		Yes
Ips latidens (LeConte)		35,39	5ag				5		
Ips paraconfusus Lanier		33,34,40, 43	5s,ag		3		5	3	Yes
Ips pini (Say)	pine engraver*	34,35,45, 51	5s,ag		3		5	3	Yes
Ips sexdentatus (Börner)	6-toothed bark beetle	49,50	5s,ag				5		No
Ips typographus (L.)	spruce bark beetle; 8-toothed engraver beetle	15,16,182	5s,ag	1			5		Yes
Lasioderma serricorne (F.)	cigarette beetle*	730,749, 750	5s	2af			5s		Yes (s,af)

Leptinotarsa decemlineata (Say)	Colorado potato beetle*	751	2s					
Limonius californicus (Mannerheim)	sugarbeet wireworm*	752	3s			3		No
Limonius canus LeConte	Pacific Coast wireworm*	484,752	4s			3		No
Limonius infuscatus Motschulsky	western field wireworm*		1	1				
Limonius subauratus LeConte	Columbia Basin wireworm*		1	1				
Lissorhoptrus oryzophilus Kuschel	rice water weevil*		1					
Melanotus depressus Melsheimer	a wireworm	120		5sa				
Monolepta albidipennis Weise			1	1				
Necrobia rufipes (De Geer)	redlegged ham beetle*	730		3af		4		Yes
Orcytes rhinoceros (L.)	coconut rhinoceros beetle	477,478, 632,730, 753	1	5af	5			Yes
Orthotomicus caelatus (Eichhoff)	a beetle	35		5ag				
Oryzaephilus mercator (Fauvel)	merchant grain beetle*	730	1	2af		4		Yes
Oryzaephilus surinamensis (L.)	sawtoothed grain beetle*	730	1	2af		4		Yes
Palorus ratzeburgi (Wissmann)	smalleyed flour beetle*	730		2af		4		Yes
Phyllophaga sp.	white grubs		1	1				
Pissodes strobi (Peck)	white pine weevil*	754,755	5s,ag			5		No
Pityogenes chalcographus (L.)		18,612	5s,ag			5		No
Pityokteines curvidens (Germar)		41,42,46	5s,ag			5	3	No
Pityokteines spinidens (Reitter)		42	5s					
Pityokteines vorontzovi (Jacobson)		42	5s,ag					
Platypus flavicornis (F.)	an ambrosia beetle (pinhole borer)	7	5s			5		No

Table 1 (continued)
STATUS OF RESEARCH, DEVELOPMENT, AND USE OF PHEROMONES AND NONPHEROMONAL ATTRACTANTS FOR ARTHROPODS

Species		References	Pheromone identified	Nonpheromonal behavior modifier identified	Annihilative trapping or bait demonstrated	Chemical used for predictive monitoring	Chemical used for survey and detection	Chemical used to disrupt mating	In operational use: attractant or pheromone
Polyphylla alba (Pallas)	a scarabid beetle in U.S.S.R.	731	3s						
Popillia japonica Newman	Japanese beetle*	454,455, 490,756	5s	5	4	5	5	2	Yes
Ptinus fur (L.)	whitemarked spider beetle*	730		2af			4		Yes
Ptinus ocellus Brown	Australian spider beetle*	730		2af			4		Yes
Ptinus tactus, see *Ptinus ocellus* Brown									
Rhabdoscelus obscurus (Boisduval)	New Guinea sugarcane weevil*	179,730		5sa		4			
Rhizotrogus aequinoctialis (Herbst)	a scarabid beetle in U.S.S.R.	869	4s						
Rhizotrogus majalis (Razoumowsky), see *Amphimallon majalis* (Razoumowsky)									
Rhyzopertha dominica (F.)	lesser grain borer*	757	5s,ag				5		
Scolytus multistriatus (Marsham)	smaller European elm bark beetle*	20,610, 611	5s,ag	1	4	1	5	2	Yes
Scolytus pygmaeus (F.)		21,22	5s,ag						

Scolytus scolytus (F.)	larger European elm bark beetle	23—28, 723	5s,ag	1			5	2	
Selatosomus latus (F.)	an elaterid beetle	870	3s	1					No
Sitophilus granarius (L.)	granary weevil*	761	2s,ag	5af,ag					Yes (af)
Sitophilus oryzae (L.)	rice weevil*	730	2s,ag	2af					Yes (af)
Sitophilus zeamais Motschulsky	maize weevil*	730,758	2s,ag	3af			5		Yes (af)
Smicronyx fulvus LeConte	sunflower seed weevil		1	1					
Stegobium paniceum (L.)	drugstore beetle*	575,730	5s	2af			5		Yes (af)
Temnochila chlorida (Mannerheim)		599	1	5k			3		
Tenebrio molitor L.	yellow mealworm*	663,759	2 antiaphrodisiac						
Thanasimus dubius (F.)		608,757	1	5k			3		
Thanasimus formicarius (L.)		17	1	5k			3		
Thanasimus rufipes (Brahm)		17	1	5k			3		
Thanasimus undatulus Say		173,603	1	5k			3		
Tomicus piniperda (L.)	a bark beetle	658, 734—736	5s,ag	1			5		No
Tribolium castaneum (Herbst)	red flour beetle*	761,763	2s,ag	5af,ag					
Tribolium confusum Jacquelin duVal	confused flour beetle*	760,762, 763	5s	5af,ag			5		Yes (af)
Trichodes ornatus Say	checkered flower beetle	480	1	5	5	1	5	1	Yes
Trogoderma glabrum (Herbst)		159,161, 531,536	5s	2af			5		Yes
Trogoderma granarium Everts	khapra beetle*	158,472, 532,535	5s	2af	1	0	5	2	Yes (s,af)
Trogoderma grassmani Beal		157,159		5sa					
Trogoderma inclusum LeConte	a dermestid beetle	157,159, 531	5s	2af			5		
Trogoderma parabile, see *Trogoderma variabile* Ballion									
Trogoderma simplex Jayne		157,159		5sa					
Trogoderma sternale Jayne		157,159		5sa					

Table 1 (continued)
STATUS OF RESEARCH, DEVELOPMENT, AND USE OF PHEROMONES AND NONPHEROMONAL ATTRACTANTS FOR ARTHROPODS

Species		References	Pheromone identified	Nonpheromonal behavior modifier identified	Annihilative trapping or bait demonstrated	Chemical used for predictive monitoring	Chemical used for survey and detection	Chemical used to disrupt mating	In operational use: attractant or pheromone
Trogoderma variabile Ballion	warehouse beetle	159,160, 533	5s	2			5		Yes
Trypodendron domesticum (L.), see *Trypodendron lineatum* (Olivier)									
Trypodendron lineatum (Olivier)	striped ambrosia beetle*	8,10,613, 614	5s,ag	1	5	1	5		Yes
Typhaea stercorea (L.)	hairy fungus beetle*	730	1	2af			4		Yes
Xyloterus domestica (L.), see *Trypodendron lineatum* (Olivier)									
Xyloterus lineatus (Olivier), see *Trypodendron lineatum* (Olivier)									
Zabrus tenebrioedes (Goeze)			1	1					
Zygogramma exclamationis (F.)	sunflower beetle*		1	1					
DIPTERA									
Aedes aegypti (L.)	yellowfever mosquito*	768	1	5	1	1	1	0	No
Anastrepha ludens (Loew)	Mexican fruit fly*	779	2s	3					
Anastrepha suspensa (Loew)	Caribbean fruit fly	769—778	4s	4	1	1	3	1	No
Anopheles sp.	malaria vectors	768	1	2af					
Callantra aequalis (Coquillett)	a tropical fruit fly	557		5sa			4		
Callantra spp.		558		5sa			4		

Ceratitis capitata (Wiedemann)	Mediterranean fruit fly*	53,466, 481, 780—782	4s,ag	5sa	5	2	5	2	Yes (sa)
Chloropidae	eye gnats	457,458	1	5sa					
Cochliomyia hominivorax (Coquerel)	screwworm*	661,783, 784	4c,2op	5af,4oa	5	1	5	0	Yes (a)
Cochliomyia macellaria (F.)	secondary screwworm*	785		4			3		
Culex tarsalis Coquillett	mosquito, encephalitis vector	871	2op,c	1					
Culicoides melleus (Coquillett)		639		5sa					
Dacus absonifacies (May)		557	1	5sa					
Dacus bancrofti (Tryon)		557	1	5sa					
Dacus bryoniae (Tryon)		557	1	5sa					
Dacus cacuminatus (Hering)		557,625	1	5sa					
Dacus choristus (May)		557	1	5sa					
Dacus correctus Bezzi	mango fruit fly	626	1	5sa			4		
Dacus cucurbitae Coquillett	melon fly*	560	4s	5sa	5	2	3s,5np	1	Yes (sa)
Dacus dorsalis Hendel	oriental fruit fly*	627,786	4s	5sa	5	2	3s,5np	1	Yes (sa)
Dacus endiandrae (Perkins & May)		557		5sa			5		
Dacus mayi Hardy		557		5sa			5		
Dacus neohumeralis Hardy		557		5sa			5		
Dacus newmani (Perkins)		557		5sa			5		
Dacus ochrosiae Malloch		560	1	5sa			4		
Dacus oleae (Gmelin)	olive fruit fly*	52,637	5s				5		
Dacus piceus Drew		628		5sa					
Dacus signatifrons (May)		557		5sa					
Dacus spp.		558		5sa					
Dacus tryoni (Froggatt)	Queensland fruit fly	561		5sa					
Dacus umbrosus F.	jack fruit fly	558,559		5sa			4		
Dacus zonatus (Saunders)		629		5sa					
Fannia canicularis (L.)	little house fly*	447,644	5s,ag	1			5		No
Fannia femoralis (Stein)		648	5s,ag	1			5		No
Fannia pusio (Wiedemann)	chicken dung fly*	649	5s,ag	1			5		No

Table 1 (continued)
STATUS OF RESEARCH, DEVELOPMENT, AND USE OF PHEROMONES AND NONPHEROMONAL ATTRACTANTS FOR ARTHROPODS

Species		References	Pheromone identified	Nonpheromonal behavior modifier identified	Annihilative trapping or bait demonstrated	Chemical used for predictive monitoring	Chemical used for survey and detection	Chemical used to disrupt mating	In operational use: attractant or pheromone
Forcipomyia sp.		456		5			3		
Glossina morsitans morsitans Westwood	a tsetse fly	653	5s,ag				5		No
Glossina palpalis palpalis Robineau-Desvoidy	a tsetse fly	787	3s						
Haematobia irritans (L.)	horn fly*	788	5s	1					No
Hippelates spp.	eye gnats	789	1	5	3	1	5	0	No
Hylemya antiqua (Meigen)	onion maggot*		1	1					
Hylemya brassicae (Bouché)	cabbage maggot*	790	1	2af		4	4		Yes
Hylemya cinerella (Fallén)		785		4			3		
Hylemya platura (Meigen)	seedcorn maggot*		1	1					
Lucilia cuprina (Wiedemann)	Australian sheep blowfly	791	2op,2c						No
Lycoriella mali (Fitch)	mushroom fly	631,634	5s				5		No
Mayetiola destructor (Say)	Hessian fly*		1	1					
Medetera bistriata Parent		607		5k					
Melanagromyza phaseoli (Tryon)	bean fly		1	1					
Musca autumnalis De Geer	face fly*	646	5s	1	0	1	5	0	No
Musca domestica (L.)	house fly*	473,642, 643	5s	3	0	3	5	0	Yes
Neophyllomyza sp.		456		5					
Orseolia oryzae (Wood-Mason)	rice gall midge of China	792	3s	1					

Pachydiplosis oryzae (Wood-Mason), see *Orseolia oryzae* (Wood-Mason)									
Ravinia effrenata (Walker)	a sarcophagid	785		4			3		
Rhagoletis cerasi (L.)	European cherry fruit fly	793	3od						
Rhagoletis cingulata (Loew)	cherry maggot/ cherry fruit fly*		1	1					
Rhagoletis pomonella (Walsh)	apple maggot*	794,795	4od	5af	0	1	4np	0	Yes (a)
Sarcophaga bullata Parker	flesh fly	492	5s				5		No
Siphonella sp.		456	1	5			3		
Stomoxys calcitrans (L.)	stable fly*	645—647, 652	5s,c				5		No
Synthesiomyia nudiseta (Wulp)	a muscid	785		4			3		
Tetanops myopaeformis (Roder)	sugarbeet root maggot*		1	1					
Tipula simplex Doane	range crane fly*	662	2ag						
ISOPTERA									
Reticulitermes flavipes (Kollar)	eastern subterranean termite	796	5t						No
Reticulitermes lucifugus (Rossi)		797,799	5t	5af	4				No
Reticulitermes santonensis Feytaud	French termite	798,799	5t	5af	4				No
THYSANOPTERA									
Liothrips setinodis (Reuter)	European		1						
Scirtothrips citri (Moulton)	citrus thrips		1	1					
SIPHONAPTERA									
Ceratophyllus gallinae (Schrank)	European chicken flea*; hen flea	800	2c						

Table 1 (continued)
STATUS OF RESEARCH, DEVELOPMENT, AND USE OF PHEROMONES AND NONPHEROMONAL ATTRACTANTS FOR ARTHROPODS

Species		References	Pheromone identified	Nonpheromonal behavior modifier identified	Annihilative trapping or bait demonstrated	Chemical used for predictive monitoring	Chemical used for survey and detection	Chemical used to disrupt mating	In operational use: attractant or pheromone
Ctenocephalides canis (Curtis)	dog flea*		1	1					
Ctenocephalides felis (Bouché)	cat flea*		1	1					
Tunga penetrans L.	chigoe*		1	1					
HOMOPTERA									
Acyrthosiphon dirhorum (Walker)	rose grass aphid	801	5a						No
Acyrthosiphon pisum (Harris)	pea aphid*	801	5a						No
Acyrthosiphon solani (Kaltenbach)	foxglove aphid*	801	5a						
Adelges japonicus (Monzen)			1	1					
Adelges tardus (Dreyfus)			1	1					
Aonidiella aurantii (Maskell)	California red scale*	205—207	5s	1			5		Yes
Aonidiella citrina (Coquillett)	yellow scale*	208	5s	1			5		Yes
Aphis gossypii Glover	cotton/melon aphid*	801	5a						No
Chrysomphalus aonidum (L.)	Florida red scale*		1	1					
Coccus hesperidum L.	brown soft scale*		1	1					
Dialeurodes citri (Ashmead)	citrus whitefly*		1	1					
Empoasca fabae (Harris)	potato leafhopper*		1	1					

Empoasca maligna Walsh	apple leafhopper*		1	1		
Entylia bactriana Germar	a myrmecophilous treehopper	802,804	2a			
Lepidosaphes beckii (Newman)	purple scale*		1	1		
Lepidosaphes newsteadi (Sulc)	Newstead's scale		1	1		
Lepidosaphes tubulorum Ferris	an Asian scale		1	1		
Lepidosaphes ulmi (L.)	oystershell scale*		1	1		
Macrosiphum avenae (F.)	English grain aphid*	801	5a			No
Macrosiphum euphorbiae (Thomas)	potato aphid*	801	5a			No
Macrosiphum rosae (L.)	rose aphid*	801	5a			No
Metalopium dirhorum, see *Acyrthosiphon dirhorum* (Walker)						
Myzus persicae (Sulzer)	green peach aphid*	801,803	5a			No
Planococcus citri (Risso)	citrus mealybug*	476	5s	1	5	Yes
Pseudaulacaspis pentagona (Targioni-Tozzetti)	white peach scale*	450	5s	1	5	
Pseudococcus comstocki (Kuwana)	Comstock mealybug*	197	5s	1	5	Yes
Psylla pyricola Forster	pear psylla*		1	1		
Publia concava, see *Publilia concava* (Say)						
Publilia concava (Say)	a myrmecophilous treehopper	802,804	2a			
Quadraspidiotus perniciosus (Comstock)	San Jose scale*	449	5s		5	Yes
Rhopalosiphum maidis (Fitch)	corn leaf aphid*	801	5a			
Rhopalosiphum padi (L.)	oatbird-cherry aphid	801	5a			No
Saissetia oleae (Olivier)	black scale*		1	1		

Table 1 (continued)
STATUS OF RESEARCH, DEVELOPMENT, AND USE OF PHEROMONES AND NONPHEROMONAL ATTRACTANTS FOR ARTHROPODS

Species		References	Pheromone identified	Nonpheromonal behavior modifier identified	Annihilative trapping or bait demonstrated	Chemical used for predictive monitoring	Chemical used for survey and detection	Chemical used to disrupt mating	In operational use: attractant or pheromone
Schizaphis graminum (Rondani)	greenbug*	801	5a						
Therioaphis trifolii (Monell)	yellow clover aphid*	801	5a						No
Trialeurodes vaporariorum (Westwood)	greenhouse whitefly*		1	1					
Unaspis citri (Comstock)	citrus snow scale*		1	1					
Vanduzeea arquata (Say)	myrmecophilous treehopper	802	2a						
HEMIPTERA									
Acrosternum hilare (Say)	green stink bug*		1	1					
Chinoaspis salicis	a scale		1	1					
Cimex lectularius L.	bed bug*	805,806	2ag,5a						
Dysdercus intermedius Distant	a pyrrhocorid bug	807	31ag,2ag, 5a						
Eurygaster integriceps Puton	a shield bug	808	5s				5		
Irbisia brachycerus (Uhler)	a black grass bug		1	1					
Irbisia solani, see *Irbisia brachycerus* (Uhler)									
Labops spp.	(black) grass bugs		1						

Leptoglossus phyllopus (L.)	leaffooted bug*	809	4ag		
Lethocercus indicus (Lepeletier & Serville)	a giant water bug	664	2ap		
Lygus elisus Van Duzee	pale legume bug*		1	1	
Lygus hesperus Knight	a western lygus bug		1	1	
Lygus lineolaris (Palisot de Beauvois)	tarnished plant bug*		1	1	
Macrosteles laevis (Ribault)	a leafhopper		1	1	
Murgantia histrionica (Hahn)	harlequin bug*		1	1	
Musgraveia (= Rhoecocoris) sulciventris (Stål)	bronze orange bug	810	2ag,ap	1	
Nezara viridula (L.)	southern green stink bug*	811,812	2a,2s		
Oncopeltus fasciatus (Dallas)	large milkweed bug*	813	2s	1	
Pentatomidae	stinkbugs	802		4ag	
Podisus maculiventris (Say)	spined soldier bug*	809	5s		
Pseudatomoscelis seriatus (Reuter)	cotton fleahopper*		1	1	
Rhodnius prolixus Stål	an assassin bug	814	2s		
Targionia vitis (Signoret)			1	1	
HYMENOPTERA					
Acanthomyops claviger (Roger)	smaller yellow ant*	815	2s		
Ancistocerus antilope antilope (Panzer)	eumenid wasp	882	3ag	alkylpyrazines	
Anisopteromalus calandrae (Howard)	a pteromalid	816	2s	1	
Anthidium spp.	megachilid bees	817	2s	1	
Anthophora acervorum (L.)	anthophorid bee	817	2s		

Table 1 (continued)
STATUS OF RESEARCH, DEVELOPMENT, AND USE OF PHEROMONES AND NONPHEROMONAL ATTRACTANTS FOR ARTHROPODS

Species		References	Pheromone identified	Nonpheromonal behavior modifier identified	Annihilative trapping or bait demonstrated	Chemical used for predictive monitoring	Chemical used for survey and detection	Chemical used to disrupt mating	In operational use: attractant or pheromone
Anthophora edwardsii	anthophorid bee	885	3s						
Apanteles glomeratus (L.)	a braconid	818	3s	1					
Apanteles liparidis (Bouché)	a braconid	819	2s	1					
Apanteles medicaginis Muesebeck	a braconid	820	2s	1					
Apanteles melanoscelus (Ratzeburg)	a braconid	819	3s	1					
Aphytis maculicornis (Masi)	an aphelinid	822	2s	1					
Aphytis mytilaspidis (LeBaron)	an aphelinid	822	2s	1					
Aphytis spp.	aphelinids	821	2s	1					
Apis cerana F.	Oriental honey bee	486	5s						
Apis dorsata F.	giant honey bee	486	5s						
Apis florea F.	dwarf honey bee	486	5s						
Apis mellifera L.	honey bee*	487	5s,5a,5m, 3w,3b,3t				5		No
Argogorytes fargeii (Shuckard)	nyssonine wasp	884	3ag						
Argogorytes mystaceus (L.)	nyssonine wasp	884	3ag						
Atta cephalotes (L.)	leaf-cutter ant	823	5s						
Atta sexdens rubropilosa Forel	leaf cutting ant	823	5t						
Atta texana (Buckley)	Texas leafcutting ant*	823	5s,5t						
Bathyplectes curculionis (Thomson)	an ichneumonid	824	2s	1					
Biosteres oophilus (Fullaway)	a braconid	837	2s	1					
Bombus sp.	bumble bees	664	4ag,2s				5		No
Brachymeria intermedia (Nees)	a chalcid	825	3s,2ag	1					

Brachymeria lasus (Walker)	a chalcid	825	3s,2ag	1		
Bracon brevicornis Wesmael	a braconid	872	2s	1		
Bruchophagus roddi (Gussakovsky)	alfalfa seed chalcid*	826		4		
Campoletis sonorensis (Cameron)	an ichneumonid	827	3s	1		
Camponotus herculeanus (L.)	carpenter ant	828	2s			
Camponotus ligniperda (Latreille)	carpenter ant	828	2s			
Cardiochiles nigriceps Viereck	a braconid	650	3s	5k		
Cephus cinctus Norton	wheat stem sawfly*		1			
Colletes inequalis Say	a colletid bee	32	5ag		5	No
Colletes similis Robertson	a colletid bee	493	5ag		5	No
Colletes thoracicus Smith	a colletid bee	32	5ag,s		5	No
Colletes validus Cresson	a colletid bee	32	5ag		5	No
Diachasma alloeum (Muesebeck)	a braconid	836	2s	1		
Diprion frutetorum (F.), see *Gilpinia frutetorum* (F.)						
Diprion similis (Hartig)	introduced pine sawfly*	405,452	5s		5	Yes
Dolichovespula maculata (L.)	baldfaced hornet*	836	1			
Eumenes fruternus Say	a eumenid wasp	882	4ag	alkylpyrazines		
Exetastes cinctipes Ratzeburg	an ichneumonid	248	1	5k		
Formica aquilonia Yarrow	an ant in U.S.S.R.	829	5a			
Gilpinia frutetorum (F.)		405	5s		3	
Habrobracon brevicormis, see *Bracon brevicornis* Wesmael						
Iridomyrmex humilis (Mayr)	Argentine ant*	873	5ag			
Itoplectis conquisitor (Say)	a wasp	494	5s		5	No
Lasius alienus (Förster)	cornfield ant*		1			
Leptothorax acerrorum (F.)	formicid ant	886	2ag	(tandem pheromone)		
Leptothorax muscorum (Nylander)	formicid ant	886	2ag	(tandem pheromone)		
Leptothorax nylanderi (Förster)	formicid ant	886	2ag	(tandem pheromone)		
Macrocentrus ancylivorus Rohwer	a braconid	830	2s	1		

Table 1 (continued)
STATUS OF RESEARCH, DEVELOPMENT, AND USE OF PHEROMONES AND NONPHEROMONAL ATTRACTANTS FOR ARTHROPODS

Species		References	Pheromone identified	Nonpheromonal behavior modifier identified	Annihilative trapping or bait demonstrated	Chemical used for predictive monitoring	Chemical used for survey and detection	Chemical used to disrupt mating	In operational use: attractant or pheromone
Macrocentrus grandii (Goidanich)	a braconid	874	2s	1					
Melissodes denticulata Smith	anthophorid bee	881	3ag,3s,3a						
Melissodes desponsa Smith	anthophorid bee	881	3ag,3s,3a						
Melittobia chalybii Ashmead	a eulophid	831	2s	1					
Meteorus pulchricornis (Wesmael)	a braconid	832	3s	1					
Microplitis croceipes (Cresson)		651		5k					
Monobia quadridens (L.)	a eumenid wasp	882	4ag	alkylpyrazines					
Monomorium pharaonis (L.)	Pharaoh ant*	833—835	5t						
Myrmecia gulosa (F.)	an ant	887,888	5agr						
Neodiprion frutetorum (F.), see *Gilpinia frutetorum* (F.)									
Neodiprion lecontei (Fitch)	redheaded pine sawfly*	405—407	5s				5		Yes
Neodiprion pinetum (Norton)	white pine sawfly*	408	5s						
Neodiprion sertifer (Geoffroy)	European pine sawfly*	405	5s				5		
Neodiprion swainei Middleton	Swaine jack pine sawfly*	405,406	5s				5		Yes
Nysson spinosus (J. Förster)	nyssonine wasp	884	4ag						
Opius alloeus (Muesebeck), see *Diachasma alloeum* (Muesebeck)									
Opius oophilus, see Biosteres oophilus (Fullaway)									

Opius vandenboschi Fullaway	a braconid	837	2s	1		
Orgilus lepidus Muesebeck	a braconid	485	2	5k		
Parancistrocerus fulvipes rufovestris (Bohart)		882	4ag	alkylpyrazines		
Phaeogenes invisor (Thunberg)	an ichneumonid	875	3s	1		
Philanthus triangulum (F.)	a philanthine wasp	884	4ag			
Pithitis smaradula (F.)	ceratinine bee	880	3a			
Praon exoletum palitans Muesebeck	a braconid	838	2s	1		
Praon palitans, see *Praon exoletum palitans* Muesebeck						
Pseudodynerus quadrisectus (Say)	a eumenid wasp	882	4ag	alkylpyrazines		
Sceliphron caementarium (Drury)	a solitary sphecid	883	4ag	geranyl acetate and 2-decen-1-ol		
Sirex notilio F.	a horntail		1			
Solenopsis invicta Buren	red imported fire ant*	839,840	3b,t			
Stenodynerus fulvipes (Saussure), see *Parancistrocerus fulvipes rufovestris* (Bohart)						
Sterictiphora geminata (Gmelin)		204	1	5sa	3	
Telenomus remus Nixon	a scelionid	841	2s	1		
Tetramorium caespitum (L.)	pavement ant*	876	5ag			
Tetrastichus incertus (Ratzeburg)	a eulophid	842	2s	1		
Trichogramma achaeae Nagaraja & Nagarkatti		640,641		5k		
Trichogramma evanescens Westwood		638		5k		
Trichogramma pretiosum Riley		641		5k		
Trissolcus grandis Thomson	a scelionid	843,844	2sr,c			
Vespa crabro germana Christ	European hornet*	879	2agr,2te,2s			
Vespa orientalis L.	oriental hornet	877	4s		5	No
Vespula atropilosa (Sladen)	a yellow jacket wasp	461,462	1	5af	4	

Table 1 (continued)
STATUS OF RESEARCH, DEVELOPMENT, AND USE OF PHEROMONES AND NONPHEROMONAL ATTRACTANTS FOR ARTHROPODS

Species		References	Pheromone identified	Nonpheromonal behavior modifier identified	Annihilative trapping or bait demonstrated	Chemical used for predictive monitoring	Chemical used for survey and detection	Chemical used to disrupt mating	In operational use: attractant or pheromone
Vespula maculata (L.), see *Dolichovespula maculata* (L.)									
Vespula pensylvanica (Saussure)	western yellowjacket*	461,462	1	5af	4	1	5	0	Yes
Vespula spp.	yellow jacket	459,460, 463	1	5af		4			
Vespula sulphurea (Saussure)	a yellow jacket wasp	461,462	1	5af			4		
Vespula vulgaris L.	a yellow jacket wasp		1						
Xylocopa hirsutissima Maidl	a carpenter bee	845	5te						
Xylocopa virginica (L.)	carpenter bee*	878	5s				5		No
Zadiprion vallicola (Rohwer)	a conifer sawfly		1						
NEUROPTERA									
Chrysopa albolineata Killington		559		5af					
Chrysopa basalis Walker	green lacewing	624		5af					
Chrysopa carnea Stephens	common green lacewing*	654,846	1	5af,k					
Chrysopa sp.		559		5af					
ORTHOPTERA									
Blaberus cranifer Burmeister		633	5ag						

Blattella germanica (L.)	German cockroach*	479, 551—556	5s,2ag	5af		5	Yes (af)
Byrsotria fumigata (Guérin-Méneville)	a cockroach	847	2s				
Ectobius pallidus (Olivier)	spotted Mediterranean cockroach*	848	2s				
Eurycotis floridana (Walker)	Florida wood roach	849	2s				
Leucophaea maderae (F.)	Madeira cockroach*	850,851	2s				
Locusta migratoria migratioides (Reiche & Fairmaire)	migratory locust	802,852	5pt				No
Nauphoeta cinerea (Olivier)	cinereous cockroach*	848	2s	pheromone is ''seducin''			
Periplaneta americana (L.)	American cockroach*	572—574, 655	5s,2ag, 2agr	5af	5np	5	Yes (af)
Schistocerca gregaria (Forskål)	desert locust	853	2op,c,3m				
Supella longipalpa (F.)	brownbanded cockroach*	854—857	2s				
TRICHOPTERA							
Sericostoma personatum (Spence)		858	2s				
MECOPTERA							
Harpobittacus australis (Klug)	a scorpion fly	664,859, 860	2s				
Harpobittacus nigriceps (Selys)	a scorpion fly	664,859, 860	2s				
Harpobittacus similis Esben-Petersen	a scorpion fly	664,859, 860	2s				
Harpobittacus tillyardi Esben-Petersen	a scorpion fly	664,859, 860	2s				

Table 1 (continued)
STATUS OF RESEARCH, DEVELOPMENT, AND USE OF PHEROMONES AND NONPHEROMONAL ATTRACTANTS FOR ARTHROPODS

Species		References	Pheromone identified	Nonpheromonal behavior modifier identified	Annihilative trapping or bait demonstrated	Chemical used for predictive monitoring	Chemical used for survey and detection	Chemical used to disrupt mating	In operational use: attractant or pheromone
ACARI									
Amblyomma americanum (L.)	lone star tick*	538,619, 620	5s				5		Yes
Amblyomma cajennense (F.)	Cayenne tick*	621		5sa					
Amblyomma hebraeum Koch			1	1					
Amblyomma maculatum Koch	Gulf Coast tick*	620	5s	4np	1	1	5	1	Yes
Amblyomma variegatum (F.)		538	5s						
Argas arboreus Kaiser	soft-bodied ticks	861	2ag						
Argas brumpti Neumann		861	2ag						
Argas cooleyi Kohls & Hoogstraal		861	3ag						
Argas persicus (Oken)	fowl tick*	861	2ag						
Boophilus annulatus (Say)	cattle tick*		1						
Boophilus microplus (Canestrini)	southern cattle tick*		1						
Dermacentor andersoni Stiles	Rocky Mountain wood tick*	618,622	5s				5		Yes
Dermacentor variabilis (Say)	American dog tick*	618	5s				5		Yes
Eutetranychus orientalis (Klein)	Oriental red mite		1	1					
Hyalomma truncatum (Koch)		538	5s						
Ixodes persulcatus Schulze		862	1	2sa					

Oligonychus pratensis (Banks)	Banks grass mite*		1	1			
Ornithodoros boliviensis Kohls & Clifford	argasid tick	861	2ag				
Ornithodoros capensis Neuman		861	2ag				
Ornithodoros concanensis Cooley & Kohls		861	3ag				
Ornithodoros coriaceus Koch		861	2ag				
Ornithodoros monbata (Murray)		861	2ag				
Ornithodoros savigryi (Audouin)		861	2ag				
Ornithodoros turicata (Dugés)	relapsing fever tick*	861	1				
Ornithonyssus sylviarum (Canestrini & Fanzago)	northern fowl mite*		1	1			
Petrobia latents (Müller)	brown wheat mite*		1	1			
Psoroptes spp.	scab mites		1	1			
Rhipicephalus appendiculatus Neumann		538	1	5sa	4		
Rhipicephalus compositus Neumann		538	5s	1			
Rhipicephalus pulchellus Neumann		538	5s		5		
Rhipicephalus sanguineus (Latreille)	brown dog tick*	623	5s		5		
Rhipicephalus simus Koch		538	5s		5		
Sarcoptes scabiei (De Geer)	itch mite*		1	1			
Tetranychus turkestani (Ugarov & Nikolski)	strawberry spider mite*		1	1			
Tetranychus urticae Koch	twospotted spider mite*	31,97	5s	3af	5	2	Yes
Trombicula spp.	chiggers		1	1	5		
Tyrophagus putrescentiae (Schrank)	mold mite*	863	1	4af	5		

REFERENCES

1. **Inscoe, M. N. and Beroza, M.,** Insect behavior chemicals active in field trials, in *Pest Management With Insect Sex Attractants and Other Behavior-Controlling Chemicals,* Beroza, M., Ed., ACS Symp. Ser. 23, American Chemical Society, Washington, D.C., 1976, 145.
2. **Rothschild, G. H. L.,** Problems in defining synergists and inhibitors of the oriental fruit moth pheromone by field experimentation, *Entomol. Expl. Appl.,* 17, 294, 1974.
3. **Baker, T. C. and Cardé, R. T.,** Analysis of pheromone-mediated behaviors in male *Grapholitha molesta,* the Oriental fruit moth (Lepidoptera: Tortricidae), *Environ. Entomol.,* 8, 956, 1979.
4. **Metcalf, R. L.,** Plants, chemicals, and insects: some aspects of coevolution, *Bull. Entomol. Soc. Am.,* 25, 30, 1979.
5. **Pitman, G. B., Hedden, R. L., and Gara, R. I.,** Synergistic effects of ethyl alcohol on the aggregation of *Dendroctonus pseudotsugae* (Coleoptera, Scolytidae) in response to pheromones, *Z. Angew. Entomol.,* 78, 203, 1975.
6. **Cade, S. C., Hrutfiord, B. F., and Gara, R. I.,** Identification of a primary attractant for *Gnathotrichus sulcatus* isolated from western hemlock logs, *J. Econ. Entomol.,* 63, 1014, 1970.
7. **Renwick, J. A. A., Vité, J. P., and Billings, R. F.,** Aggregation pheromones in the ambrosia beetle *Platypus flavicornis, Naturwissenschaften,* 64, 226, 1977.
8. **Klimetzek, D., Vité, J. P., and Mori, K.,** Zur Wirkung und Formulierung des Populationslockstoffes des Nutzholzborkenkafers *Trypodendron (= Xyloterus) lineatum, Z. Angew. Entomol.,* 89, 57, 1980.
9. **Vité, J. P. and Bakke, A.,** Synergism between chemical and physical stimuli in host colonization by an ambrosia beetle, *Naturwissenschaften,* 66, 528, 1979.
10. **Borden, J. H., Handley, J. R., Johnston, B. D., MacConnell, J. G., Silverstein, R. M., Slessor, K. N., Swigar, A. A., and Wong, D. T. W.,** Synthesis and field testing of 4,6,6-lineatin, the aggregation pheromone of *Trypodendron lineatum* (Coleoptera: Scolytidae), *J. Chem. Ecol.,* 5, 681, 1979.
11. **Brand, J. M. and Barras, S. J.,** The major volatile constituents of a Basidiomycete associated with the southern pine beetle, *Lloydia (J. Natural Prod.),* 40, 398, 1977.
12. **Byrne, K. J., Swigar, A. A., Silverstein, R. M., Borden, J. J., and Stokkink, E.,** Sulcatol: population aggregation pheromone in the scolytid beetle, *Gnathotrichus sulcatus, J. Insect Physiol.,* 20, 1895, 1974.
13. **Stoakley, J. T., Bakke, A., Renwick, J. A. A., and Vité, J. P.,** The aggregation pheromone system of the larch bark beetle, *Ips cembrae (Heer), Z. Angew. Entomol.,* 86, 174, 1978.
14. **Bakke, A.,** Aggregation pheromone componentsof the bark beetle *Ips acuminatus, Okikos,* 31, 184, 1978.
15. **Krawielitzki, S., Klimetzek, D., Bakke, A., Vité, J. P., and Mori, K.,** Field and laboratory response of *Ips typographus* to optically pure pheromonal components, *Z. Angew. Entomol.,* 83, 300, 1977.
16. **Bakke, A., Frøyen, P., and Skattelbøl, L.,** Field response to a new pheromonal compound isolated from *Ips typographus, Naturwissenschaften,* 64, 98, 1977.
17. **Bakke, A. and Kvamme, T.,** Kairomone response by the predators *Thanasimus formicarius* and *Thanasimus rufipes* to the synthetic pheromone of *Ips typographus, Norw. J. Entomol.,* 25, 41, 1978.
18. **Francke, W., Heemann, V., Gerken, B., Renwick, J. A. A., and Vité, J. P.,** 2-Ethyl-1,6-dioxaspiro[4.4]noname, principal aggregation pheromone of *Pityogenes chalcographus* (L.), *Naturwissenschaften,* 64, 590, 1977.
19. **Renwick, J. A. A. and Pittman, G. B.,** An attractant isolated from female Jeffrey pine beetles, *Dendroctonus jeffreyi, Environ. Entomol.,* 8, 40, 1979.
20. **Pearce, G. T., Gore, W. E., Silverstein, R. M., Peacock, J. W., Cuthbert, R. A., Lanier, G. N., and Simeone, J. B.,** Chemical attractants for the smaller European elm bark beetle *Scolytus multistriatus* (Coleoptera, Scolytidae), *J. Chem. Ecol.,* 1, 115, 1975.
21. **Vité, J. P., Lühl, R., Gerken, B., and Lanier, G. N.,** Ulmensplintkafer: Anlockversuche mit synthetischen Pheromonen in Oberrheintal, *Z. Pflanzenkr. Pflanzenschutz,* 83, 166, 1976.
22. **van Deventer, P. and Minks, A. K.,** Enkele waarnemingen over de schorskever, *Scolytus pygmaeus* (F.) (Coleoptera, Scolytidae), *Entomol. Ber. (Amsterdam),* 37, 138, 1977.
23. **Blight, M. M., Mellon, F. A., Wadhams, L. J., and Wenham, M. J.,** Volatiles associated with *Scolytus scolytus* beetles in English elm, *Experientia,* 33, 845, 1977.
24. **Blight, M. M., Wadhams, L. J., and Wenham, M. J.,** Volatiles associated with unmated *Scolytus scolytus* beetles on English elm: differential production of α-multistriatin and 4-methyl-3-heptanol, and their activities in a laboratory bioassay, *Insect Biochem.,* 8, 135, 1978.
25. **Blight, M. M., Wadhams, L. J., and Wenham, M. J.,** Chemically-mediated behavior in the large elm bark beetle, *Scolytus scolytus, Bull. Entomol. Soc. Am.,* 25, 122, 1979.
26. **Blight, M. M., Wadhams, L. J., and Wenham, M. J.,** The stereoisomeric composition of the 4-methyl-3-heptanol produced by *Scolytus scolytus* and the preparation and biological activity of the four synthetic stereoisomers, *Insect Biochem.,* 9, 525, 1979.

27. **Blight, M. M., Wadhams, L. J., Wenham, M. J., and King, C. J.**, Field attraction of *Scolytus scolytus* to the enantiomers of 4-methyl-3-heptanol, the major component of the aggregation pheromone, *Forestry*, 52, 83, 1979.
28. **Blight, M. M., King, C. J., Wadhams, L. J., and Wenham, M. J.**, Attraction of *Scolytus scolytus* (F.) to the components of Multilure, the aggregation pheromone of *S. multistriatus* (Marsham) (Coleoptera: Scolytidae), *Experientia*, 34, 1119, 1978.
29. **Borden, J. H. and McLean, J. A.**, Secondary attraction in *Gnathotrichus retusus* and cross-attraction of *G. sulcatus* (Coleoptera: Scolytidae), *J. Chem. Ecol.*, 5, 79, 1979.
30. **Johnston, B. D. and Slessor, K. N.**, Facile syntheses of the enantiomers of sulcatol, *Can. J. Chem.*, 57, 233, 1979.
31. **Regev, S. and Cone, W. W.**, Analyses of pharate female twospotted spider mites for nerolidol and geraniol: evaluation for sex attraction of males, *Environ. Entomol.*, 5, 133, 1976.
32. **Hefetz, A., Batra, S. W. T., and Blum, M. S.**, Linalool, neral and geranial in the mandibular glands of *Colletes bees* — an aggregation pheromone, *Experientia*, 35, 319, 1979.
33. **Young, J. C., Brownlee, R. G., Rodin, J. O., Hildebrand, D. N., Silverstein, R. M., Wood, D. L., Birch, M. C., and Browne, L. E.**, Identification of linalool produced by two species of bark beetles of the genus *Ips*, *J. Insect Physiol.*, 19, 1615, 1973.
34. **Birch, M. C. and Wood, D. L.**, Mutual inhibition of the attractant pheromone response by two species of *Ips* (Coleoptera: Scolytidae), *J. Chem. Ecol.*, 1, 101, 1975.
35. **Furniss, M. M. and Ladd Livingston, R.**, Inhibition of ipsenol of pine engraver attraction in northern Idaho, *Environ. Entomol.*, 8, 369, 1979.
36. **Young, J. C., Silverstein, R. M., and Birch, M. C.**, Aggregation pheromone of the beetle *Ips confusus*: isolation and identification, *J. Insect Physiol.*, 19, 2273, 1973.
37. **Birch, M. C., Tilden, P. E., Wood, D. L., Browne, L. E., Young, J. C., and Silverstein, R. M.**, Biological activity of compounds isolated from air condensates and frass of the bark beetle, *Ips confusus*, *J. Insect Physiol.*, 23, 1373, 1977.
38. **Vité, J. P. and Renwick, J. A. A.**, Population aggregating pheromone in the bark beetle, *Ips grandicollis*, *J. Insect Physiol.*, 17, 1699, 1971.
39. **Wood, D. L., Stark, R. W., Silverstein, R. M., and Rodin, J. O.**, Unique synergistic effects produced by the principal sex attractant compounds of *Ips confusus* (LeConte) (Coleptera: Scolytidae), *Nature (London)*, 215, 206, 1967.
40. **Silverstein, R. M., Rodin, J. O., and Wood, D. L.**, Sex attractants in frass produced by male *Ips confusus* in ponderosa pine, *Science*, 154, 509, 1966.
41. **Harring, C. M., Vité, J. P., and Hughes, P. R.**, "Ipsenol", der Populationslockstoff des krummzahnigen Tannenborkenkafers, *Naturwissenschaften*, 62, 488, 1975.
42. **Harring, C. M.**, Aggregation pheromones of the European fir engraver beetles *Pityokteines curvidens*, *P. spinidens*, and *P. vorontzovi* and the role of juvenile hormone in pheromone biosynthesis, *Z. Angew. Entomol.*, 85, 281, 1978.
43. **Vité, J. P., Ohloff, G., and Billings, R. F.**, Pheromonal chirality and integrity of aggregation response in southern species of the bark beetle *Ips* sp., *Nature (London)*, 272, 817, 1978.
44. **Vité, J. P., Hedden, R., and Mori, K.**, *Ips grandicollis:* field response to the optically pure pheromone, *Naturwissenschaften*, 63, 43, 1976.
45. **Birch, M. C., Light, D. M., and Mori, K.**, Selective inhibition of response of *Ips pini* to its pheromone by the (*S*)-(−)-enantiomer of ipsenol, *Nature (London)*, 270, 738, 1977.
46. **Harring, C. M. and Mori, K.**, *Pityokteines curvidens* Germ. (Coleoptera: Scolytidae) :aggregation in response to optically pure ipsenol, *Z. Angew. Entomol.*, 82, 327, 1976/77.
47. **Renwick, J. A. and Vité, J. P.**, Pheromones and host volatiles that govern aggregation of the six-spined engraver beetle, *Ips calligraphus*, *J. Insect Physiol.*, 18, 1215, 1972.
48. **Bakke, A.**, Aggregation pheromone in the bark beetle *Ips duplicatus* (Sahlberg), *Norw. J. Entomol.*, 22, 67, 1975.
49. **Vité, J. P., Bakke, A., and Hughes, P. R.**, Ein Populationslockstoff des zwölfzähnigen Kiefernborkenkafers *Ips sexdentatus*, *Naturwissenschaften*, 61, 365, 1974.
50. **Vité, J. P.**, Chemische und biotechnische Schadlingsbekampfung, *Forst. Holzwirt*, 31, 101, 1976.
51. **Light, D. M. and Birch, M. C.**, Inhibition of the attractive pheromone response in *Ips paraconfusus* by (*R*)-(−)-ipsdienol, *Naturwissenschaften*, 66, 159, 1979.
52. **Rossi, R., Carpita, A., and Vita, G.**, (*Z*)-6-Nonen-1-ol and related compounds as attractants of the olive fruit fly, *Dacus oleae* (Gmelin) (Diptera: Trypetidae), *Gazz. Chim. Ital.*, 108, 709, 1978.
53. **Jacobson, M., Ohinata, K., Chambers, D. L., Jones, W. A., and Fujimoto, M. S.**, Insect sex attractants. XIII. Isolation, identification, and synthesis of sex pheromones of the male Mediterranean fruit fly, *J. Med. Chem.*, 16, 248, 1973.
54. **Ohinata, K., Jacobson, M., Nakagawa, S., Fujimoto, M., and Higa, H.**, Mediterranean fruit fly: laboratory and field evaluations of synthetic sex pheromones, *J. Environ. Sci. Health*, 12A, 67, 1977.

55. **Roelofs, W., Kochansky, J., Anthon, E., Rice, R., and Cardé, R.,** Sex pheromone of the peach twig borer moth *(Anarsia lineatella), Environ. Entomol.,* 4, 580, 1975.
56. **Formigoni, A., Patuzzo, L., and Leri, G.,** Esperienze quadriennali di lotta contro la tignola orientale del pesco (*Grapholitha molesta* (Busck) e l'anarsia *Anarsia lineatella* Zell.) con l'ausilio di trappole a feromoni (Orfamone ed Anamone), *Inf. Fitopatol.,* 27, 77, 1977.
57. **Hrdy, I., Marek, J., and Krampl, F.,** Sexual pheromone activity of 8-dodecenyl and 11-tetradecenyl acetates for males of several lepidopteran species in field trials, *Acta Entomol. Bohemoslov.,* 76, 65, 1979.
58. **Baker, T. C., Cardé, R. T., and Roelofs, W. L.,** Behavioral responses of male *Argyrotaenia velutinana* (Lepidoptera: Tortricidae) to components of its sex pheromone, *J. Chem. Ecol.,* 2, 333, 1976.
59. **Cardé, R. T., Baker, T. C., and Roelofs, W. L.,** Behavioural role of individual components of a multichemical attractant system in the Oriental fruit moth, *Nature,* 253, 348, 1975.
59a. **Cardé, R. T., Baker, T. C., and Roelofs, W. L.,** Ethological function of components of a sex attractant system for Oriental fruit moth males, *Grapholitha molesta* (Lepidoptera: Tortricidae), *J. Chem. Ecol.,* 1, 475, 1975.
60. **Cardé, A. M., Baker, T. C., and Cardé, R. T.,** Identification of a four-component sex pheromone of the female Oriental fruit moth, *Grapholitha molesta* (Lepidoptera: Tortricidae), *J. Chem. Ecol.,* 5, 423, 1979.
61. **Kochansky, J. P., Roelofs, W. L., and Sivapalan, P.,** Sex pheromone of the tea tortrix moth (*Homona coffearia* Neitner), *J. Chem. Ecol.,* 4, 623, 1978.
62. **Bierl, B. A., Beroza, M., Staten, R. T., Sonnet, P. E., and Adler, V. E.,** The pink bollworm sex attractant, *J. Econ. Entomol.,* 67, 211, 1974.
63. **Steck, W., Underhill, E. W., Chisholm, M. D., Bailey, B. K., Loeffler, J., and Devlin, C. G.,** Sex attractants for males of 12 moth species found in western Canada, *Can. Entomol.,* 109, 157, 1977.
64. **Struble, D. L. and Swailes, G. E.,** A sex attractant for adult males of the pale western cutworm, *Agrotis orthogonia* (Lepidoptera: Noctuidae), *Can. Entomol.,* 110, 769, 1978.
65. **Steck, W. F., Underhill, E. W., Chisholm, M. D., and Byers, J. R.,** Sex attractants for *Agrotis venerabilis* and *Euxoa albipennis* based on (*Z*)-5-decenyl acetate and (*Z*)-7-dodecenyl acetate, *Environ. Entomol.,* 8, 1126, 1979.
66. **Steck, W. F., Underhill, E. W., Chisholm, M. D., and Gerber, H. S.,** Sex attractant for male alfalfa looper moths, *Autographa californica, Environ. Entomol.,* 8, 373, 1979.
67. **Roelofs, W. L. and Comeau, A.,** Sex attractants in Lepidoptera, in *Chemical Releasers in Insects,* Vol. 3, Tahori, A. S., Ed., Proc. IUPAC 2nd Int. Congr. Pesticide Chem., Tel Aviv, Israel, Gordon and Breach, New York, 1971, 91.
68. **Underhill, E. W., Chisholm, M. D., and Steck, W.,** Olefinic aldehydes as constituents of sex attractants for noctuid moths, *Environ. Entomol.,* 6, 333, 1977.
69. **Steck, W., Bailey, B. K., Chisholm, M. D., and Underhill, E. W.,** A sex attractant for males of the cutworm *Euxoa pleuritica* (Lepidoptera: Noctuidae), *Can. Entomol.,* 110, 775, 1978.
70. **Weatherston, J., Davidson, L. M., and Simonini, D.,** Attractants for several male forest Lepidoptera, *Can. Entomol.,* 106, 781, 1974.
71. **Weatherston, J., Hedlin, A. F., Ruth, D. S., MacDonald, L. M., Leznoff, C. C., and Fyles, T. M.,** Chemical and field studies on the sex pheromones of the cone and seed moths *Barbara colfaxiana* and *Laspeyresia youngana, Experientia,* 33, 723, 1977.
72. **Dix, M. E., Tagestad, A., and Jacobson, M.,** Synthetic attractants for male Lepidoptera, *Proc. North Cent. Branch Entomol. Soc. Am.,* 31, 29, 1976.
73. **Dix, M. E., Tagestad, A. D., and Jacobson, M.,** Trapping male *Petrova metallica* (Busck) with synthetic attractants, *Proc. North Cent. Branch Entomol. Soc. Am.,* 32, 63, 1977.
74. **Ando, T., Yoshida, S., Tatsuki, S., and Takahashi, N.,** Sex attractants for male Lepidoptera, *Agric. Biol. Chem.,* 41, 1485, 1977.
75. **McLaughlin, J. R., Mitchell, E. R., Chambers, D. L., and Tumlinson, J. H.,** Perception of *Z*-7-dodecen-1-ol and modification of the sex pheromone response of male loopers, *Environ. Entomol.,* 3, 677, 1974.
76. **Tumlinson, J. H., Mitchell, E. R., Browner, S. M., Mayer, M. S., Green, N., Hines, R., and Lindquist, D. A.,** *cis*-7-Dodecen-1-ol, a potent inhibitor of the cabbage looper sex pheromone, *Environ. Entomol.,* 1, 354, 1972.
77. **Arn, H., Schwarz, C., Limacher, H., and Mani, E.,** Sex attractant inhibitors of the codling moth *Laspeyresia pomonella* L., *Experientia,* 30, 1142, 1974.
78. **Beroza, M., Gentry, C. R., Blythe, J. L., and Muschik, G. M.,** Isomer content and other factors influencing captures of Oriental fruit moth by synthetic pheromone traps, *J. Econ. Entomol.,* 66, 1307, 1973.
79. **Biwer, G., Descoins, C., and Gallois, M.,** Etude des constituants volatils presents dans la glande productrice de pheromone de la femelle vierge de *Grapholitha molesta* Busck, Lepidoptere Tortricidae (Olethreutinae), *C. R. Acad. Sci. Ser. D,* 288, 413, 1979.

80. **Roelofs, W. L., Hill, A. S., Berisford, C. W., and Godbee, J. F.**, Sex pheromone of the subtropical pine tip moth *Rhyacionia subtropica*, *Environ. Entomol.*, 8, 894, 1979.
81. **Saglio, P., Priesner, E., Descoins, C., and Gallois, M.**, Un attractif sexuel de synthese pour la Pyrale de la Vigne *Sparganothis pilleriana* (Schiffermuller), *C. R. Acad. Sci. Ser. D*, 284, 2007, 1977; also in *Les Pheromones Sexuelles des Lepidopteres, Compte Rendu Reunion Oct. 1976*, Publ. INRA, Centre de Recherches de Bordeaux 1976, 109.
82. **Arn, H., Roehrich, R., Descoins, C., and Rauscher, S.**, Performance of five sex attractant formulations for the grape moth, *Eupoecilia ambiguella* Hb., in European vineyards, *Mitt. Schweiz. Entomol. Ges.*, 52, 45, 1979.
83. **Saglio, P., Descoins, C., Gallois, M., Jaouen, D., and Mercier, J.**, Etude de la pheromone sexuelle de la Cochylis de la vigne *Eupoecilia (Clysia) ambiguella* Hb. Lepidoptere Tortricoidea Cochylidae, *Ann. Zool. Ecol. Anim.*, 9, 553, 1977; also in *Les Pheromones Sexuelles des Lepidopteres, Compte Rendu Reunion Oct. 1976*, Publ. INRA, Centre de Recherches de Bordeaux 1976, 118.
84. **Vu, M. H., Ando, T., Yoshida, S., Takahashi, N., Tatsuki, S., Katagiri, K., Yamane, A., and Ikeda, T.**, 5,7-Dodecadien-1-ol: candidate for the sex pheromone of the pine moth, *Agric. Biol. Chem.*, 43, 1615, 1979.
85. **Vu, M. H., Ando, T., Takahashi, N., Tatsuki, S., Yamane, A., Ikeda, T., and Yamazaki, S.**, Identification of the female sex pheromone of the pine moth, *Agric. Biol. Chem.*, 44, 231, 1980.
86. **Chambon, J. P. and d'Aguilar, J.**, Remarques sur la specificite des quelques pheromones sexuelles de syntheses: analyse faunistique des Tortricides en vergers, *Ann. Zool. Ecol. Anim.*, 6, 423, 1974.
87. **Lewis, T., Wall, C., Macaulay, E. D. M., and Greenway, A. R.**, The behavioural basis of a pheromone monitoring system for pea moth, *Cydia nigricana*, *Ann. Appl. Biol.*, 80, 257, 1975.
88. **Roelofs, W., Comeau, A., Hill, A., and Milicevic, G.**, Sex attractant of the codling moth: characterization with electroantennogram technique, *Science*, 174, 297, 1971.
89. **Beroza, M., Bierl, B. A., and Moffitt, H. R.**, Sex pheromones: (*E,E*)-8,10-dodecadien-1-ol in the codling moth, *Science*, 183, 89, 1974.
90. **Cardé, R. T., Baker, T. C., and Castrovillo, P. J.**, Disruption of sexual communication in *Laspeyresia pomonella* (codling moth), *Grapholitha molesta* (Oriental fruit moth) and *G. prunivora* (lesser appleworm) with hollow fiber attractant sources, *Entomol. Exp. Appl.*, 22, 280, 1977.
91. **Audemard, H., Beauvais, F., and Descoins, C.**, La lutte contre le Carpocapse (*Laspeyresia pomonella* L.) avec une pheromone sexuelle de synthese par le methode de confusion des males: premier essai en verger commercial de pommiers, *Rev. Zool. Agric. Pathol. Veg.*, 76, 15, 1977.
92. **Moffitt, H. R.**, Control of the codling moth through mating disruption with the sex pheromone, *Mitt. Biol. Bundesdnst. Land Forstwirtsch. Berlin Pahlem*, 180, 42, 1978.
93. **Fluri, P., Mani, E., Wildbolz, T., and Arn, H.**, Untersuchungen über das Paarungsverhalten des Apfelwicklers (*Laspeyresia pomonella* L.) and über den Einfluss von künstlichem Sexuallockstoff auf die Kopulationshäufigkeit, *Mitt. Schweiz, Entomol. Ges.*, 47, 253, 1974.
94. **Charmillot, P. J., Rosset, S., and Rhyn, D.**, Comportement sexuel du carpocapse (*Laspeyresia pomonella* L.): influence de l'attractif synthetique, *Mitt. Schweiz. Entomol. Ges.*, 49, 143, 1976.
95. **Charmillot, P. J.**, Reduction des captures de carpocapse (*Laspeyresia pomonella* L.) par inhibition des males due a la diffusion d'attractif sexuel synthetique en verger, *Mitt. Schweiz. Entomol. Ges.*, 51, 5, 1978.
96. **Brakefield, J. J. and Bartlett, D. H.**, The use of pheromones in trapping moth pests in U.K. top fruit orchards, *Proc. 8th British Insecticide Fungicide Conf.*, Vol. 1, British Crop Protection Council, London, 1975, 277.
97. **Regev, S. and Cone, W. W.**, Evidence of farnesol as a male sex attractant of the twospotted spider mite, *Tetranychus urticae* Koch (Acarina, Tetranychidae), *Environ. Entomol.*, 4, 307, 1975.
98. **Klun, J. A., Neal, J. W., Jr., and Bierl-Leonhardt, B. A.**, unpublished results.
99. **Underhill, E. W., Arthur, A. P., Chisholm, M. D., and Steck, W. F.**, Sex pheromone components of the sunflower moth *Homeosoma electellum*, (*Z*)-9, (*E*)-12-tetradecadienol and (*Z*)-9-tetradecenol, *Environ. Entomol.*, 8, 740, 1979.
100. **Lee, K. R., Park, M. -t., Koh, J. -B., Lee, J. -J., and Ji, Y. -D.**, Studies on the sex pheromones during the metamorphsis of the fall webworm *Hyphantria cunea* Drury, *Korean J. Entomol.*, 8, 45, 1978; *Chemoreception Abstr.*, 8(1), 82, 1980.
101. **Struble, D. L. and Swailes, G. E.**, A sex attractant for the adult males of the army cutworm, *Euxoa auxiliaris*: a mixture of *Z*-5-tetradecen-1-yl acetate and *E*-7-tetradecen-1-yl acetate, *Environ. Entomol.*, 6, 719, 1977.
102. **Steck, W. F., Bailey, B. K., Underhill, E. W., and Chisholm, M. D.**, A sex attractant for the Great Dart, *Eurois occulta*: a mixture of (*Z*)-9-tetradecen-1-ol acetate and (*Z*)-11-hexadecen-1-ol acetate, *Environ. Entomol.*, 5, 523, 1976.
103. **Klun, J. A., Lynch, R. E., Schwarz, M., and Bierl-Leonhardt, B. A.**, Sex pheromone of the lesser cornstalk borer, *Elasmopalpus lignosellus*, in preparation.

104. **Struble, D. L., Swailes, G. E., Steck, W. F., Underhill, E. W., and Chisholm, M. D.**, A sex attractant for *Leucania commoides*: a mixture of *Z*-9-tetradecen-1-yl acetate, *Z*-11-hexadecen-1-yl acetate, and *Z*-11-hexadecen-1-ol, *Can. Entomol.*, 109, 1393, 1977.
105. **Arsura, E., Capizzi, A., Piccardi, P., and Spinelli, P.**, (*Z*)-9-Tetradecen-1-ol and (*Z*)-9-tetradecenyl acetate: a potent attractant system for male *Sesamia cretica* Led. (Lep., Noctuidae), *Experientia*, 33, 1423, 1977.
106. **Voerman, S.**, Synthesis, purification, and field screening of potential insect sex pheromones, in *Chemical Ecology: Odour Communication in Animals*, Ritter, F. J., Ed., Elsevier/North-Holland, Amsterdam, 1979, 353.
107. **Ando, T., Yoshida, S., Tatsuki, S., and Takahashi, N.**, Lepidopterous sex attractants, *Agric. Biol. Chem.*, 39, 1163, 1975.
108. **Steck, W., Underhill, E. W., and Chisholm, M. D.**, Attraction and inhibition in moth species responding to sex-attractant lures containing *Z*-11-hexadecen-1-yl acetate, *J. Chem. Ecol.*, 3, 603, 1977.
109. **Struble, D. L., Swailes, G. E., Steck, W. F., Underhill, E. W., and Chisholm, M. D.**, A sex attractant for the adult males of variegated cutworm, *Peridroma saucia, Environ. Entomol.*, 5, 988, 1976.
110. **Ando, R., Kuroko, H., Nakagaki, S., Saito, O., Oku, T., and Takahashi, N.**, Two-component sex attractants for male moths of the subfamily Tortricinae (Lepidoptera), *Agric. Biol. Chem.*, 42, 1081, 1978.
111. **Roelofs, W., Hill, A., Cardé, R., Tette, J., Madsen, H., and Vakenti, J.**, Sex pheromones of the fruittree leafroller moth, *Archips argyrospilus, Environ. Entomol.*, 3, 747, 1974.
112. **Roelofs, W., Hill, A., Cardé, A., Cardé, R., Madsen, H., and Vakenti, J.**, Sex pheromone of the European leafroller, *Archips rosanus, Environ. Entomol.*, 5, 362, 1976.
113. **AliNiazee, M. T.**, Field studies on sex pheromone trapping of the filbert leafroller, *Archips rosanus*, in Oregon, *Ann. Entomol. Soc. Am.*, 69, 820, 1976.
114. **Madsen, H. F. and Peters, F. E.**, Pest management: monitoring populations of *Archips argyrospilus* and *Archips rosanus* (Lepidoptera: Tortricidae) with sex pheromone traps, *Can. Entomol.*, 108, 1281, 1976.
115. **Weatherston, J., Grant, G. G., MacDonald, L. M., Frech, D., Werner, R. A., Leznoff, C. C., and Fyles, T. M.**, Attraction of various tortricine moths to blends containing *cis*-11-tetracedenal, *J. Chem. Ecol.*, 4, 543, 1978.
116. **Comeau, A. and Roelofs, W. L.**, Sex attraction specificity in the Tortricidae, *Entomol. Exp. Appl.*, 16, 191, 1973.
117. **Sanders, C. J.**, Disruption of mating behavior of the spruce budworm in air permeated by synthetic sex attractant, *Can. For. Serv. Bi-Mon. Res. Notes*, 31, 10, 1975.
118. **Reissig, W. H., Novak, M., and Roelofs, W. L.**, Orientation disruption of *Argyrotaenia velutinana* and *Choristoneura rosaceana* male moths, *Environ. Entomol.*, 7, 631, 1978.
119. **Hill, A. S. and Roelofs, W. L.**, Sex pheromone components of the obliquebanded leafroller moth, *Choristoneura rosaceana, J. Chem. Ecol.*, 5, 3, 1979.
120. **Weires, R. W.**, Evidence of an attractant for the click beetle, *Melanotus depressus*, in the pheromones of two *Platynota* spp. (Lepidopter Tortricidae), *Environ. Entomol.*, 5, 920, 1976.
121. **Roelofs, W. L. and Comeau, A.**, Lepidopterous sex attractants discovered by field screening tests, *J. Econ. Entomol.*, 63, 969, 1970.
122. **Frerot, B., Gallois, M., and Einhorn, J.**, La pheromone sexuelle produite par la femelle vierge de *Pandemis heparana* (Den. et Schiff.) (Lepidoptere Tortricidae, Tortricinae), *C. R. Acad. Sci. Ser. D*, 288, 1611, 1979.
123. **Hill, A. S., Cardé, R. T., Bode, W. M., and Roelofs, W. L.**, Sex pheromone components of the variegated leafroller moth, *Platynota flavedana, J. Chem. Ecol.*, 3, 369, 1977.
124. **Hill, A., Cardé, R., Comeau, A., Bode, W., and Roelofs, W.**, Sex pheromones of the tufted apple bud moth *(Platynota idaeusalis), Environ. Entomol.*, 3, 249, 1974.
125. **Hill, A. S. and Roelofs, W. L.**, Sex pheromone components of the omnivorous leafroller moth, *Platynota stultana, J. Chem. Ecol.*, 1, 91, 1975.
126. **Baker, J. L., Hill, A. S., Cardé, R. T., Kurokawa, A., and Roelofs, W. L.**, Sex pheromone field trapping of the omnivorous leafroller, *Platynota stultana, Environ. Entomol.*, 4, 90, 1975.
127. **Hall, D. R. and Read, J. S.**, Sex attractants for two zygaenid moths, *J. Entomol. Soc. South Afr.*, 42, 115, 1979.
128. **Benz, G. and von Salis, G.**, Use of synthetic sex attractant of larch bud moth *Zeiraphera diniana* (Gn.) in monitoring traps under different conditions, and antagonistic action of *cis*-isomere, *Experientia*, 29, 729, 1973.
129. **Sanders, C. J., Bartell, R. J., and Roelofs, W. L.**, Field trials for synergism and inhibition of *trans*-11-tetradecenal, sex pheromone of the eastern spruce budworm, *Bi-Mon. Res. Notes Can. For. Serv.*, 28, 9, 1972.

130. **Weatherston, J. and MacLean, W.**, The occurrence of (*E*)-11-tetradecen-1-ol, a known sex attractant inhibitor, in the abdominal tips of virgin female eastern spruce budworm, *Choristoneura fumiferana* (Lepidoptera, Tortricidae), *Can. Entomol.*, 106, 281, 1974.

131. **Wiesner, C. J., Silk, P. J., Tan, S. -H., Palaniswamy, P., and Schmidt, J. O.**, Components of the sex pheromone gland of the eastern spruce budworm, *Choristoneura fumiferana* (Lepidoptera: Tortridicae), *Can. Entomol.*, 111, 1311, 1979.

132. **Sower, L. L., Vick, K. W., and Tumlinson, J. H.**, (*Z,E*)-9,12-Tetradecadien-1-ol: a chemical released by female *Plodia interpunctella* that inhibits the sex pheromone response of male *Cadra cautella*, *Environ. Entomol.*, 3, 120, 1974.

133. **Read, J. S. and Haines, C. P.**, The functions of the female sex pheromones of *Ephestia cautella* (Walker) (Lepidoptera, Phycitidae), *J. Stored Prod. Res.*, 12, 49, 1976.

134. **Read, J. S. and Beevor, P. S.**, Analytical studies on the sex pheromone complex of *Ephestia cautella* (Walker) (Lepidoptera: Phycitidae), *J. Stored Prod. Res.*, 12, 55, 1976.

135. **Sower, L. L. and Fish, J. C.**, Rate of release of the sex pheromone of the female Indian meal moth, *Environ. Entomol.*, 4, 168, 1975.

136. **Nesbitt, B. F., Beevor, P. S., Hall, D. R., and Lester, R.**, Female sex pheromone components of the cotton bollworm, *Heliothis armigera*, *J. Insect. Physiol.*, 25, 535, 1979.

136a. **Neumark, S., Teich, I., and Green, N.**, Hexalure, pink bollworm attractant: masking effect of *cis*-7-hexadecen-l-ol, *Environ. Lett.*, 5, 1, 1973.

137. **Bariola, L. A. and Green, N.**, Trapping pink bollworms with hexalure: inhibitory effects of chemical by-products, *Southwest. Entomol.*, 2, 207, 1977.

138. **Sato, R., Yaginuma, K., and Kumakura, M.**, Sex attractant of the stone leek miner, *Acrolepia alliella* Semenov et Kuznesov: (*Z*)-11-hexadecenyl acetate, (*Z*)-11-hexadecenal, and *(Z)-11-hexadecenol*, *Jpn. J. Appl. Entomol. Zool.*, 23, 115, 1979.

139. **Nesbitt, B. F., Beevor, P. S., Hall, D. R., Lester, R., Davies, J. C., and Seshu Reddy, K. V.**, Components of the sex pheromone of the female spotted stalk borer, *Chilo partellus* (Swinhoe) (Lepidoptera: Pyralidae): identification and preliminary field trials, *J. Chem. Ecol.*, 5, 153, 1979.

140. **Kanno, K., Tatsuki, S., Uchiumi, K., Kurihara, M., Fukami, J., Fujimoto, Y., and Tatsuno, T.**, Disruption of sex attraction in the rice stem borer moth, *Chilo suppressalis* Walker, with components of the sex pheromone and related chemicals, *Appl. Entomol. Zool.*, 13, 321, 1978.

141. **McDonough, L. M. and Kamm., J. A.**, Sex pheromone of the cranberry girdler, *Chrysoteuchia topiaria* (Zeller) (Lepidoptera: Pyralidae), *J. Chem. Ecol.*, 5, 211, 1979.

142. **Kamm, J. A. and McDonough, L. M.**, Field tests with the sex pheromone of the cranberry girdler, *Environ. Entomol.*, 8, 773, 1979.

143. **Steck, W., Underhill, E. W., Bailey, B. K., and Chisholm, M. D.**, A sex attractant for male moths of the glassy cutworm *Crymodes devastator* (Brace): a mixture of *Z*-11-hexadecen-l-yl acetate, *Z*-11-hedecenal and *Z*-7-dodencen-l-yl acetate, *Environ. Entomol.*, 6, 270, 1977.

144. **Klun, J. A., Plimmer, J. R., Bierl-Leonhardt, B. A., Sparks, A. N., and Chapman, O. L.**, Trace chemicals: the essence of sexual communication systems in *Heliothis* species, *Science*, 204, 1328, 1979.

145. **Klun, J. A., Bierl-Leonhardt, B. A., Plimmer, J. R., Sparks, A. N., Primiani, M., Chapman, O. L., Lepone, G., and Lee, G. H.**, Sex pheromone chemistry of the female tobacco budworm moth, *Heliothis virescens*, *J. Chem. Ecol.*, 6, 177, 1980.

146. **Hartsack, A. W., Jr., Lopez, J. D., Klun, J. A., Witz, J. A., Shaver, T. N., and Plimmer, J. R.**, New trap designs and pheromone bait formulatons for *Heliothis*, *Proc. 1980 Beltwide Cotton Prod. Res. Conf.*, National Cotton Council, Memphis, 1980, 132.

147. **Takahashi, S., Kawaradani, M., Sato, Y., and Sakai, M.**, Sex pheromone components of *Leucania separata* Wlk. and *Leucania loreyi* DuPonchel, *Jpn. J. Appl. Entomol. Zool.*, 23, 78, 1979.

148. **Struble, D. L., Arn, H., Buser, H. R., Städler, E., and Freuler, J.**, Identification of 4 sex pheromone components isolated from calling females of *Mamestra brassicae*, *Z. Naturforsch.*, 35c, 45, 1980.

149. **Descoins, C., Priesner, E., Gallois, M., Arn, H., and Martin, G.**, Sur la secretion pheromonale des femelles vierges de *Mamestra brassicae* L. et de *Mamestra oleracea* L. (Lepidopteres Noctuidae, Hadeninae), *C. R. Acad. Sci. Ser. D.*, 286, 77, 1978.

150. **Ando, T., Koshihara, T., Yamada, H., Vu, M. H., Takahashi, N., and Tamaki, Y.**, Electroantennogram activities of sex pheromone analogues and their synergistic effect on field attraction in the diamondback moth, *Appl. Entomol. Zool.*, 14, 362, 1979.

151. **McDonough, L. M., Kamm, J. A., and Bierl-Leonhardt, B. A.**, Sex pheromone of the armyworm, *Pseudaletia unipuncta* (Haworth) (Lepidoptera: Noctuidae), *J. Chem. Ecol.*, 6, 565, 1980.

152. **Struble, D. L. and Swailes, G. E.**, A sex attractant for the clover cutworm, *Scotogramma trifolii* (Rottenberg), a mixture of *Z*-11-hexadecen-l-ol acetate and *Z*-11-hexadecen-l-ol, *Environ. Entomol.*, 4, 632, 1975.

153. **Struble, D. L. and Swailes, G. E.**, Sex attractant for clover cutworm, *Scotogramma trifolii:* field tests with various ratios of *Z*-11-hexadecen-l-yl acetate and *Z*-11-hexadecen-l-ol, and with various quantities of attractant on two types of carriers, *Can. Entomol.*, 109, 369, 1977.
154. **Underhill, E. W., Steck, W. F., and Chisholm, M. D.**, Sex pheromone of the clover cutworm moth, *Scotogramma trifolii:* isolation, identification and field studies, *Environ. Entomol.*, 5, 307, 1976.
155. **Kuwahara, Y. and Yen, L. T. M.**, Identification of bombykol in *Bombyx mandarina* females (Lepidoptera: Bombycidae), *Appl. Entomol. Zool.*, 14, 114, 1979.
156. **Butenandt, A., Beckmann, R., Stamm, D., and Hecker, E.**, Über den Sexual-Lockstoff des Seidenspinners *Bombyx mori.* Reindarstellung und Konstitution, *Z. Naturforsch. Teil. B*, 14, 283, 1959.
156a. **Kasang, G., Schneider, D., and Schäfer, W.**, The silkworm moth *Bombyx mori.* Presence of the (*E,E*)-stereoisomer of bombykol in the female pheromone gland, *Naturwissenschaften*, 65, 337, 1978.
157. **Rodin, J. O., Silverstein, R. M., Burkholder, W. E., and Gorman, J. E.**, Sex attractant of female dermestid beetle *Trogoderma inclusum* Le Conte, *Science*, 165, 904, 1969.
158. **Levinson, H. Z. and Levinson, A. R.**, The pheromone activity of *Z*-(−)-methyl-8-hexadecen-l-ol for male khapra beetles, *Naturwissenschaften*, 61, 685, 1974.
159. **Greenblatt, R. E., Burkholder, W. E., Cross, J. H., Cassidy, R. F., Silverstein, R. M., Levinson, A. R., and Levinson, H. Z.**, Chemical basis for interspecific responses to sex pheromones of *Trogoderma* species (Coleoptera: Dermestidae), *J. Chem. Ecol.*, 3, 337, 1977.
160. **Cross, J. H., Byler, R. C., Silverstein, R. M., Greenblatt, R. E., Gorman, J. E., and Burkholder, W. E.**, Sex pheromone components and calling behavior of the female dermestid beetle, *Trogoderma variabile* Ballion (Coleoptera: Dermestidae), *J. Chem. Ecol.*, 3, 115, 1977.
161. **Yarger, R. G., Silverstein, R. M., and Burkholder, W. E.**, Sex pheromone of the female dermestid beetle *Trogoderma glabrum*, *J. Chem. Ecol.*, 1, 323, 1975.
162. **Nielsen, D. G. and Purrington, F. F.**, Sex attractants: a new warning system to time clearwing borer control practices, *Ohio Agr. Res. Dev. Cent. Res. Circ.*, 236, 7, 1978.
163. **Barry, M. W., Nielsen, D. G., Purrington, F. F., and Tumlinson, J. H.**, Attractivity of pheromone blends to male peachtree borer, *Synanthedon exitiosa*, *Environ. Entomol.*, 7, 1, 1978.
164. **Sharp, J. L., McLaughlin, J. R., James, J., Eichlin, T. D., and Tumlinson, J. H.**, Seasonal occurrence of male Sesiidae in North Central Florida determined with pheromone trapping methods, *Fla. Entomol.*, 61, 245, 1978.
165. **Nielsen, D. G., Purrington, F. F., and Shambaugh, G. F.**, EAG and field responses of sesiid males to sex pheromones and related compounds, in *Pheromones of the Sesiidae*, Neal, J. W., Jr., Ed., SEA, Agr. Res. Results ARR-NE-6, U.S. Department of Agriculture, Washington, D.C., 1979, 11.
166. **Nielsen, D. G. and Purrington, F. F.**, Field attraction of male *Podosesia syringae* and *P. aureocincta* to blends of synthetic compounds, *Environ. Entomol.*, 7, 708, 1978.
167. **Underhill, E. W., Steck, W., Chisholm, M. D., Worden, H. A., and Howe, J. A. G.**, A sex attractant for the cottonwood crown borer, *Aegeria tibialis* (Lepidoptera: Sesiidae), *Can. Entomol.*, 110, 495, 1978.
168. **Nielsen, D. G., Purrington, F. F., Campbell, R. L., Wilmot, T. R., Capizzi, J., and Tumlinson, J. H.**, Sex attractants for sequoia pitch moth and strawberry crown moth, *Environ. Entomol.*, 7, 544, 1978.
169. **Campbell, R. L.**, A sex pheromone for *Vespamima sequoiae*, *Proc. Wash. State Entomol. Soc.*, 38, 499, 1976: and *Bull. Oreg. Entomol. Soc.*, 61, 499, 1976.
170. **Vité, J. P., Pitman, G. B., Fentiman, A. F., Jr., and Kinzer, G. W.**, 3-Methyl-2-cyclohexen-l-ol isolated from *Dendroctonus*, *Naturwissenschaften*, 59, 469, 1972.
171. **Furniss, M. M., Baker, B. H., and Hostetler, B. B.**, Aggregation of spruce beetles (Coleoptera) to seudenol and repression of attraction by methylcyclohexenone in Alaska, *Can. Entomol.*, 108, 1297, 1976.
172. **Baker, B. H., Hostetler, B. B., and Furniss, M. M.**, Response of eastern larch beetle (Coleoptera: Scolytidae) in Alaska to its natural attractant and to Douglas-fir beetle pheromones, *Can. Entomol.*, 109, 289, 1977.
173. **Dyer, E. D. A. and Hall, P. M.**, Effect of the living host tree (*Picea*) on the response of *Dendroctonus rufipennis* (Coleoptera: Scolytidae) and a predator *Thanasimus undatulus* (Coleoptera: Cleridae) to frontalin and seudenol, *Can. Entomol.*, 112, 167, 1980.
174. **Tumlinson, J. H., Hardee, D. D., Gueldner, R. C., Thompson, A. C., Hedin, P. A., and Minyard, J. P.**, Sex pheromones produced by male boll weevil: isolation, identification, and synthesis, *Science*, 166, 1010, 1969.
175. **Mitchell, E. R., Tumlinson, J. H., and Davich, T. B.**, unpublished data, cited by Tumlinson, J. H., in *Pest Management with Insect Sex Attractants and Other Behavior-Controlling Chemicals*, Beroza, M., Ed., ACS Symp. Ser. 23, American Chemical Society, Washington, D.C., 1976, 63.
176. **Huddleston, P. M., Mitchell, E. B., and Wilson, N. M.**, Disruption of boll weevil communication, *J. Econ. Entomol.*, 70, 83, 1977.

177. **Hedin, P. A., McKibben, G. H., Mitchell, E. B., and Johnson, W. L.**, Identification and field evaluation of the compounds comprising the sex pheromone of the female boll weevil, *J. Chem. Ecol.*, 5, 617, 1979.
178. **Patrick, D. R.**, Boll weevil sex-lure traps attract *Dectes texanus texanus* (Coleoptera, Cerambycidae), *J. Ga. Entomol. Soc.*, 9, 17, 1974.
179. **Chang, V. C. S. and Curtis, G. A.**, Pheromone production by the New Guinea sugarcane weevil, *Environ. Entomol.*, 1, 476, 1972.
180. **Polles, S. G., Payne, J. A., and Jones, R. L.**, Attraction of the pecan weevil to its natural pheromone and grandlure, *Pecan South*, 4, 26, 1977.
181. **Hedin, P. A., Payne, J. A., Carpenter, T. L., and Neel, W.**, Sex pheromones of the male and female pecan weevil, *Curculio caryae:* behavioral and chemical studies, *Environ. Entomol.*, 8, 521, 1979.
182. **Bakke, A.**, Spruce bark beetle, *Ips typographus:* pheromone production and field response to synthetic pheromones, *Naturwissenschaften*, 63, 92, 1976.
183. **Renwick, J. A. A. and Vité, J. P.**, Systems of chemical communication in *Dendroctonus, Contrib. Boyce Thompson Inst.*, 24, 283, 1970.
184. **Renwick, J. A. A. and Vité, J. P.**, Bark beetle attractants: mechanism of colonization by *Dendroctonus frontalis, Nature (London)*, 224, 1222, 1969.
185. **Pitman, G. B. and Vité, J. P.**, Aggregation behavior of *Dendroctonus ponderosae* (Coleoptera: Scolytidae) in response to chemical messengers, *Can. Entomol.*, 101, 143, 1969.
186. **Renwick, J. A. A., Hughes, P. R., and Vité, J. P.**, The aggregation pheromone system of a *Dendroctonus* bark beetle in Guatemala, *J. Insect. Physiol.*, 21, 1097, 1975.
187. **Renwick, J. A. A., Pitman, G. B., and Vité, J. P.**, 2-Phenylethanol isolated from bark beetles, *Naturwissenschaften*, 63, 198, 1976.
188. **Nesbitt, B. F., Beevor, P. S., Hall, D. R., Lester, R., Sternlicht, M., and Goldenberg, S.**, Identification and synthesis of the female sex pheromone of the citrus flower moth, *Prays citri, Insect Biochem.*, 7, 355, 1977.
189. **Butler, L. I., Halfhill, J. E., McDonough, L. M., and Butt, B. A.**, Sex attractant of the alfalfa looper *Autographa californica* and the celery looper *Anagrapha falcifera* (Lepidoptera: Noctuidae), *J. Chem. Ecol.*, 3, 65, 1977.
190. **Gothilf, S., Kehat, M., Jacobson, M., and Galun, R.**, Screening pheromone analogues by EAG technique for biological activity on males of *Earias insulana, Heliothis armigera,* and *Spodoptera littoralis, Environ. Entomol.*, 7, 31, 1978.
191. **Beevor, P. S., Hall, D. R., Nesbitt, B. F., Dyck, V. A., Arida, G., Lippold, P. C., and Oloumi-Sadeghi, H.**, Field trials of the synthtic sex pheromones of the striped rice borer, *Chilo suppressaliso* (Walker) (Lepidoptera: Pyralidae), and of related compounds, *Bull. Entomol Res.*,67, 439, 1977.
192. **Beevor, P. S. and Campion, D. C.**, The field use of 'inhibitory' components of leipdopterous sex pheromones and pheromone mimics, in *Chemical Ecology: Odour Communication in Animals,* Ritter, F. J., Ed., Elsevier/North-Holland, Amsterdam, 1979, 313.
193. **Jacobson, M., Landis, B. J., Hendricks, D. E., Priesner, E., and Bestmann, H. J.**, Science concentrates, *Chem. Eng. News*, 50(49), 19, 1972.
194. **Mitchell, E. R., Jacobson, M., and Baumhover, A. H.**, *Heliothis* spp.: disruption of pheromonal communication with (*Z*)-9-tetradecen-1-ol formate, *J. Chem. Ecol.*, 4, 577, 1975.
195. **Mitchell, E. R., Baumhover, A. H., and Jacobson, M.**, Reduction of mating potential of male *Heliothis* spp. and *Spodoptera frugiperda* in field plots treated with disruptants, *Environ. Entomol.*, 5, 484, 1976.
196. **Mitchell, E. R., Tumlinson, J. H., and Baumhover, A. H.**, *Heliothis virescens*: attraction of males to blends of (*Z*)-9-tetradecen-l-ol formate and (*Z*)-9-tetradecenal, *J. Chem. Ecol.*, 4, 709, 1978.
197. **Bierl-Leonhardt, B. A., Moreno, D. S., Schwarz, M., Forster, H. S., Plimmer, J. R., and DeVilbiss, E. D.**, Identification of the pheromone of the Comstock mealybug, *Life Sci.*, 27, 399, 1980.
198. **Voerman, S. and Minks, A. K.**, Sex pheromones of summerfruit tortrix moth, *Adoxophyes orana.* II. Compounds influencing their attractant activity, *Environ. Entomol.*, 2, 751, 1973.
199. **Roelofs, W. L., Tette, J. P., Taschenberg, E. F., and Comeau, A.**, Sex pheromone of the grape berry moth: identification by classical and electroantennogram methods, and field tests, *J. Insect. Physiol.*, 17, 2235, 1971.
199a. **Struble, D. L., Buser, H. R., Arn, H., and Swailes, G. E.**, Identification of sex pheromone components of rebacked cutworm, *Euxoa ochrogaster,* and modification of sex attractant blend for adult males, *J. Chem. Ecol.*, 6, 573, 1980.
200. **Struble, D. L. and Swailes, G. E.**, Sex attractant for adult males of the redbacked cutworm, *Euxoa ochrogaster, Environ. Entomol.*, 7, 384, 1978.
201. **Wakamura, S.**, Sex attractant pheromone of the common cutworm moth, *Agrotis fucosa* Butler (Lepidoptera: Noctuidae): isolation and identification, *Appl. Entomol., Zool.*, 13, 290, 1978.

202. **Wakamura, S.,** Sex attractant pheromone of the common cutworm moth, *Agrotis fucosa* Butler: field evaluation, *Appl. Entomol. Zool.,* 15, 167, 1980.
203. **Bestmann, H. J., Vostrowsky, O., Koschatzky, K.-H., Platz, H., Brosche, T., Kantardjiew, I., Rheinwald, M., and Knauf, W.,** (*Z*)-5-Decenyl acetate, a sex attractant for the male turnip moth *Agrotis segetum* (Lepidoptera), *Angew. Chem.,* 90, 815, 1978; *Angew. Chem.,* (English ed.), 17, 768, 1978.
204. **Kovaleva, A. S., Ivanov, L. L., Pyatnova, I. B., Kovalev, B. G., Minyailo, V. A., and Minyailo, A. K.,** Results of testing synthetic compounds for their ability to attract insects in nature, *Khemoretseptsiya Nasekomykh,* 2, 159, 1975.
205. **Roelofs, W. L., Gieselmann, M. J., Cardé, A. M., Tashiro, H., Moreno, D. S., Henrick, C. A., and Anderson, R. J.,** Sex pheromone of the California red scale, *Aonidiella aurantii, Nature (London),* 267, 698, 1977.
206. **Roelofs, W., Gieselmann, M., Cardé, A., Tashiron, H., Moreno, D. S., Henrick, C. Ca., and Anderson, R. J.,** Identification of the California red scale sex pheromone, *J. Chem. Ecol.,* 4, 211, 1978.
207. **Gieselmann, M. J., Henrick, C. A., Anderson, R. J., Moreno, D. S., and Roelofs, W. L.,** Responses of male California red scale to sex pheromone isomers, *J. Insect Physiol.,* 26, 179, 1980.
208. **Gieselmann, M. J., Moreno, D. S., Fagerlund, J., Tashiro, H., and Roelofs, W. L.,** Identification of the sex pheromone of the yellow scale, *J. Chem. Ecol.,* 5, 27, 1979.
209. **Tamaki, Y., Noguchi, H., Sugie, H., Sato, R., and Kariya, A.,** Minor components of the female sex-attractant pheromone of the smaller tea tortrix moth (Lepidoptera: Tortricidae): isolation and identification, *Appl. Entomol. Zool.,* 14, 101, 1979.
210. **Cardé, R. T., Cardé, A. M., Hill, A. S., and Roelofs, W. L.,** Sex pheromone specificity as a reproductive isolating mechanism among the sibling species *Archips argyrospilus* and *Archips mortuanus* and other sympatric tortricine moths (Lepidoptera: Tortridicae), *J. Chem. Ecol.,* 3, 71, 1977.
211. **Roelofs, W. L. and Comeau, A.,** Sex pheromone perception, *Nature (London),* 220, 600, 1968.
212. **Roelofs, W. L. and Comeau, A.,** Sex pheromone perception: synergists and inhibitors for the redbanded leaf roller attractant, *J. Insect Physiol.,* 17, 435, 1971.
213. **Roelofs, W., Hill, A., and Cardé, R.,** Sex pheromone components of the redbanded leafroller, *Argyrotaenia velutinana* (Lepidoptera: Tortricidae), *J. Chem. Ecol.,* 1, 83, 1975.
214. **Roelofs, W. L. and Tette, J. P.,** Sex pheromone of the oblique-banded leaf roller moth, *Nature (London),* 226, 1172, 1970.
214a. **Biwer, G., Descoins, C., Gallois, M., Priesner, E., Chambon, J. -P., Genestier, G., and Martinez, M.,** Etude de la secretion pheromonale de la femelle vierge de la Tordeuse des cereales: *Cnephasia pumicana* Zeller (Lepidoptere, Tortricidae), *Ann. Zool. Ecol. Anim.,* 10, 129, 1977.
215. **Nesbitt, B. F., Beevor, P. S., Cole, R. A., Lester, R., and Poppi, R. G.,** Sex pheromones of two noctuid moths, *Nature (London), New Biol.,* 244, 208, 1973.
216. **Rauscher, S. and Arn, H.,** Mating suppression in tethered females of *Eupoecilia ambiguella* by evaporation of (*Z*)-9-dodecenyl acetate in the field. Pheromonal control of European grape berry moth, *Entomol. Exp. Appl.,* 25, 16, 1979.
217. **Arn, H., Rauscher, S., and Schmid, A.,** Sex attractant formulations and traps for the grape moth *Eupoecilia ambiguella* Hb., *Mitt. Schweiz. Entomol. Ges.,* 52, 49, 1979.
218. **Steck, W. F., Struble, D. L., Lilly, C. E., Chisholm, M. D., Undergill, E. W., and Swailes, G. E.,** A sex attractant for males of the early cutworm, *Euxoa tristicula* (Lepidoptera: Noctuidae), *Can. Entomol.,* 111, 337, 1979.
219. **Roelofs, W. L., Cardé, R. T., and Tette, J. P.,** Oriental fruit moth attractant synergists, *Environ. Entomol;,* 2, 252, 1973.
220. **Roelofs, W. L. and Cardé, R. T.,** Oriental fruit moth and lesser appleworm attractant mixtures refined, *Environ. Entomol.,* 3, 586, 1974.
221. **Frerot, B., Priesner, E., and Gallois, M.,** A sex attractant for the green budworm moth *Hedya nubiferana, Z. Naturforsch.,* 34c, 1248, 1979.
222. **Staten, R. T., Beroza, M., Bierl, B. A., and Adler, V. E.,** Pink bollworm: field, laboratory, and electroantennogram data on hexalure and inhibitors of its attraction of the moth, *J. Econ. Entomol.,* 66, 1263, 1973.
223. **Neumark, S., Jacobson, M., and Teich, I.,** Field evaluation of propylure, hexalure, and their formulations with deet, dodecyl acetate, and an antioxidant as attractants for male pink bollworm moths, *Environ. Lett.,* 7, 21, 1974.
224. **Smith, R. G., Daterman, G. E., Daves, G. D., Jr., McMurtrey, K. D., and Roelofs, W. L.,** Sex pheromone of the European pine shoot moth: chemical identification and field tests, *J. Insect Physiol.,* 20, 661, 1974.
225. **Kaae, R. S., Shorey, H. H., McFarland, S. U., and Gaston, L. K.,** Sex pheromones of Lepidoptera. XXXVII. Roles of sex pheromones and other factors in reproductive isolation among ten species of Noctuidae, *Ann. Entomol. Soc. Am.,* 66, 444, 1973.

226. **Hill, A. S., Rings, R. W., Swier, S. R., and Roelofs, W. L.,** Sex pheromone of the black cutworm moth, *Agrotis ipsilon, J. Chem. Ecol.,* 5, 439, 1979.
227. **McLaughlin, J. R. and Shorey, H. H.,** unpublished, cited by Mayer, M. S. and McLaughlin, J. R., in *An Annotated Compendium of Insect Sex Pheromones,* Univ. Fla. Agr. Exp. Sta. Monogr. Ser. No. 6, University of Florida, Gainesville, 1975.
228. **McLaughlin, J. R., Mitchell, E. R., Beroza, M., and Bierl, B. A.,** Effect of *E-Z* concentration of 7-dodecenyl acetate on captures of four noctuid species in pheromone traps, *J. Ga. Entomol. Soc.,* 10, 338, 1975.
229. **Kaae, R. S., Shorey, H. H., and Gaston, L. K.,** Pheromone concentration as a mechanism for reproductive isolation between two lepidopterous species, *Science,* 179, 487, 1973.
230. **Kaae, R. S., McLaughlin, J. R., Shorey, H. H., and Gaston, L. K.,** Sex pheromones of Lepidoptera. XXXII. Disruption of intraspecific pheromone communication in various species of Lepidoptera by permeation of the air with looplure or hexalure, *Environ. Entomol.,* 1, 651, 1972.
231. **Ghizdavu, I., Hodosan, F., Popovici, N., Botar, A., and Barabas, A.,** A sex attractant for *Autographa (Plusia) gamma* L. moth, *Rev. Roum. Biol. Ser. Biol. Anim.,* 24, 87, 1979.
232. **Tumlinson, J. H., Mitchell, E. R., Browner, S. M., and Lindquist, D. A.,** A sex pheromone for the soybean looper, *Environ. Entomol.,* 1, 466, 1972.
233. **Daterman, G. E., Daves, G. D., and Jacobson, M.,** Inhibition of pheromone perception in European pine shoot moth by synthetic acetates, *Environ. Entomol.,* 1, 382, 1972.
234. **Jacobson, M. and Jennings, D. T.,** Attraction of *Rhyacionia neomexicana* (Dyar) to synthetic pheromones, *Environ. Sci. Health,* 13, 429, 1978.
235. **Mitchell, E. R. and Tumlinson, J. H.,** An attractant for males of *Spodoptera dolichos* (Lepidoptera: Noctuidae), *Ann. Entomol. Soc. Am.,* 66, 917, 1973.
236. **Mitchell, E. R., Copeland, W. W., Sparks, A. N., and Sekul, A. A.,** Fall armyworm: disruption of pheromone communication with synthetic acetates, *Environ. Entomol.,* 3, 778, 1974.
237. **Berger, R. S.,** Isolation, identification, and synthesis of the sex attractant of the cabbage looper, *Trichoplusia ni, Ann. Entomol. Soc. Am.,* 59, 767, 1966.
238. **Mitchell, E. R., Webb, J. C., Baumhover, A. H., Hines, R. W., Stanley, J. W., Endris, R. G., Lindquist, D. A., and Masuda, S.,** Evaluation of cylindrical electric grids as pheromone traps for loopers and tobacco hornworms, *Environ. Entomol.,* 1, 365, 1972.
239. **Read, J. S., Warren, F. L., and Hewitt, P. H.,** Identification of the sex pheromone of the false codling moth *(Argyroploce leucotreta), Chem. Commun.,* 1968, 792.
240. **Read, J. S., Hewitt, P. H., Warren, F. L., and Myberg, A. D.,** Isolation of the sex pheromone of the moth *Argyroploce leucotreta, J. Insect Physiol.,* 20, 441, 1974.
241. **Meng, H. -T., Hu, J. -H., and Li, P. -S.,** Field trials for trapping the Oriental fruit moth and summer fruit tortrix with the synthetic pheromone, *cis*-8-dodecenyl acetate, *Acta Entomol. Sin.,* 21, 7, 1978.
242. **Sziraki, G.,** Examinations on tortricid moths trapped by synthetic attractants (Lepidoptera), *Folia Entomol. Hung.,* 31, 259, 1978.
243. **Frilli, F.,** L'impiego di attrattivi sessuali di sintesi per il controllo di *Cydia molesta, Entomologica (Bari),* 10, 31, 1974.
244. **Beauvais, F., Biwer, G., and Charmillot, P. J.,** Essais de piegeage sexuel de *Grapholitha janthinana* Dup. par un melange d'acetoxy-l dodecene 8Z et 8E (8Z DDA et 8E DDA), *Ann. Zool. Ecol. Anim.,* 9, 457, 1977; also in Les Pheromones Sexuelles des Lepidopteres, Compte Rendu Reunion, Oct. 1976, Publ. INRA, Centre de Recherches de Bordeaux, 1976, 83.
245. **Alford, D. V.,** Observations on the specificity of pheromone-baited traps for *Cydia funebrana* (Treitschke) (Lepidoptera: Tortricidae), *Bull. Entomol. Res.,* 68, 97, 1978.
245a. **Persoons, C. J., Ritter, F. J., Hainaut, D., and Demoute, J. P.,** Sex pheromone of the false codling moth *Cryptophlebia (= Argyroploce) leucotreta* (Lepidoptera: Tortricidae), *trans*-8-dodecenyl acetate. A corrected structure, *Meded. Fac. Landbouwwet. Rijksuniv. Gent,* 41/2, 937, 1976.
246. **Hainaut, D., Lhoste, J., Ritter, F. J., and Persoons, C. J.,** *Trans*- and *cis*-isomers of 8-dodecenyl acetate useful as attractants for *Cryptophlebia leucotreta,* French Patent 2317272, July 8, 1975.
247. **Persoons, C. J., Ritter, F. J., and Nooyen, W. J.,** Sex pheromone of the false codling moth *Cryptophlebia leucotreta* (Lepidoptera: Tortricidae). Evidence for a two-component system, *J. Chem. Ecol.,* 3, 717, 1977.
248. **Hrdy, I. and Sedivy, J.,** Males of *Exetastes cinctipes* (Hymenoptera: Ichneumonidae) attracted to 8-dodecenyl and 11-tetradecenyl acetates, *Acta Entomol. Bohemoslov.,* 76, 59, 1979.
249. **Granges, J. and Baggiolini, M.,** Une pheromone sexuelle synthetique attractive pour le carpocapse des prunes *(Grapholitha funebrana* Tr. [Lep. Tortricidae]), *Rev. Suisse Vitic. Arboric. Hortic.,* 3, 93, 1971.
250. **Arn, H., Delley, B., Baggiolini, M., and Charmillot, P. J.,** Communication disruption with sex attractant for control of the plum fruit moth, *Grapholitha funebrana:* a two-year field study, *Entomol. Exp. Appl.,* 19, 139, 1976.

251. **Mani, E., Arn, H., Wildbolz, T., and Hauri, H.**, Ein Feldversuch zur Bekampfung des Pflaumenwicklers mit der Desorientierungsmethode bei hoher Populationsdichte, *Mitt. Schweiz. Entomol. Ges.*, 51, 307, 1978.
252. **Iacob, M.**, Combaterea daunatorilor *Grapholitha molesta* Busck si *Grapholitha funebrana* Tr. prin metoda dezorientarii masculilor cu feromoni sexuali de sinteza, *An. Inst. Cercet. Prot. Plant. Inst. Cent. Cerc. Agric. Bucharest*, 14, 107, 1978.
253. **Baggiolini, M., Delley, B., Arn, H., and Charmillot, P. J.**, Inhibition des captures des males de *Grapholitha funebrana* Tr. par diffusion d'attractif sexuel synthetique en verger de pruniers, *Mitt. Schweiz. Entomol. Ges.*, 50, 213, 1977.
254. **Roelofs, W. L., Comeau, A., and Selle, R.**, Sex pheromone of the Oriental fruit moth, *Nature (London)*, 224, 723, 1969.
255. **Beroza, M., Muschik, G. M., and Gentry, C. R.**, Small proportion of opposite geometric isomer increases potency of synthetic pheromone of Oriental fruit moth, *Nature (London) New Biol.*, 244, 149, 1973.
256. **Gentry, C. R., Beroza, M., Blythe, J. L., and Bierl, B. A.**, Efficacy trials with the pheromone of the Oriental fruit moth and data on the lesser appleworm, *J. Econ. Entomol.*, 67, 607, 1974.
257. **Gentry, C. R., Beroza, M., and Blythe, J. L.**, Pecan bud moth captures in Georgia in traps baited with the pheromone of the oriental fruit moth, *Environ. Entomol.*, 4, 227, 1975.
258. **Gentry, C. R., Yonce, C. E., and Bierl-Leonhardt, B. A.**, Oriental fruit moth: mating disruption trials with pheromone, this volume.
259. **Kwangtung Institute of Analysis,** Isolation, identification, and synthesis of the sex pheromone of *Argyroploce (Eucosma) schistaceana* Snellen (Lepidoptera: Tortricidae), *Kexue Tongbao (K'o Hsueh T'ung Pao)*, 23, 438, 1978.
260. **Marks, R. J.**, Field studies with the synthetic sex pheromone and inhibitor of the red bollworm *Diparposis castanea* Hmps. (Lepidoptera, Noctuidae) in Malawi, *Bull. Entomol. Res.*, 66, 243, 1976.
261. **Marks, R. J., Nesbitt, B. F., Hall, D. R., and Lester, R.**, Mating disruption of the red bollworm of cotton *Diparopsis castanea* Hampson (Lepidoptera: Noctuidae) by ultra-low-volume spraying with a microencapsulated inhibitor of mating, *Bull. Entomol. Res.*, 68, 11, 1978.
262. **Minks, A. K., Voerman, S., and van de Vrie, M.**, A sex attractant for the cherry-bark tortrix moth, *Enarmonia formosana, Entomol. Exp. Appl.*, 19, 301, 1976.
263. **Wakamura, S., Kitamura, C., and Takahashi, S.**, Disruption of mating behaviour in the almond moth, *Cadra cautella* Walker (Lep., Pyralidae), with *cis*-9, *trans*-12-tetradecadien-1-ol acetate and *cis*-9-tetradecen-1-ol acetate, *Jpn. J. Appl. Entomol. Zool.*, 19, 227, 1975.
264. **Taschenberg, E. F. and Roelofs, W. L.**, Pheromone communication disruption of the grape berry moth with microencapsulated and hollow fiber systems, *Environ. Entomol.*, 5, 688, 1976.
265. **Sower, L. L., Daterman, G. E., Sartwell, C., and Cory, H. T.**, Attractants for the Western pine shoot borer, *Eucosma sonomana*, and *Rhyacionia zozana* determined by field screening, *Environ. Entomol.*, 8, 265, 1979.
266. **Sartwell, C., Daterman, G. E., Sower, L. L., Overhulser, D. L., and Koerber, T. W.**, Mating disruption with synthetic sex attractants controls damage by *Eucosma sonomana* (Lepidoptera: Tortricidae, Olethreutinae) in *Pinus ponderosa* plantations. I. Manually applied polyvinyl chloride formulation, *Can. Entomol.*, 112, 159, 1980.
267. **Arn, H., Rauscher, S., Buser, H. -R., and Roelofs, W. L.**, Sex pheromone of *Eupoecilia ambiguella*: *cis*-9-dodecenyl acetate as a major component, *Z. Naturforsch.*, 31c, 499, 1976.
268. **Noguchi, H., Tamaki, Y., and Yushima, T.**, Sex pheromone of the tea tortrix moth: isolation and identification, *Appl. Entomol. Zool.*, 14, 225, 1979.
269. **Taschenberg, E. F., Cardé, R. T., and Roelofs, W. L.**, Sex pheromone mass trapping and mating disruption for control of redbanded leafroller and grape berry moths in vineyards, *Environ. Entomol.*, 3, 239, 1974.
270. **Lange, R. and Hoffmann, D.**, Essigsäure-*cis*-dodecen-(9)-yl-ester, ein Inhibitor des Sexualpheromons des Kiefernknospentriebwicklers, *Naturwissenschaften*, 59, 217, 1972.
271. **Mitchell, E. R.**, Inhibition of pheromone perception by male cabbage loopers and beet armyworms: proximity vs. atmospheric permeation, *Environ. Entomol.*, 5, 770, 1976.
272. **Sekul, A. A. and Sparks, A. N.**, Sex attractant of the fall armyworm moth, *U.S. Dep. Agric. Tech. Bull.*, 1542, 1976.
273. **Mitchell, E. R., and Doolittle, R. E.**, Sex pheromones of *Spodoptera exigua, S. eridania,* and *S. frugiperda*: bioassay for field activity, *J. Econ. Entomol.*, 69, 324, 1976.
274. **Marks, R. J.**, The influence of behaviour modifying chemicals on mating success of the red bollworm *Diparopsis castanea* Hmps. (Lepidoptera, Noctuidae) in Malawi, *Bull. Entomol. Res.*, 66, 279, 1976.
275. **Steck, W. and Bailey, B. K.**, Pheromone traps for moths: evaluation of cone trap designs and design parameters, *Environ. Entomol.*, 7, 449, 1978.
276. **Berisford, C. W. and Hedden, R. L.**, Suppression of male *Rhyacionia frustrana* response to live females by the sex pheromone of *R. buoliana, Environ. Entomol.*, 7, 532, 1978.

277. **Hill, A. S., Berisford, C. W., Brady, U. E., and Roelofs, W. L.,** Sex pheromone of the pitch pine tip moth, *Rhyacionia rigidana, Environ. Entomol.,* 5, 959, 1976.
278. **Baltensweiler, W., Priesner, E., Arn, H., and Delucchi, V.,** Unterschiedliche Sexuallockstoffe bei Larchen- und Arvenform des Grauen Lärchenwicklers *(Zeiraphera diniana* Gn., Lep., Tortricidae), *Mitt. Schweiz. Entomol. Ges.,* 51, 133, 1979.
279. **Vrkoc, J., Skuhravy, V., and Baltensweiler, W.,** Freilanduntersuchungen zur Sexuallockstoff-Reaktion der Fichtenform des grauen Lärchenwicklers, *Zeiraphera diniana* Gn. (Lep., Tortricidae), *Anz. Schadlingskd. Pflanz. Umweltschutz,* 52, 129, 1979.
280. **Wall, C., Greenway, A. R., and Burt, P. E.,** Electroantennographic and field responses of the pea moth *Cydia nigricana* to sex attractants and related compounds, *Physiol. Entomol.,* 1, 151, 1976.
281. **Macaulay, E. D. M. and Lewis, T.,** Attractant traps for monitoring pea moth, *Cydia nigricana* (Fabr.), *Ecol. Entomol.,* 2, 279, 1977.
282. **Roelofs, W. L., Reissig, W. H., and Weires, R. W.,** Sex attractant for the spotted tentiform leaf miner moth, *Lithocolletis blancardella, Environ. Entomol.,* 6, 373, 1977.
283. **Voerman, S. and Herrebout, W. M.,** A sex attractant for the leaf miner moth *Lithocolletis corylifoliella* and its influence on that of *L. blancardella* (Lep., Gracillariidae), *Entomol. Exp. Appl.,* 23, 96, 1978.
284. **Nesbitt, B. F., Beevor, P. S., Cole, R. A., Lester, R., and Poppi, R. G.,** The isolation and identification of the female sex pheromones of the red bollworm moth, *Diparopsis castanea, J. Insect Physiol.,* 21, 1091, 1975.
285. **Marks, R. J.,** Female sex pheromone release and the timing of male flight in the red bollworm *Diparopsis castanea* Hmps. (Lepidoptera, Noctuidae), measured by pheromone traps, *Bull. Entomol. Res.,* 66, 219, 1976.
286. **Buser, H. -R., and Arn, H.,** Analysis of insect pheromones by quadrupole mass fragmentography and high-resolution gas chromatography, *J. Chromatogr.,* 106, 83, 1975.
287. **Roelofs, W., Kochansky, J., Cardé, R., Arn, H., and Rauscher, S.,** Sex attractant of the grape vine moth, *Lobesia botrana, Mitt. Schweiz. Entomol. Ges.,* 46, 71, 1973.
288. **Roehrich, R., Carles, J. -P., Darrioumerle, Y., Pargade, P., and Lalanne Cassou, B.,** Essais en vignoble de pheromones de synthese pour la capture des males de l'eudemis *(Lobesia botrana* Schiff.), *Ann. Zool. Ecol. Anim.,* 8, 473, 1976.
289. **Roehrich, R., Carles, J. -P., and Tresor, C.,** Essai preliminaire de protection du vignoble contre *Lobesia botrana* Schiff. au moyen de la pheromone sexuelle de synthese (Methode de confusion), *Rev. Zool. Agric. Pathol. Veg.,* 76, 25, 1977; also in Les Pheromones Sexuelles des Lepidopteres, Compte Rendu Reunion, Oct. 1976, Publ. INRA, Centre de Recherches de Bordeaux, 1976, 183.
290. **Biwer, G., Lalanne-Cassou, B., Descoins, C., and Samain, D.,** Sur le piegeage sexuel de *Sterrha biselata* (Hufn.) (Lepidoptere Geometridae, Sterrhinae) par l'acetoxy-1 dodecadiene-7 E, 9 Z pheromone sexuelle de *Lobesia botrana* (Schiff.) (Lepidoptere Tortricidae, Olethreutinae), *C. R. Acad. Sci. Ser. D,* 280, 1469, 1975.
291. **Hathaway, D. O., McGovern, T. P., Beroza, M., Moffitt, H. R., McDonough, L. M., and Butt, B. A.,** An inhibitor of sexual attraction of male codling moths to a synthetic sex pheromone and virgin females in traps, *Environ. Entomol.,* 3, 522, 1974.
292. **Hathaway, D. O., McDonough, L. M., George, D. A., and Moffitt, H. R.,** Antipheromone of the codling moth: potential for control by air permeation, *Environ. Entomol.,* 8, 318, 1979.
293. **Berisford, C. W.,** Inhibition of male attraction of the Nantucket pine tip moth by a synthetic pheromone for the pitch pine tip moth, *Environ. Entomol.,* 6, 245, 1977.
294. **Bacon, O. G., Seiber, J. N., and Kennedy, G. G.,** Evaluation of survey trapping techniques for potato tuberworm moths with chemical baited traps, *J. Econ. Entomol.,* 69, 569, 1976.
295. **Voerman, S., Minks, A. K., and Goewie, E. A.,** Specificity of the pheromone system of *Adoxophyes orana* and *Clepsis spectrana, J. Chem. Ecol.,* 1, 423, 1975.
296. **Cardé, R. T. and Roelofs, W. L.,** Attraction of redbanded leafroller moths, *Argyrotaenia velutinana,* to blends of (*Z*)- and (*E*)-11-tridecenyl acetates, *J. Chem. Ecol.,* 3, 143, 1977.
297. **Roelofs, W. L., Kochansky, J. P., Cardé, R. T., Kennedy, G. G., Henrick, C. A., Labovitz, J. N., and Corbin, V. L.,** Sex pheromone of the potato tuberworm moth, *Phthorimaea operculella, Life Sci.,* 17, 699, 1975.
298. **Persoons, C. J., Voerman, S., Verwiel, P. E. J., Ritter, F. J., Nooyen, W. J., and Minks, A. K.,** Sex pheromone of the potato tuberworm moth, *Phthorimaea operculella:* isolation, identification and field evaluation, *Entomol. Exp. Appl.,* 20, 289, 1976.
299. **Voerman, S. and Rothschild, G. H. L.,** Synthesis of the two components of the sex pheromone system of the potato tuberworm moth, *Phthorimaea operculella* (Zeller) (Lepidoptera: Gelechiidae) and field experience with them, *J. Chem. Ecol.,* 4, 531, 1978.
300. **Thal, J.,** Zur Wirkung von Sexualpheromonen der Kartoffelmotte, *Phthorimaea operculella* Zeller (Lepid., Gelechiidae) im Vergleich mit verschiedenen Lichtfallen, *Anz. Schadlingskd. Pflanz. Umweltschutz,* 51, 107, 1978.

301. **Negishi, T., Ishiwatari, T., Uchida, M., and Asano, S.,** A new synergist, 11-methyl-*cis*-9,12-trideca-dienyl acetate, for the smaller tea tortrix attractant, *Appl. Entomol. Zool.,* 14, 478, 1979.
302. **Negishi, T., Ishiwatari, T., Asano, S., and Fujikawa, H.,** Mass trapping for the smaller tea tortrix control, *Appl. Entomol. Zool.,* 15, 113, 1980.
303. **Sugie, H., Yaginuma, K., and Tamaki, Y.,** Sex pheromone of the Asiatic leafroller, *Archippus breviplicanus* Walsingham (Lepidoptera: Tortricidae): isolation and identification, *Appl. Entomol. Zool.,* 12, 69, 1977.
304. **Klun, J. A., Bierl-Leonhardt, B. A., Schwarz, M., Litsinger, J. A., Barrion, A. T., Chiang, H. C., and Jiang, Z.,** Sex pheromone of the Asian corn borer moth, *Life Sci.,* 27, 1603, 1980.
305. **Klun, J. A. and Junk, G. A.,** Iowa European corn borer sex pheromone: isolation and identification of four C_{14} esters, *J. Chem. Ecol.,* 3, 447, 1977.
306. **Beroza, M., Staten, R. T., and Bierl, B. A.,** Tetradecyl acetate and related compounds as inhibitors of attraction of the pink bollworm moth to the sex lure hexalure, *J. Econ. Entomol.,* 64, 580, 1971.
307. **Subchev, M. A. and Kuznetsov, N. V.,** Preliminary identification of the sex pheromone of *Rhynchopacha* sp. (Lepidoptera: Gelechiidae) by field screening tests, *Dokl. Bolg. Akad. Nauk,* 27, 993, 1974.
308. **Herrebout, W. N.,** Sex attractants for Yponomeuta species, *Neth. J. Zool.,* 28 (Abstr.), 276, 1978.
309. **Fujimura, T.,** Mating behavior and sex attractant of the spotted cutworm moth, *Amathes c-nigrum* L. (Lepidoptera: Noctuidae), *Jpn. J. Appl. Entomol. Zool.,* 20, 133, 1976.
310. **Hirai, Y.,** *cis*-7-Tetradecen-1-ol acetate: a sex pheromone component of the spotted cutworm moth, *Amathes c-nigrum* L., in *Proc. Internatl. Symp. on Insect Pheromones and Their Applications,* Japan Plant Protection Association, Tokyo, 1977, 165.
311. **Bestmann, H. J., Vostrowsky, O., Platz, H., Brosche, T., Koschatzky, K. H., and Knauf, W.,** Pheromone. XXIII. (*Z*)-7-Tetradecenylacetat, ein Sexuallockstoff für Männchen von *Amathes c-nigrum* (Noctuidae, Lepidoptera), *Tetrahedron Lett.,* 497, 1979.
312. **Roelofs, W. L., Cardé, R., Benz, G., and von Salis, G.,** Sex attractant of the larch bud moth found by electroantennogram method, *Experientia,* 27, 1438, 1971.
313. **Arn, H., Städler, E., and Rauscher, S.,** The electroantennographic detector — a selective and sensitive tool in the gas chromatographic analysis of insect pheromones, *Z. Naturforsch.,* 30c, 722, 1975.
314. **Tamaki, Y., Noguchi, H., Yushima, T., Hirano, C., Honma, K., and Sugawara, H.,** Sex pheromone of the summer fruit tortrix: isolation and identification, *Kontyu,* 39, 338, 1971.
315. **Meijer, G. M., Ritter, F. J., Persoons, C. J., Minks, A. K., and Voerman, S.,** Sex pheromones of summer fruit tortrix moth *Adoxophyes orana:* two synergistic isomers, *Science,* 175, 1469, 1972.
316. **Sato, R. and Tamaki, Y.,** Isolating factors between the smaller tea tortrix and the summer-fruit tortrix (Lepidoptera: Tortricidae). IV. Role of the pheromonal components, (*Z*)-9- and (*Z*)-11-tetradecen-1-ol acetates, *Appl. Entomol. Zool.,* 12, 50, 1977.
317. **Tamaki, Y., Noguchi, H., Yushima, T., and Hirano, C.,** Two sex pheromones of the smaller tea tortrix: isolation, identification, and synthesis, *Appl. Entomol. Zool.,* 6, 139, 1971.
318. **Tamaki, Y., Ishiwatari, T., and Osakabe, M.,** Inhibition of the sexual behavior of the smaller tea tortrix moth by the sex pheromone and its components, *Jpn. J. Appl. Entomol. Zool.,* 19, 187, 1975.
319. **Negishi, T., Ishiwatari, T., and Asano, S.,** A potent synergist for the sex pheromone of the smaller tea tortrix, *Appl. Entomol. Zool.,* 12, 178, 1977.
320. **Tamaki, Y.,** Multicomponent sex pheromone of Lepidoptera with special reference to *Adoxophyes* sp., in *Chemical Ecology: Odour Communication in Animals,* Ritter, F. J., Ed., Elsevier/North Holland, Amsterdam, 1979, 169.
321. **Brady, U. E.,** Response of male *Sitotroga* to the sex pheromones of *Sitotroga* and *Pectinophora* females and to propylure and deet. Deet analysis in *Sitotroga* females, *J. Ga. Entomol. Soc.,* 4, 11, 1969.
322. **Roelofs, W. L. and Comeau, A.,** Sex pheromone specificity: taxonomic and evolutionary aspects in Lepidoptera, *Science,* 165, 398, 1969.
323. **Milaire, H. G.,** Role et portee du piegeage avec des substances attractives sexuelles dans les cultures fruitieres, *Pomol. Fr.,* 17, 13, 1975.
324. **Minks, A. K., Roelofs, W. L., Ritter, F. J., and Persoons, C. J.,** Reproductive isolation of two tortricid moth species by different ratios of a two-component sex attractant, *Science,* 180, 1073, 1973.
325. **Chisholm, M. D., Underhill, E. W., and Steck, W. F.,** Field trapping of the diamondback moth *Plutella xylostella* using synthetic sex attractants, *Environ. Entomol.,* 8, 516, 1979.
326. **Kondratyev, Y. A., Lebedeva, E. V., and Pyatnova, Y. B.,** Progress in insect pheromone investigation, in *Proc. VIII Internatl. Congr. Plant Protection,* Moscow, 1975, Vol. 3, 1975, 44.
327. **Brady, U. E.,** Isolation, identification and stimulatory activity of a second component of the sex pheromone system (complex) of the female almond moth, *Cadra cautella* (Walker), *Life Sci.,* 13, 227, 1973.
328. **Read, J. S. and Haines, C. P.,** Secondary pheromone components and synergism in stored-products Phycitinae, *J. Chem. Ecol.,* 5, 251, 1979.

329. **Roelofs, W., Cardé, A., Hill, A., and Cardé, R.**, Sex pheromone of the threelined leafroller, *Pandemis limitata*, *Environ. Entomol.*, 5, 649, 1976.

330. **Underhill, E. W., Steck, W. F., and Chisholm, M. D.**, A sex pheromone mixture for the bertha armyworm moth *Mamestra configurata*: (*Z*)-9-tetradecen-1-ol acetate and (*Z*)-11-hexadecen-1-ol acetate, *Can. Entomol.*, 109, 1335, 1977.

331. **Steck, W., Underhill, E. W., Chisholm, M. D., Peters, C. C., Philip, H. G., and Arthur, A. P.**, Sex pheromone traps in population monitoring of adults of the bertha armyworm, *Mamestra configurata* (Lepidoptera: Noctuidae), *Can. Entomol.*, 111, 91, 1979.

332. **Ando, T., Kishino, K. -I., Tatsuki, S., Nakajima, H., Yoshida, S., and Takahashi, N.**, Identification of the female sex pheromone of the rice green caterpillar, *Agric. Biol. Chem.*, 41, 1819, 1977.

332a. **Ando, T., Kishino, K. -I., Tatsuki, S., and Takahashi, N.**, Sex pheromone of the rice green caterpillar: chemical identification of three components and field tests with synthetic pheromones, *Agric. Biol. Chem.*, 44, 765, 1980.

333. **Roelofs, W. L., Lagier, R. F., and Hoyt, S. C.**, Sex pheromone of the moth, *Pandemis pyrusana*, *Environ. Entomol.*, 6, 353, 1977.

334. **Priesner, E., Bogenschutz, H., Altenkirch, W., and Arn, H.**, A sex attractant for the pine beauty moth, *Panolis flammea*, *Z. Naturforsch.*, 33c, 1000, 1978.

335. **Knauf, W., Bestmann, H. J., Koschatzky, K. H., Süss, J., and Vostrowsky, O.**, Untersuchungen über die Lockwirkung synthetischer Sex-Pheromone bei *Tortrix viridana* (Eichenwickler) und *Panolis flammea* (Kieferneule), *Z. Angew. Entomol.*, 88, 307, 1979.

336. **Bestmann, H. J., Brosche, T., Koschatzky, K. H., Michaelis, K., Platz, H., Vostrowsky, O., and Knauf, W.**, Pheromone. XXX. Identifizierung eines Pheromonkomplexes aus der Graseule *Scotia exclamationis* (Lepidoptera), *Tetrahedron Lett.*, 21, 747, 1980.

337. **Campion, D. G.**, Some observations on the use of pheromone traps as a survey tool for *Spodoptera littoralis*, *Centre Overseas Pest Res. Misc. Rep.*, No. 4, 1972.

338. **Beevor, P. S., Hall, D. R., Lester, R., Poppi, R. G., Read, J. S., and Nesbitt, B. F.**, Sex pheromones of the armyworm moth, *Spodoptera exempta* (Wlk.), *Experientia*, 31, 22, 1975.

339. **Campion, D. G., Odiyo, P. O., Mushi, A. M., Hall, D. R., Lester, R., and Nesbitt, B. F.**, Field tests with the synthetic sex pheromone of the African armyworm *Spodoptera exempta* (Walk.), *Centre Overseas Pest Res. Misc. Rep.*, No. 25, 1976.

340. **Mitchell, E. R.**, Recent advances in the use of sex pheromones for control of insect pests, in *Proc. 4th Annu. Controlled-Release Pestic. Symp.*, Goulding, R. L., Ed., Oregon State University, Corvallis, Ore., 1977.

341. **Jones, R. L. and Sparks, A. N.**, (*Z*)-9-Tetradecen-1-ol acetate, a secondary sex pheromone of the fall armyworm, *Spodoptera frugiperda* (J. E. Smith), *J. Chem. Ecol.*, 5, 721, 1979.

342. **Kehat, M., Gothilf, S., Dunkelblum, E., and Greenberg, S.**, Communication disruption of the cotton leafworm, *Spodoptera littoralis*, by using its sex pheromone or other related compounds emitted at different distances from the attractant, *Environ. Entomol.*, 8, 973, 1979.

343. **Tamaki, Y., Osawa, T., Yushima, T., and Noguchi, H.**, Sex pheromone and related compounds secreted by the virgin females of *Spodoptera litura* (F.), *Jpn. J. Appl. Entomol. Zool.*, 20, 81, 1976.

344. **Miyashita, K., Kawasaki, K., Nakamura, K., Uesumi, Y., and Sugiura, T.**, Mating suppression of *Spodoptera litura* F. (Lepidoptera: Noctuidae) in greenhouses by a component of its sex pheromone, *Appl. Entomol. Zool.*, 11, 364, 1976.

345. **Kawasaki, K. and Miyashita, K.**, Mating suppression by individual components of the sex pheromone of *Spodoptera litura* F. (Lepidoptera: Noctuidae) under field conditions, *App. Entomol. Zool.*, 11, 320, 1976.

346. **Hirano, C.**, (*Z*)-9-Tetradecenyl acetate as a potent inhibitor of attraction of *Spodoptera litura* (Lepidoptera: Noctuidae) to synthetic sex pheromone, *Prot. Ecol.*, 1, 171, 1979.

347. **Campion, D. G.**, Sex pheromones and their uses for control of insects of the genus *Spodoptera*, *Meded. Fac. Landbouwwet. Rijksniv. Gent*, 40, 283, 1975.

348. **Blair, B. W. and Tannock, J.**, A possible pheromone for *Spodoptera triturata* (Walker) (Lepidoptera, Noctuidae), *Rhod. J. Agric. Res.*, 15, 225, 1977.

349. **Blair, B. W. and Tannock, J.**, A further note on the possible pheromone for *Spodoptera triturata* (Walker) (Lepidoptera, Noctuidae), *Rhod. J. Agric. Res.*, 16, 221, 1978.

350. **Minks, A. K. and Voerman, S.**, Sex pheromones of the summerfruit tortrix moth, *Adoxophyes orana*: trapping performance in the field, *Entomol. Exp. Appl.*, 16, 541, 1973.

351. **Minks, A. K., Voerman, S., and Klun, J. A.**, Disruption of pheromone communication with microencapsulated antipheromones against *Adoxophyes orana*, *Entomol. Exp. Appl.*, 20, 163, 1976.

352. **Persoons, C. J., Minks, A. K., Voerman, S., Roelofs, W. L., and Ritter, F. J.**, Sex pheromones of the moth, *Archips podana*: isolation, identification and field evaluation of two synergistic geometrical isomers, *J. Insect Physiol.*, 20, 1181, 1974.

353. **Boness, M.**, Versuche zur Bekampfung des Fruchtschalenwicklers *Archips podana* mit pheromonen, *Z. Angew. Entomol.*, 82, 104, 1976.

354. **Boness, M.,** Erfahrungen mit Sexualpheromonen von Lepidopteren, *Anz. Schädlingskd. Pflanz. Umweltschutz,* 51, 161, 1978.
355. **Miller, J. R., Baker, T. C., Cardé, R. T., and Roelofs, W. L.,** Reinvestigation of oak leaf roller sex pheromone components and the hypothesis that they vary with diet, *Science,* 192, 140, 1976.
356. **Hill, A. S., Cardé, R. T., Kido, H., and Roelofs, W. L.,** Sex pheromone of the orange tortrix moth, *Argyrotaenia citrana* (Lepidoptera: Tortricidae), *J. Chem. Ecol.,* 1, 215, 1975.
357. **Daterman, G. E ., Robbins, R. G., Eichlin, T. D., and Pierce, J.,** Forest Lepidoptera attracted by known sex attractants of western spruce budworms, *Choristoneura* spp. (Lepidoptera: Tortricidae), *Can. Entomol.,* 109, 875, 1977.
358. **Maini, S.,** Prova preliminare con attrattivi sessuali nei confronti della Piralide, *Inf. Fitopatol.,* 23(9), 11, 1973.
359. **Roelofs, W. L. and Arn, H.,** Sex attractant of the red-banded leaf roller moth, *Nature (London),* 219, 513, 1968.
360. **Taschenberg, E. F. and Roelofs, W. L.,** Male redbanded leafroller moth orientation disruption in vineyards, *Environ. Entomol.,* 7, 103, 1978.
361. **Anglade, P.,** Emploi de pheromones sexuelles synthetiques pour l'attraction des males de la Pyrale du maïs *(O. nubilalis* Hbn.), *Rev. Zool. Agric. Pathol. Veg.,* 73, 37, 1974.
362. **Struble, D. L. and Lilly, C. E.,** An attractant for the beet webworm, *Loxostege sticticalis* (Lepidoptera: Pyralidae), *Can. Entomol.,* 109, 261, 1977.
363. **Klun, J. A. and Brindley, T. A.,** *cis*-11-Tetradecenyl acetate, a sex stimulant of the European corn borer, *J. Econ. Entomol.,* 63, 779, 1970.
364. **Klun, J. A. and Robinson, J. F.,** Inhibition of European corn borer mating by *cis*-11-tetradecenyl acetate, a borer sex stimulant, *J. Econ. Entomol.,* 63, 1281, 1970.
365. **Klun, J. A. and Robinson, J. F.,** European corn borer moth: sex attractant and sex attraction inhibitors, *Ann. Entomol. Soc. Am.,* 64, 1083, 1971.
366. **Klun, J. A. and Robinson, J. F.,** Olfactory discrimination in the European corn borer and several pheromonally analogous moths, *Ann. Entomol. Soc. Am.,* 651337, 1972.
367. **Klun, J. A., Chapman, O. L., Mattcs, K. C., Wojtkowski, P. W., Beroza, M., and Sonnet, P. E.,** Insect sex pheromones: minor amount of opposite geometrical isomer critical to attraction, *Science,* 181, 661, 1973.
368. **Kochansky, J., Cardé, R. T., Liebherr, J., and Roelofs, W. L.,** Sex pheromone of the European corn borer, *Ostrinia nubilalis* (Lepidoptera: Pyralidae), in New York, *J. Chem. Ecol.,* 1, 225, 1975.
369. **Stockel, J. and Anglade, P.,** Influence de la concentration en isomeres *cis* et *trans* dans la pheromone sexuelle d'*Ostrinia nubilalis* Hb. sur le comportement d'orientation de cette espece et sur celui de *Pyrausta aurata* Sc., *C. R. Acad. Sci. Ser. D,* 285, 61, 1977.
370. **Baker, J. L., Hill, A. S., and Roelofs, W. L.,** Seasonal variations in pheromone trap catches of male omnivorous leafroller moths, *Platynota stultana, Environ. Entomol.,* 7, 399, 1978.
371. **Loughner, G.,** Citrus pest management: OLR in the San Joaquin, *Citrograph,* 62, 171, 1977.
372. **Klun, J. A., Chapman, O. L., Mattes, K. C., and Beroza, M.,** European corn borer and redbanded leafroller: disruption of reproduction behavior, *Environ. Entomol.,* 4, 871, 1975.
373. **Sanders, C. J.,** Disruption of sex attraction in the eastern spruce budworm, *Environ. Entomol.,* 5, 868, 1976.
374. **Sanders, C. J., Daterman, G. E., Shepherd, R. F., and Cerezke, H.,** Sex attractants for two species of western spruce budworm, *Choristoneura biennis* and *C. viridis* (Lepidoptera: Tortricidae), *Can. Entomol.,* 106, 157, 1974.
375. **Bartell, R. J. and Lawrence, L. A.,** Reduction in responsiveness of male light-brown apple moths, *Epiphyas postvittana,* to sex pheromone following pulsed pre-exposure to pheromone components, *Physiol. Entomol.,* 2, 89, 1977.
376. **Hall, D. R., Lester, R., and Langmaid, J. R.,** unpublished observations in Reference 127.
377. **Bode, W. M., Asquith, D., and Tette, J. P.,** Sex attractants and traps for tufted apple budmoth and redbanded leafroller males, *J. Econ. Entomol.,* 66, 1129, 1973.
378. **Doolittle, R. E., Roelofs, W. L., Solomon, J. D., Cardé, R. T., and Beroza, M.,** (*Z,E*)-3,5-Tetradecadien-l-ol acetate: sex attractant for the carpenterworm moth, *Prionoxystus robiniae* (Peck) (Lepidoptera: Cossidae), *J. Chem. Ecol.,* 2, 399, 1976.
379. **Solomon, J. D., Dix, M. E., and Doolittle, R. E.,** Attractiveness of the synthetic carpenterworm sex attractant increased by isomeric mixtures and prolonged by preservatives, *Environ. Entomol.,* 7, 39, 1978.
380. **Tagestad, A. D. and Stein, J. D.,** Response of the carpenterworms, *Prionoxystus robiniae* and *Acossus centerensis* (Lepidoptera: Cossidae) to a synthetic sex attractant in North Dakota, *Proc. North Cent. Branch Entomol. Soc. Am.,* 30, 101, 1975.
381. **Doolittle, R. E., Tagestad, A., and McKnight, M. E.,** Trapping carpenterworms and aspen carpenterworms with sex attractants in North Dakota, *Environ. Entomol.,* 5, 267, 1976.

382. **Tamaki, Y. and Yushima, T.,** Sex pheromone of the cotton leafworm, *Spodoptera littoralis*, *J. Insect Physiol.*, 20, 1005, 1974.
383. **Tamaki, Y., Noguchi, H., and Yushima, T.,** Sex pheromone of *Spodoptera litura* (F.) (Lepidoptera: Noctuidae): isolation, identification, and synthesis, *Appl. Entomol. Zool.*, 8, 200, 1973.
384. **Tamaki, Y. and Yushima, T.,** Biological activity of the synthesized sex pheromone and its geometrical isomers of *Spodoptera litura* (F.) (Lepidoptera: Noctuidae), *Appl. Entomol. Zool.*, 9, 73, 1974.
385. **Ishiwatari, T., Negishi, T., and Asano, S.,** Promising sex-pheromonal activity of the combination of *cis*-9, *trans*-12-tetradecadienyl acetate and *cis*-11-tetradecenyl acetate for the smaller tea tortrix, *Appl. Entomol. Zool.*, 12, 211, 1977.
386. **Kuwahara, Y., Hara, H., Ishii, S., and Fukami, H.,** The sex pheromone of the Mediterranean flour moth, *Agric. Biol. Chem.*, 35, 447, 1971.
387. **Dahm, K. H., Richter, I., Meyer, D., and Röller, H.,** The sex attractant of the Indian-meal moth *Plodia interpunctella* (Hübner). II, *Life Sci.*, 10, 531, 1979.
388. **Mitchell, E. R., Chalfant, R. B., and Lupo, T. J., Jr.,** Lesser cornstalk borer: effects of 9,12-(*Z,E*)-tetradecadien-l-ol acetate on pheromone communication, *J. Ga. Entomol. Soc.*, 11, 211, 1976.
389. **Wakamura, S., Kitamura, C., and Takahashi, S.,** Effects of the synthetic sex pheromone, *cis*-9, *trans*-12-tetradecadien-l-ol acetate on the reproduction of the almond moth, *Cadra cautella* Wlk. (Lepidoptera: Pyralidae), *Jpn. J. Appl. Entomol. Zool.*, 19, 182, 1975.
390. **Brady, U. E., Tumlinson, J. H., Brownlee, R. G., and Silverstein, R. M.,** Sex stimulant and attractant in the Indian meal moth and in the almond moth, *Science*, 171, 802, 1971.
391. **Kuwahara, Y., Kitamura, C., Takahashi, S., Hara, H., Ishii, S., and Fukami, H.,** Sex pheromone of the almond moth and the Indian meal moth: *cis*-9, *trans*-12-tetradecadienyl acetate, *Science*, 171, 801, 1971.
392. **Takahashi, F., Masui, A., Kuwahara, Y., Ishii, S., and Fukami, H.,** Studies on sex pheromone of Pyralididae. IV. The male response to the female sex pheromone of the almond moth, *Cadra cautella* Walker (Phycitinae), *Botyu Kagaku*, 37, 56, 1972.
393. **Haines, C. P.,** The use of synthetic sex pheromones for pest management in stored-product situations, *Pestic. Sci.*, 7, 647, 1976.
394. **McLaughlin, J. R. and Hagstrum, D. W.,** Effects of a dark environment and air permeation with synthetic sex pheromone on mating in the almond moth, *Environ. Entomol.*, 5, 1057, 1976.
395. **Sower, L. L. and Whitmer, G. P.,** Population growth and mating success of Indian meal moths and almond moths in the presence of synthetic sex pheromone, *Environ. Entomol.*, 6, 17, 1977.
396. **Reichmuth, C., Schmidt, H. -U., Levinson, A. R., and Levinson, H. Z.,** Die Fängigkeit pheromonbeköderter Klebefallen fur Speichermotten *(Ephestia elutella* Hbn.) in unterscheidlich dicht befallenen Getreidelagern, *Z. Angew. Entomol.*, 86, 417, 1978.
397. **Adler, V. E., Jacobson, M., Edmiston, J. F., and Fleming, M. H.,** (*Z,E*)-9,12-Tetradecadien-l-ol acetate: a potent attractant for male *Ephestiodes infimella*, *J. Econ.Entomol.*, 69, 706, 1976.
398. **Sower, L. L., Turner, W. K., and Fish, J. C.,** Population-density-dependent mating frequency among *Plodia interpunctella* (Lepidoptera: Phycitidae) in the presence of synthetic sex pheromone with behavioral observations, *J. Chem. Ecol.*, 1, 335, 1975.
399. **Mitchell, E. R.,** Disruption of pheromonal communication among coexistent pest insects with multichemical formulations, *BioScience*, 25, 493, 1975.
400. **Hirano, C.,** Efficiency of water-pan trap baited with synthetic sex pheromone in attracting male *Spodoptera litura* (Lepidoptera: Noctuidae), *Jpn. J. Appl. Entomol. Zool.*, 21, 216, 1977.
401. **Oyama, M.,** Mating suppression of *Spodoptera litura* (F.) by (*Z,E*)-9,12-tetradecadienyl acetate, *Appl. Entomol. Zool.*, 12, 281, 1977.
402. **Oyama, M. and Wakamura, S.,** Effective range for mating suppression of *Spodoptera litura* (F.) by evaporating (*Z,E*)-9,12-tetradecadienyl acetate, *Appl. Entomol. Zool.*, 12, 372, 1977.
403. **Voerman, S., Minks, A. K., and Houx, N. W. H.,** Sex pheromones of summerfruit tortrix moth, *Adoxophyes orana:* investigations on compounds modifying their attractancy, *Environ. Entomol.*, 3, 701, 1974.
404. **Ritter, F. J. and Persoons, C. J.,** Recent development in insect pheromone research, in particular in the Netherlands, *Neth. J. Zool.*, 25, 261, 1975.
405. **Jewett, D. M., Matsumura, F., and Coppel, H. C.,** Sex pheromone specificity in the pine sawflies: interchange of acid moieties in an ester, *Science*, 192, 51, 1976.
406. **Jewett, D. M., Matsumura, F., and Coppel, H. C.,** Preparation and use of sex attractants for four species of pine sawflies, *J. Chem. Ecol.*, 4, 277, 1978.
407. **Matsumura, F., Tai, A., Coppel, H. C., and Imaida, M.,** Chiral specificity of the sex pheromone of the red-headed pine sawfly, *Neodiprion lecontei*, *J. Chem. Ecol.*, 5, 237, 1979.
408. **Kraemer, M., Coppel, H. C., Matsumura, F., Kikukawa, T., and Mori, K.,** Field responses of the white pine sawfly, *Neodiprion pinetum*, to optical isomers of sawfly sex pheromones, *Environ. Entomol.*, 8, 519, 1979.

409. **Struble, D. L., Swailes, G. E., and Ayre, G. L.**, A sex attractant for males of the darksided cutworm, *Euxoa messoria* (Lepidoptera: Noctuidae), *Can. Entomol.*, 109, 975, 1977.
410. **Green, N., Jacobson, M., and Keller, J. C.**, Hexalure, an insect sex attractant discovered by empirical screening, *Experientia*, 25, 682, 1969.
411. **McLaughlin, J. R., Shorey, H. H., Gaston, L. K., Kaae, R. S., and Stewart, F. D.**, Sex pheromones of Lepidoptera. XXXI. Disruption of sex pheromone communication in *Pectinophora gossypiella* with hexalure, *Environ. Entomol.*, 1, 645, 1972.
412. **Sower, L. L., Vick, K. W., and Long, J. S.**, Isolation and preliminary biological studies of the female-produced sex pheromone of *Sitotroga cerealella* (Lepidoptera: Gelechiidae), *Ann. Entomol. Soc. Am.*, 66, 184, 1973.
413. **Struble, D. L., Jacobson, M., Green, N., and Warthen, J. D.**, Bertha armyworm (Lepidoptera: Noctuidae): detection of a sex pheromone and the stimulatory effect of some synthetic chemicals, *Can. Entomol.*, 107, 355, 1975.
414. **Rothschild, G. H. L.**, Attractants for *Heliothis armigera* and *H. punctigera*, *J. Aust. Entomol. Soc.*, 17, 389, 1978.
415. **Bestmann, H. J., Vostrowsky, O., Koschatzky, K. H., Platz, H., and Szymanska, A.**, Pheromone XVII. (*Z*)-11-Hexadecenylacetat, ein Sexuallockstoff des Pheromonsystems der Kohleule *Mamestra brassicae*, *Tetrahedron Lett.*, 605, 1978.
416. **Hirai, Y., Kimura, H., Kawasaki, K., and Tamaki, Y.**, (*Z*)-11-Hexadecenyl acetate: a sex-pheromone component of the cabbage armyworm moth, *Mamestra brassicae Linné*, *Appl. Entomol. Zool.*, 13, 136, 1978.
417. **Kovalev, B. G., Nedopekina, S. F., Lebedeva, K. V., and Kost, A. N.**, Isolation and identification of the sex pheromone of females of the cabbage moth *Mamestra brassicae* L. (Lepidoptera: Noctuidae), *Bioorg. Khim.*, 5, 912, 1979; *Sov. J. Bioorg. Chem.*, 5, 685, 1979.
418. **Chisholm, M. D., Steck, W. F., Arthur, A. P., and Underhill, E. W.**, Evidence for *cis*-11-hexadecen-1-ol acetate as a major component of the sex pheromone of the bertha armyworm, *Mamestra configurata* (Lepidoptera: Noctuidae), *Can. Entomol.*, 107, 361, 1975.
419. **Tamaki, Y., Kawasaki, K., Yamada, H., Koshihara, T., Osaki, N., Ando, T., Yoshida, S., and Kakinohana, H.**, (*Z*)-11-Hexadecenal and (*Z*)-hexadecenyl acetate: sex-phermone components of the diamondback moth (Lepidoptera: Plutellidae), *Appl. Entomol. Zool.*, 12, 208, 1977.
420. **Chow, Y. S., Lin, Y. M., and Hsu, C. L.**, Sex pheromone of the diamondback moth (Lepidoptera: Plutellidae), *Bull. Inst. Zool. Acad. Sin.*, 16, 99, 1977.
421. **Chow, Y. S., Hsu, C. L., and Lin, Y. M.**, Field attraction experiment of the sex pheromone of the diamondback moth, *Plutella xylostella* (L.) in Taiwan, *Natl. Sci. Counc. Mon.*, 6, 651, 1978.
422. **Koshihara, Y., Yamada, H., Tamaki, Y., and Ando, T.**, Field attractiveness of the synthetic sex-pheromone of the diamondback moth, *Plutella xylostella* (L.), *Appl. Entomol. Zool.*, 13, 138, 1978.
423. **Nesbitt, B. F., Beevor, P. S., Hall, D. R., Lester. R., and Dyck, V. A.**, Identification of the female sex pheromone of the purple stem borer moth, *Sesamia inferens*, *Insect Biochem.*, 6, 105, 1976.
23a. **Hirano, C., Muramoto, H., and Horiike, M.**, Sex pheromone produced by female sweet potato leaf folder moth, *Naturwissenschaften*, 63, 439, 1976.
424. **Kochansky, J., Tette, J., Taschenberg, E. F., Cardé, R. T., Kaissling, K. E., and Roelofs, W. L.**, Sex pheromone of the moth, *Antheraea polyphemus*, *J. Insect Physiol.*, 21, 1977, 1975.
425. **Kochansky, J. P., Cardé, R. T., Taschenberg, E. F., and Roelofs, W. L.**, Rhythms of male *Antheraea polyphemus* attraction and female attractiveness, and an improved pheromone synthesis, *J. Chem. Ecol.*, 3, 419, 1977.
426. **Oetting, R. D.**, Immature stages and biology of *Chionodes psiloptera* a pest in bluegrass seed fields (Lepidoptera, Gelechiidae), *Pan Pac. Entomol.*, 53, 258, 1977.
427. **Rothschild, G. H. L.**, Attractants for monitoring *Pectinophora scutigera* and related species in Australia, *Environ. Entomol.*, 4, 983, 1975.
428. **Hummel, H. E., Gaston, L. K., Shorey, H. H., Kaae, R. S., Byrne, K. J., and Silverstein, R. M.**, Clarification of the chemical status of the pink bollworm sex pheromone, *Science*, 181, 873, 1973.
429. **Shorey, H. H., Gaston, L. K., and Kaae, R. S.**, Air-permeation with gossyplure for control of the pink bollworm, in *Pest Management with Insect Sex Attractants and Other Behavior-Controlling Chemicals*, Beroza, M., Ed., ACS Symp. Ser. 23, American Chemical Society, Washington, D. C., 1976, 67.
430. **Smith, R. L., Flint, H. M., and Forey, D. E.**, Air permeation with gossyplure: feasibility studies on chemical confusion for control of the pink bollworm, *J. Econ. Entomol.*, 71, 257, 1978.
431. **Flint, H. M., Balasubramanian, M., Campero, J., Strickland, G. R., Ahmad, Z., Barral, J., Barbosa, S., and Khail, A. F.**, Pink bollworm: response of native males to ratios of *Z,Z*- and *Z,E*-isomers of gossyplure in several cotton growing areas of the world, *J. Econ. Entomol.*, 72, 758, 1979.
432. **Vick, K. W., Su, H. C. F., Sower, L. L., Mahany, P. G., and Drummond, P. C.**, (*Z,E*)-7,11-Hexadecadien-1-ol acetate: the sex pheromone of the Angoumois grain moth, *Sitotroga cerealella*, *Exper-*

433. **Vick, K. W., Coffelt, J. S., and Sullivan, M. A.,** Disruption of pheromone communication in the Angoumois grain moth with synthetic female sex pheromone, *Environ. Entomol.,* 7, 528, 1978.
434. **Duckworth, W. D. and Eichlin, T. D.,** Two new species of clearwing moths (Sesiidae) from eastern North America clarified by sex pheromones, *J. Lepid. Soc.,* 31, 191, 1977.
435. **Duckworth, W. D. and Eichlin, T. D.,** A new species of clearwing moth from southcentral Texas, *Pan Pac. Entomol.,* 53, 175, 1977.
436. **Brennan, B. M.,** Attraction of *Ereunetis simulans* (Lepidoptera: Tineidae) to the sex attractant of peachtree borer, *Sanninoidae exitiosa* (Lepidoptera: Sessiidae), *Proc. Hawaii. Entomol. Soc.,* 22, 457, 1977.
437. **Greenfield, M. D., and Karandinos, M. G.,** A new species of *Paranthrene* (Lepidoptera: Sesiidae), *Proc. Entomol. Soc. Wash.,* 81, 499, 1979.
438. **Nielsen, D. G., Purrington, F. F., Tumlinson, J. H., Doolittle, R. E., and Yonce, C. E.,** Response of male clearwing moths to caged virgin females, female extracts, and synthetic sex attractants, *Environ. Entomol.,* 4, 451, 1975.
439. **Tumlinson, J. H., Yonce, C. E., Doolittle, R. E., Heath, R. R., Gentry, C. R., and Mitchell, E. R.,** Sex pheromones and reproductive isolation of the lesser peachtree borer and the peachtree borer, *Science,* 185, 614, 1974.
440. **McLaughlin, J. R., Doolittle, R. E., Gentry, C. R., Mitchell, E. R., and Tumlinson, J. H.,** Response to pheromone traps and disruption of pheromone communication in the lesser peachtree borer and the peachtree borer (Lepidoptera: Sesiidae), *J. Chem. Ecol.,* 2, 73, 1976.
441. **Yaginuma, K., Kumakura, M., Tamaki, Y., Yushima, T., and Tumlinson, J. H.,** Sex attractant for the cherry tree borer, *Synanthedon hector* Butler (Lepidoptera: Sesiidae), *Appl. Entomol. Zool.,* 11, 266, 1976.
442. **Uchida, M., Mori, K., and Matsui, M.,** Synthesis of (*Z,Z*)-3,13-octadecadienyl acetate and its (*E,Z*)-isomer, the attractant for the cherrytree borer, *Agric. Biol. Chem.,* 42, 1067, 1978.
443. **Ebata, T. and Mori, K.,** A convenient synthesis of a mixture of (*Z,Z*)-3,13-octadecadienyl acetate and its (*E,Z*)-isomer, the attractant for the cherrytree borer, *Agric. Biol. Chem.,* 43, 1567, 1979.
444. **Voerman, S., Minks, A. K., Vanwetswinkel, G., and Tumlinson, J. H.,** Attractivity of 3,13-octadecadien-1-ol acetate to the male clearwing moth *Synanthedon myopaeformis* (Borkhausen) (Lepidoptera, Sesiidae), *Entomol. Exp. Appl.,* 23, 301, 1978.
445. **Karandinos, M. G., Tumlinson, J. H., and Eichlin, T. D.,** Field evidence of synergism and inhibition in the Sesiidae sex pheromone system, *J. Chem. Ecol.,* 3, 57, 1977.
446. **Tamaki, Y., Yushima, T., Oda, M., Kida, K., Kitamura, K., Yabuki, S., and Tumlinson, J. H.,** Attractiveness of 3,13-octadecadienyl acetates for males of clearwing moths, *Jpn. J. Appl. Entomol. Zool.,* 21, 106, 1977.
447. **Uebel, E. C., Sonnet, P. E., Menzer, R. E., Miller, R. W., and Lusby, W. R.,** Mating-stimulant pheromone and cuticular lipid constituents of the little house fly, *Fannia canicularis* (L.), *J. Chem. Ecol.,* 3, 269, 1977.
448. **Oleshchenko, I. N., Palchkov, V. A., Ivashchenko, I. I., Filin-Koldakov, B. V., Adamenko, E. A., and Pleshkova, V. S.,** Attractant for the male click beetles *Agriotes litigiosus* Rossi, U.S.S.R. Patent 351, 517, September 21, 1972; *Chem. Abstr.,* 78, 54047, 1973.
449. **Gieselmann, M. J., Rice, R. E., Jones, R. A., and Roelofs, W. L.,** Sex pheromone of the San Jose scale, *J. Chem. Ecol.,* 5, 891, 1979.
450. **Heath, R. R., McLaughlin, J. R., Tumlinson, J. H., Ashley, T. R., and Doolittle, R. E.,** Identification of the white peach scale sex pheromone. An illustration of micro techniques, *J. Chem. Ecol.,* 5, 941, 1979.
451. **George, D. A., McDonough, L. M., Hathaway, D. O., and Moffitt, H. R.,** Inhibitors of sexual attraction of male codling moths, *Environ. Entomol.,* 4, 606, 1975.
452. **Longhurst, C., Baker, R., and Mori, K.,** Response of the sawfly *Diprion similis* to chiral sex pheromones, *Experientia,* 36, 946, 1980.
453. **Chow, Y. S., Young, L. C. T., and Lin, Y. M.,** *cis*-9-Hexadecenyl propionate, a sex stimulant pheromone of the diamondback moth, presented at the XV Int. Congr. Entomol., Washington, D. C., August 19 to 27, 1976.
454. **McGovern, T. P., Beroza, M., Ladd, T. L., Jr., Ingangi, J. C., and Jurimas, J. P.,** Phenethyl propionate, a potent new attractant for Japanese beetles, *J. Econ. Entomol.,* 63, 1727, 1970.
455. **McGovern, T. P. and Ladd, T. L., Jr.,** Japanese beetle: new synthetic attractants, *J. Econ. Entomol.,* 74, 194, 1981.
456. **Sugawara, R. and Muto, T.,** Attraction of several dipterous insects to aliphatic esters (Diptera: Milichiidae, Chloropidae and Ceratopogonidae), *Appl. Entomol. Zool.,* 9, 11, 1974.
457. **Fluno, J. A., Davis, H. G., and Rogoff, W. M.,** Chloropid flies attracted to synthetic chemical lures, *Proc. Entomol. Soc. Wash.,* 74, 443, 1972.

458. **Rogoff, W. M., Davis, H. G., McGovern, T. P., Ilcken, E. H., and Kreasky, J. B.,** Synthetic attractants screened in the field as lures for Chloropidae, *Ann. Entomol. Soc. Am.*, 66, 262, 1973.
459. **Davis, H. G., Eddy, G. W., McGovern, T. P., and Beroza, M.,** 2,4-Hexadienyl butyrate and related compounds highly attractive to yellow jackets (*Vespula* spp.), *J. Med. Entomol.*, 4, 275, 1967.
460. **Davis, H. G., Eddy, G. W., McGovern, T. P., and Beroza, M.,** Heptyl butyrate, a new synthetic attractant for yellow jackets, *J. Econ. Entomol.*, 62, 1245, 1969.
461. **Reierson, D. A. and Wagner, R. E.,** Trapping yellowjackets with a new standard plastic wet trap, *J. Econ. Entomol.*, 68, 395, 1975.
462. **Reierson, D. A. and Wagner, R. E.,** Trapping to determine the sympatry and seasonal abundance of various yellowjackets, *Environ. Entomol.*, 7, 418, 1978.
463. **Davis, H. G., Peterson, R. J., Rogoff, W. M., McGovern, T. P., and Beroza, M.,** Octyl butyrate, an effective attractant for the yellowjacket, *Environ. Entomol.*, 1, 673, 1972.
464. **Henderson, H. E., Warren, F. L., Augustyn, O. P. H., Burger, B. V., Schneider, D. F., Boshoff, P. R., Spies, H. S. C., and Geertsema, H.,** Sex-pheromones. *cis*-Dec-5-en-1-yl 3-methylbutanoate as the pheromone from the pine emperor moth (*Nudaurelia cytherea cytherea* Fabr.), *Chem. Commun.*, 686, 1972.
464a. **Henderson, H. E., Warren, F. L., Augustyn, O. P. H., Burger, B. V., Schneider, D. F., Boshoff, P. R., Spies, H. S. C., and Geertsema, H.,** Isolation and structure of the sex-pheromone of the moth, *Nudaurelia cytherea cytherea*, *J. Insect Physiol.*, 19, 1257, 1973.
465. **Tashiro, H., Gertler, S. I., Beroza, M., and Green, N.,** Butyl sorbate as an attractant for the European chafer, *J. Econ. Entomol.*, 57, 230, 1964.
466. **Ohinata, K., Jacobson, M., Nakagawa, S., Urago, T., Fujimoto, M., and Higa, H.,** Methyl (*E*)-6-nonenoate: a new Mediterranean fruit fly male attractant, *J. Econ. Entomol.*, 72, 648, 1979.
467. **Fukui, H., Matsumura, F., Ma, M. C., and Burkholder, W. E.,** Identification of the sex pheromone of the furniture carpet beetle, *Anthrenus flavipes* LeConte, *Tetrahedron Lett.*, 3563, 1974.
468. **Levinson, H. Z., Levinson, A. R., Jen, T. I., Williams, J. L. D., Kahn, G., and Francke, W.,** Production site, partial composition and olfactory perception of a pheromone in the male hide beetle, *Naturwissenschaften*, 65, 543, 1978.
469. **Fukui, H., Matsumura, F., Barak, A. V., and Burkholder, W. E.,** Isolation and identification of a major sex-attracting component of *Attagenus elongatulus* (Casey) (Coleoptera: Dermestidae), *J. Chem. Ecol.*, 3, 539, 1977.
470. **Horler, D. F.,** (−) Methyl *n*-tetradeca-*trans*-2,4,5-trienoate, an allenic ester produced by the male dried bean beetle, *Acanthoscelides obtectus* (Say), *J. Chem. Soc. C*, 859, 1970.
471. **Halstead, D. G. H.,** Preliminary biological studies on the pheromone produced by male *Acanthoscelides obtectus* (Say) (Coleoptera, Bruchidae), *J. Stored Prod. Res.*, 9, 109, 1973.
472. **Levinson, A. R., Levinson, H. Z., Schwaiger, H., Cassidy, R. F., Jr., and Silverstein, R. M.,** Olfactory behavior and receptor potentials of the khapra beetle, *Trogoderma granarium* (Coleoptera: Dermestidae) induced by the major components of its sex pheromone, certain analogues, and fatty acid esters, *J. Chem. Ecol.*, 4, 95, 1978.
473. **Uebel, E. C., Schwarz, M., Lusby, W. R., and Sonnet, P. E.,** Cuticular non-hydrocarbons of the female house fly and their evaluation as mating stimulants, *Lloydia*, 41, 63, 1978.
474. **Chapman, O. L., Mattes, K. C., Sheridan, R. S., and Klun, J. A.,** Stereochemical evidence of dual chemoreceptors for an achiral sex pheromone in Lepidoptera, *J. Am. Chem. Soc.*, 100, 4878, 1978.
475. **Chapman, O. L., Klun, J. A., Mattes, K. C., Sheridan, R. S., and Maini, S.,** Chemoreceptors in Lepidoptera: stereochemical differentiation of dual receptors for an achiral pheromone, *Science*, 201, 926, 1978.
476. **Bierl-Leonhardt, B. A., Moreno, D. S., Schwarz, M., Fargerlund, J. A., and Plimmer, J. R.,** Isolation, identification and synthesis of the sex pheromone of the citrus mealybug, *Planococcus citri* (Risso), *Tetrahedron Lett.*, 22, 389, 1981.
477. **Barber, I. A., McGovern, T. P., Beroza, M., Hoyt, C. P., and Walker, A.,** Attractant for the coconut rhinoceros beetle, *J. Econ. Entomol.*, 64, 1041, 1971.
478. **Maddison, P. A., Beroza, M., and McGovern, T. P.,** Ethyl chrysanthemumate as an attractant for the coconut rhinoceros beetle, *J. Econ. Entomol.*, 66, 591, 1973.
479. **Sugawara, R., Kurihara, S., and Muto, T.,** Attraction of the German cockroach to cyclohexyl alkanoates and *n*-alkyl cyclohexaneacetates, *J. Insect Physiol.*, 21, 957, 1975.
480. **Davis, H. G., Eves, J. D., and McDonough, L. M.,** Trap and synthetic lure for the checkered flower beetle, a serious predator of alfalfa leafcutting bees, *Environ. Entomol.*, 8, 147, 1979.
481. **Beroza, M., Green, N., Gertler, S. I., Steiner, L. F., and Miyashita, D. H.,** New attractants for the Mediterranean fruit fly, *J. Agric. Food Chem.*, 9, 361, 1961.
482. **Nakagawa, S., Cunningham, R. T., and Urago, T.,** The repellent effect of high trimedlure concentrations in plastic traps to Mediterranean fruit fly in Hawaii, *J. Econ. Entomol.*, 64, 762, 1971.

483. **McGovern, T. P., Fiori, B., Beroza, M., and Ingangi, J. C.**, Propyl 1,4-benzodioxan-2-carboxylate, a new attractant for the European chafer, *J. Econ. Entomol.*, 63, 168, 1970.
484. **Butler, L. I., McDonough, L. M., Onsager, J. A., and Landis, B. J.**, Sex pheromones of the Pacific Coast Wireworm, *Limonius canus*, *Environ. Entomol.*, 4, 229, 1975.
485. **Hendry, L. B., Greany, P. D., and Gill, R. J.**, Kairomone mediated host-finding behavior in the parasitic wasp *Orgilus lepidus*, *Entomol. Exp. Appl.*, 16, 471, 1973.
486. **Sannasi, A., and Rajulu, G. S.**, 9-Oxodec-*trans*-2-enoic acid in the Indian honeybees, *Life Sci., II*, 10, 195, 1971.
487. **Gary, N. E.**, Chemical mating attractants in the queen honeybee, *Science*, 136, 773, 1962.
488. **Barak, A. V. and Burkholder, W. E.**, Pheromone studies with *Attagenus elongatulus* Casey, (Coleoptera: Dermestidae), *Proc. North Cent. Branch Entomol. Soc. Am.*, 28, 175, 1973.
489. **Silverstein, R. M., Rodin, J. O., Burkholder, W. E., and Gorman, J. E.**, Sex attractant of the black carpet beetle, *Science*, 157, 85, 1967.
490. **Tumlinson, J. H., Klein, M. G., Doolittle, R. E., Ladd, T. L., and Proveaux, A. T.**, Identification of the female Japanese beetle sex pheromone: inhibition of male response by an enantiomer, *Science*, 197, 789, 1977.
491. **Adis, J. and Kramer, E.**, Formaldehyd-Lösung attrahiert *Carabus problematicus* (Coleoptera: Carabidae), *Entomol. Ger.*, 2, 121, 1975.
492. **Girard, J. E., Germino, F. J., Budris, J. P., Vita, R. A., and Garrity, M. P.**, Pheromone of the male flesh fly, *Sarcophaga bullata*, *J. Chem. Ecol.*, 5, 125, 1979.
493. **Bergström, G. and Tengö, J.**, Linalool in mandibular gland secretion of *Colletes* bees (Hymenoptera: Apoidea), *J. Chem. Ecol.*, 4, 437, 1978.
494. **Robacker, D. C. and Hendry, L. B.**, Neral and geranial: components of the sex pheromone of the parasitic wasp, *Itoplectis conquisitor*, *J. Chem. Ecol.*, 3, 563, 1977.
495. **Leyrer, R. L. and Monroe, R. E.**, Isolation and identification of the scent of the moth, *Galleria mellonella*, and a revaluation of its sex pheromone, *J. Insect Physiol.*, 19, 2267, 1973.
496. **Finn, W. E. and Payne, T. L.**, Attraction of greater wax moth females to male-produced pheromones, *Southwest. Entomol.*, 2, 62, 1977.
497. **Dahm, K. H., Meyer, D., Finn, W. E., Reinhold, V., and Roller, H.**, The olfactory and auditory mediated sex attraction in *Achroia grisella* (Fabr.), *Naturwissenschaften*, 58, 265, 1971.
498. **Röller, H., Biemann, K., Bjerke, J. S., Norgard, D. W., and McShan, W. H.**, Sex pheromones of pyralid moths. I. Isolation and identification of the sex-attractant of *Galleria mellonella* L. (greater waxmoth), *Acta Entomol. Bohemoslov.*, 65, 208, 1968.
499. **Kamm, J. A. and McDonough, L. M.**, Sex attractant for a plume moth, *Oidaematophorus monodactylus* (L.) (Lepidoptera: Pterophoridae), *Pan Pac. Entomol.* 55, 284, 1979.
500. **Sternlicht, M., Goldenberg, S., Nesbitt, B. F., Hall, D. R., and Lester, R.**, Field evaluation of the synthetic female sex pheromone of the citrus flower moth, *Prays citri* (Mill.) (Lepidoptera: Yponomeutidae), and related compounds, *Phytoparasitica*, 6, 101, 1978.
501. **Campion, D. G., McVeigh, L. J., Polyrakis, J., Michaelakis, S., Stravrakis, G. N., Beevor, P. S., Hall, D. R., and Nesbitt, B. F.**, Laboratory and field studies of the female sex pheromone of the olive moth, *Prays oleae*, *Experientia*, 35, 1146, 1979.
502. **Renou, M., Descoins, C., Priesner, E., Gallois, M., and Lettere, M.**, Le tetradecene-7 *Z* al-1, constituant principal de la secretion pheromonale de la Teigne de l olivier: *Prays oleae* Bern. (Lepidoptere Hyponomeutidae), *C. R. Acad. Sci. Ser. D*, 288, 1559, 1979.
503. **Gothilf, S., Kehat, M., Jacobson, M., and Galun, R.**, Sex-attractants for male *Heliothis armigera* (Hbn.), *Experientia*, 34, 853, 1978.
504. **Roelofs, W. L., Hill, A. S., Cardé, R. T., and Baker, T. C.**, Two sex pheromone components of the tobacco budworm moth, *Heliothis virescens*, *Life Sci.*, 14, 1555, 1974.
505. **Tumlinson, J. H., Hendricks, D. E., Mitchell, E. R., Doolittle, R. E., and Brennan, M. M.**, Isolation, identification, and synthesis of the sex pheromone of the tobacco budworm, *J. Chem. Ecol.*, 1, 203, 1975.
506. **Hendricks, D. E.**, Tobacco budworm: disruption of courtship behavior with a component of the synthetic sex pheromone, *Environ. Entomol.*, 5, 978, 1976.
507. **Weatherston, J., Percy, J. E., and MacDonald, L. M.**, Field testing of *cis*-11-tetradecenal as attractant or synergist in Tortricinae, *Experientia*, 32, 178, 1976.
508. **Sanders, C. J. and Weatherston, J.**, Sex pheromone of the eastern spruce budworm (Lepidoptera: Tortricidae): optimum blend of *trans*- and *cis*-11-tetradecenal, *Can. Entomol.*, 108, 1285, 1976.
509. **MacDonald, L. M. and Weatherston, J.**, Gas chromatography and the structural elucidation of lepidopteran pheromones, *J. Chromatogr.*, 118, 195, 1976.
510. **Sanders, C. J.**, The role of the sex pheromone in the management of spruce budworm, in *Chemical Ecology: Odour Communication in Animals*, Ritter, F. J., Ed., Elsevier/North-Holland, Amsterdam, 1979, 281.

511. **Schmidt, J. O. and Seabrook, W. D.**, Mating of caged spruce budworm moths in pheromone environments, *J. Econ. Entomol.*, 72, 509, 1979.
512. **Schmidt, J. O., Seabrook, W. D., Ross, R. J., Darvesh, S., and Lonergan, G.**, Spruce budworm: roles of pheromone components and analogues in male disruption and attraction, *Experientia*, 36, 222, 1980.
513. **Gothilf, S., Kehat, M., Dunkelblum, E., and Jacobson, M.**, Efficacy of *Z*-11 hexadecenal and *Z*-11 tetradecenal as sex attractants for *Heliothis armigera* on 2 different dispensers, *J. Econ. Entomol.*, 72, 718, 1979.
514. **Tagestad, A. D.**, Spruce budworm detection in North Dakota shelterbelts and nurseries with a synthetic attractant, *North Dakota Farm Res.*, 33 (2), 17, 1975.
515. **Weatherston, J., Roelofs, W., Comeau, A., and Sanders, C. J.**, Studies of physiologically active arthropod secretions. X. Sex pheromone of the eastern spruce budworm, *Choristoneura fumiferana* (Lepidoptera: Tortricidae), *Can. Entomol.*, 103, 1741, 1971.
516. **Klun, J. A., Plimmer, J. R., Bierl-Leonhardt, B. A., Sparks, A. N., Primiani, M., Chapman, O. L., Lee, G. H., and Lepone, G.**, Sex pheromone chemistry of female corn earworm moth, *Heliothis zea*, *J. Chem. Ecol.*, 6, 165, 1980.
517. **Nesbitt, B. F., Beevor, P. S., Hall, D. R., Lester, R., and Dyck, V. A.**, Identification of the female sex pheromones of the moth, *Chilo suppressalis*, *J. Insect Physiol.*, 21, 1883, 1975.
518. **Ohta, K., Tatsuki, S., Uchiumi, K., Kurihara, M., and Fukami**, J. -I., Structures of sex pheromones of rice stem borer, *Agric. Biol. Chem.*, 40, 1897, 1976.
519. **Tatsuki, S., Ohta, K., Uchiumi, K., Kurihara, M., Fukami, J. -I., and Kishino, K.**, Field attractiveness of the synthetic sex pheromones of the rice stem borer moth, *Chilo suppressalis* Walker (Lepidoptera: Pyralidae), *Botyu Kagaku*, 42, 1, 1977.
520. **Kurihara, M., Tatsuki, S., Uchiumi, K., and Fukami, J. -I.**, Laboratory attraction test for sex pheromone of rice stem borer, *Chilo suppressalis* Walker (Lepidoptera: Pyralidae), *Jpn. J. Appl. Entomol. Zool.*, 22, 92, 1978.
521. **Tatsuki, S., Kurihara, M., Uchiumi, K., Fukami, J. -I., Fujimoto, Y., Tatsuno, T., and Kishino, K. -I.**, Factors improving field trapping of male rice stem borer moth, *Chilo suppressalis* Walker (Lepidoptera: Pyralidae), by using synthetic sex attractant, *Appl. Entomol. Zool.*, 14, 95, 1979.
522. **Piccardi, P., Capizzi, A., Cassani, G., Spinelli, P., Arsura, E., and Massardo, P.**, A sex pheromone component of the Old World bollworm *Heliothis armigera*, *J. Insect Physiol.*, 23, 1443, 1977.
523. **Sekul, A. A., Sparks, A. N., Beroza, M., and Bierl, B. A.**, A natural inhibitor of the corn earworm moth sex attractant, *J. Econ. Entomol.*, 68, 603, 1975.
524. **Klun, J. A., Haynes, K. F., Bierl-Leonhardt, B. A., Birch, M. C., and Plimmer, J. R.**, Sex pheromone of the female artichoke plume moth *Platyptilia carduidactyla*, *Environ. Entomol.*, in press.
525. **Kasang, G., Kaissling, K. E., Vostrowsky, O., and Bestmann, H. J.**, Bombykal, a second pheromone component of the silkworm moth *Bombyx mori* L., *Angew. Chem.*, 90, 74, 1978; *Angew. Chem. Int. Ed. Eng.*, 17, 60, 1978.
526. **Kaissling, K. E., Kasang, G., Bestmann, H. J., Stransky, W., and Vostrowsky, O.**, A new pheromone of the silkworm moth *Bombyx mori*. Sensory pathway and behavioral effect, *Naturwissenschaften*, 65, 382, 1978.
527. **Starratt, A. N., Dahm, K. H., Allen, N., Hildebrand, J. G., Payne, T. L., and Röller, H.**, Bombykal, a sex pheromone of the sphinx moth *Manduca sexta*, *Z. Naturforsch.*, 34c, 9, 1979.
528. **Kehat, M., Gothilf, S., Dunkelblum, E., and Greenberg, S.**, Field evaluation of *(E,E)*-10,12-hexadecadienal, a component of the female sex pheromone of the spiny bollworm, *Earias insulana* (Boisd.) (Lepidoptera, Noctuidae), *Phytoparasitica*, 7, 99, 1979.
529. **Hall, D. R., Beevor, P. S., Lester, R., and Nesbitt, B. F.**, *(E,E)*-10,12-Hexadecadienal: a component of the female sex pheromone of the spiny bollworm, *Earias insulana* (Boisd.) (Lepidoptera: Noctuidae), *Experientia*, 36, 152, 1980.
530. **Coffelt, J. A., Vick, K. W., Sonnet, P. E., and Doolittle, R. E.**, Isolation, identification, and synthesis of a female sex pheromone of the navel orangeworm, *Amyelois transitella* (Lepidoptera: Pyralidae), *J. Chem. Ecol.*, 5, 955, 1979.
531. **Cross, J. H., Byler, R. C., Cassidy, R. F., Jr., Silverstein, R. M., Greenblatt, R. E., Burkholder, W. E., Levinson, A. R., and Levinson, H. Z.**, Porapak-Q collection of pheromone components and isolation of (*Z*)- and (*E*)-14-methyl-8-hexadecenal, sex pheromone components, from the females of four species of *Trogoderma* (Coleoptera: Dermestidae), *J. Chem. Ecol.*, 2, 457, 1976.
532. **Rossi, R. and Niccoli, A.**, Relationship between chirality and biological activity. Biological response of *Trogoderma granarium* to optically active synthetic sex attractants, *Naturwissenschaften*, 65, 259, 1978.
533. **Silverstein, R. M. and Young, J. C.**, Insects generally use multicomponent pheromones, in *Pest Management with Insect Sex Attractants and Other Behavior-Controlling Chemicals*, Beroza, M., Ed., ACS Symp. Ser. 23, American Chemical Society, Washington, D. C., 1976, 1.

534. **Rossi, R. and Niccoli, A.**, Feromoni degli insetti. Sintesi ed attivita' biologica di componenti insaturi, otticamente attivi, del feromone del *Trogoderma granarium* Everts e di un loro analogo strutturale, *Redia*, 60, 525, 1977.
535. **Rossi, R., Salvadori, P. A., Carpita, A., and Niccoli, A.**, Chiraliy influences the biological activity of the sex pheromones of the khapra beetle, *Naturwissenschaften*, 66, 211, 1979.
536. **Shapas, T. J. and Burkholder, W. E.**, Patterns of sex pheromone release from adult females, and effects of air velocity and pheromone release rates on theoretical communication distances in *Trogoderma glabrum*, *J. Chem. Ecol.*, 4, 395, 1978.
537. **Mori, K., Suguro, T., and Uchida, M.**, Synthesis of optically active forms of *(Z)*-14-methylhexadec-8-enal. The pheromone of female dermestid beetle, *Tetrahedron*, 34, 3119, 1978.
538. **Wood, W. F., Leahy, M. G., Galun, R., Prestwich, G. D., Meinwald, J., Purnell, R. E., and Payne, R. C.**, Phenols as pheromones of ixodid ticks: a general phenomenon?, *J. Chem. Ecol.*, 1, 501, 1975.
539. **Francke, W. and Heemann, V.**, Lockversuche bei *Xyloterus domesticus* L., und *X. lineatus* Oliv. (Coleoptera: Scolytidae) mit 3-Hydroxy-3-methylbutan-2-on, *Z. Angew. Entomol.*, 75, 67, 1974.
540. **Tamaki, Y., Honma, K., and Kawasaki, K.**, Sex pheromone of the peach fruit moth, *Carposina niponensis* Walsingham (Lepidoptera: Carposinidae): isolation, identification and synthesis, *Appl. Entomol. Zool.*, 12, 60, 1977.
541. **Honma, K., Kawasaki, K., and Tamaki, Y.**, Sex-pheromone activities of synthetic 7-alken-11-ones for male peach fruit moths, *Carposina niponensis* Walsingham (Lepidoptera: Carposinidae), *Jpn. J. Appl. Entomol. Zool.*, 22, 87, 1978.
542. **Shirasaki, S., Yamada, M., Sato, R., Yaginuma, K., and Tamaki, Y.**, Field tests on attractiveness of the synthetic sex pheromone of the peach fruit moth, *Carposina niponensis* Walsingham (Lepidoptera: Carposinidae), *Jpn. J. Appl. Entomol. Zool.*, 23, 240, 1979.
543. **Daterman, G. E., Peterson, L. J., Robbins, R. G., Sower, L. L., Daves, G. D., Jr., and Smith, R. G.**, Laboratory and field bioassay of the Douglas-fir tussock moth pheromone, *(Z)*-6-heneicosene-11-one, *Environ. Entomol.*, 5, 1187, 1976.
544. **Grant, G. G.**, Interspecific pheromone responses of tussock moths and some isolating mechanisms of eastern species, *Environ. Entomol.*, 6, 739, 1977.
545. **Grant, G. G.**, Field trials on disruption of pheromone communication of tussock moths, *J. Econ. Entomol.*, 71, 453, 1978.
546. **Grant, G. G. and Frech, D.**, Mating disruption of tussock moths by atmospheric permeation with synthetic sex pheromone, *Can. For. Serv. Bi-Mon. Res. Notes*, 32, 25, 1976.
547. **Grant, G. G. and Frech, D.**, Disruption of pheromone communication of the rusty tussock moth, *Orgyia antiqua* (Lepidoptera: Lymantriidae), with *(Z)*-6-heneicosen-11-one, *Can. Entomol.*, 112, 221, 1980.
548. **Smith, R. G., Daterman, G. E., and Daves, G. D., Jr.**, Douglas-fir tussock moth: sex pheromone identification and synthesis, *Science*, 188, 63, 1975.
549. **Sower, L. L. and Daterman, G. E.**, Evaluation of synthetic sex pheromone as a control agent for Douglas-fir tussock moths, *Environ. Entomol.*, 6, 889, 1977.
550. **Sower, L. L., Daterman, G. E., Orchard, R. D., and Sartwell, C.**, Reduction of Douglas-fir tussock moth *Orgyia pseudotsugata* reproduction with synthetic sex pheromone, *J. Econ. Entomol.*, 72, 739, 1979.
551. **Nishida, R., Fukami, H., and Ishii, S.**, Sex pheromone of the German cockroach *(Blatella germanica* L.) responsible for male wing-raising: 3,11-dimethyl-2-nonacosanone, *Experientia*, 30, 978, 1974.
552. **Nishida, R., Fukami, H., and Ishii, S.**, Female sex pheromone of the German cockroach, *Blatella germanica* (L.) (Orthoptera, Blattellidae), responsible for male wing-raising. I, *Appl. Entomol. Zool.*, 10, 10, 1975.
553. **Nishida, R., Kuwahara, Y., Fukami, H., and Ishii, S.**, Female sex pheromone of the German cockroach, *Blatella germanica* (L.) (Orthoptera: Blattellidae), responsible for male wing-raising. IV. The absolute configuration of the pheromone, 3,11-dimethyl-2-nonacosanone, *J. Chem. Ecol.*, 5, 289, 1979.
554. **Burgstahler, A. W., Weigel, L. O., Sanders, M. E., Shaefer, C. G., Bell, W. J., and Vuturo, S. B.**, Synthesis and activity of 29-hydroxy-3,11-dimethyl-2-nonacosanone, component B of the German cockroach sex pheromone, *J. Org. Chem.*, 42, 566, 1977.
555. **Nishida, R., Sato, T., Kuwahara, Y., Fukami, H., and Ishii, S.**, Female sex pheromone of the German cockroach, *Blattella germanica* (L.)(Orthoptera: Blattellidae), responsible for male wing-raising. II. 29-Hydroxy-3,11-dimethyl-2-nonacosanone, *J. Chem. Ecol.*, 2, 449, 1976.
556. **Nishida, R., Sato, T., Kuwahara, Y., Fukami, H., and Ishii, S.**, Female sex pheromone of the German cockroach, *Blattella germanica* (L.), responsible for male wingraising. III. Synthesis of 29-hydroxy-3,11-dimethyl-2-nonacosanone, sex pheromone of the German cockroach, *Agric. Biol. Chem.*, 40, 1407, 1976.
557. **Hooper, G. H. S. and Drew, R. A. I.**, Effect of height of trap on capture of tephritid fruit flies with cuelure and methyl eugenol in different environments, *Environ. Entomol.*, 8, 786, 1979.

558. **Drew, R. A. I.,** The responses of fruit fly species (Diptera: Tephritidae) in the South Pacific area to male attractants, *J. Aust. Entomol. Soc.*, 13, 267, 1974.
559. **Umeya, K. and Hirao, J.,** Attraction of the jackfruit fly, *Dacus umbrosus* F. (Diptera: Tephritidae) and lace wing, *Chrysopa* sp. (Neuroptera: Chrysopidae) by lure traps baited with methyl eugenol and cue-lure in the Philippines, *Appl. Entomol. Zool.*, 10, 60, 1975.
560. **Beroza, M., Alexander, B. H., Steiner, L. F. Mitchell, W. C., and Miyashita, D. H.,** New synthetic lures for the male melon fly, *Science*, 131, 1044, 1960.
561. **Fletcher, B. S.,** Storage and release of a sex pheromone by the Queensland fruit fly, *Dacus tryoni* (Diptera: Trypetidae), *Nature (London)*, 219, 631, 1968.
562. **Rudinsky, J. A., Kline, L. N., and Diekman, J. D.,** Response-inhibition by four analogues of MCH, an antiaggregative pheromone of the Douglas-fir beetle, *J. Econ. Entomol.*, 68, 527, 1975.
563. **Rudinsky, J. A.,** Multiple functions of the Douglas fir beetle pheromone 3-methyl-2-cyclohexen-1-one, *Environ. Entomol.*, 2, 579, 1973.
564. **Rudinsky, J. A., Morgan, M., Libbey, L. M., and Michael, R. R.,** Sound production in Scolytidae: 3-methyl-2-cyclohexen-1-one released by the female Douglas fir beetle in response to male sonic signal, *Environ. Entomol.*, 2, 505, 1973.
565. **Furniss, M. M., Daterman, G. E., Kline, L. N., McGregor, M. D., Trostle, G. C., Pettinger, L. F., and Rudinsky, J. A.,** Effectiveness of the Douglas-fir beetle antiaggregative pheromone methylcyclohexenone at three concentrations and spacings around felled host trees, *Can. Entomol.*, 106, 381, 1974.
566. **Hedden, R. L. and Pitman, G. B.,** Attack density regulation: a new approach to the use of pheromones in Douglas-fir beetle population management, *J. Econ. Entomol.*, 71, 633, 1978.
567. **Rudinsky, J. A., Sartwell, C., Jr., Graves, T. M., and Morgan, M. E.,** Granular formulation of methylcyclohexenone: an antiaggregative pheromone of the Douglas fir and spruce bark beetles (Col., Scolytidae), *Z. Angew. Entomol.*, 75, 254, 1974.
568. **Kline, L. N., Schmitz, R. F., Rudinsky, J. A., and Furniss, M. M.,** Repression of spruce beetle (Coleoptera) attraction by methylcyclohexenone in Idaho, *Can. Entomol.*, 106, 485, 1974.
569. **Rudinsky, J. A. and Ryker, L. C.,** Field bioassay of male Douglas-fir beetle compound 3-methylcyclohex-3-en-1-one, *Experientia*, 35, 1302, 1979.
570. **Payne, T. L. and Richerson, J. V.,** Management implications of inhibitors for *Dendroctonus frontalis* (Col. Scolytidae), *Mitt. Schweiz. Entomol. Ges.*, 52, 323, 1979.
571. **Evans, D. A. and Higgs, M. D.,** Microscale structural studies of components of the frass of the wood-boring beetle, *Hylotrupes bajalus* (L), in *Proc. Symp. Pheromones Def. Secret. Soc. Insects (Dijon 1974)*, Noirot, D., Howse, P. E., and LeMasne, G., Eds., IUSSI, French Section, Dijon, 1975, 87.
572. **Persoons, C. J., Verwiel, P. E. J., Ritter, F. J., Talman, E., Nooijen, P. J. F., and Nooijen, W. J.,** Sex pheromones of the American cockroach, *Periplaneta americana:* a tentative structure of periplanone-B, *Tetrahedron Lett.*, 2055, 1976.
573. **Persoons, C. J., Verwiel, P. E. J., Talman, E., and Ritter, F. J.,** Sex pheromone of the American cockroach, *Periplaneta americana.* Isolation and structure elucidation of periplanone-B, *J. Chem. Ecol.*, 5, 221, 1979.
574. **Persoons, C. J. and Ritter, F. J.,** Pheromones of cockroaches, in *Chemical Ecology: Odour Communication in Animals*, Ritter, F. J., Ed., Elsevier/North-Holland, Amsterdam, 1979, 225.
575. **Kuwahara, Y., Fukami, H., Howard, R., Ishii, S., Matsumura, F., and Burkholder, W. E.,** Chemical studies on the Anobiidae: sex pheromone of the drugstore beetle, *Stegobium paniceum* (L.) (Coleoptera), *Tetrahedron*, 34, 1769, 1978.
576. **Bierl, B. A., Beroza, M., and Collier, C. W.,** Potent sex attractant of the gypsy moth: its isolation, identification, and synthesis, *Science*, 170, 87, 1970.
577. **Bierl, B. A., Beroza, M., and Collier, C. W.,** Isolation, identification, and synthesis of the gypsy moth sex attractant, *J. Econ. Entomol.*, 65, 659, 1972.
578. **Beroza, M., Hood, C. S., Trefrey, D., Leonard, D. E., Knipling, E. F., Klassen, W., and Stevens, L. J.,** Large field trial with microencapsulated sex pheromone to prevent mating of the gypsy moth, *J. Econ. Entomol.*, 67, 659, 1974.
579. **Schwalbe, C. P., Paszek, E. C., Webb, R. E., Bierl-Leonhardt, B. A., Plimmer, J. R., McComb, C. W., and Dull, C. W.,** Field evaluation of controlled release formulations of disparlure for gypsy moth mating disruption, *J. Econ. Entomol.*, 72, 322, 1979.
580. **Webb, R. E., McComb, C. W., Plimmer, J. R., Bierl-Leonhardt, B. A., Schwalbe, C. P., and Altman, R. M.,** Disruption along the "leading edge" of the infestation, *The Gypsy Moth: Research Toward Integrated Pest Management, U.S. Dep. Agric. Tech Bull.*, 1584, in press, 560.
581. **Plimmer, J. R.,** Disruption of mating. III. Forest insects: Gypsy moth, this volume.
582. **Beroza, M., Katagiri, K., Iwata, Z., Ishizuka, H., Suzuki, S., and Bierl, B. A.,** Disparlure and analogues as attractants for two Japanese lymantriid moths, *Environ. Entomol.*, 2, 966, 1973.

583. **Schönherr, J.**, Die Wirkung von Disparlure auf die Nonne, *Lymantria monacha* L., *Z. Angew. Entomol.*, 71, 260, 1972.
584. **Bierl, B. A., Beroza, M., Adler, V. E., Kasang, G., Schröter, H., and Schneider, D.**, The presence of disparlure, the sex pheromone of the gypsy moth, in the female nun month, *Z. Naturforsch.*, 30c, 672, 1975.
585. **Beroza, M., Punjabi, A. A., and Bierl, B. A.**, Disparlure and analogues as attractants for *Lymantria obfuscata*, *J. Econ. Entomol.*, 66, 1215, 1973.
586. **Punjabi, A. A., Beroza, M., and Bierl, B. A.**, Trapping of *Lymantria obfuscata* Walker with disparlure, *Indian J. Entomol.*, 36, 125, 1974.
587. **Capek, M.**, Vplyv orientacie feromonovych lapakov na ich atraktivitu pre mnisku vel'kohlavu *(Lymantria dispar* L.), *Lesnictvi*, 25, 301, 1979.
588. **Minyailo, V. A., Kovalev, B. G., Kirov, E. I., and Minyailo, A. K.**, On the attractiveness of disparlure, the sex pheromone of the gypsy moth *Porthetria dispar* (Lepidoptera, Orgyidae), for males of *Zanclognatha lunaris* (Lepidoptera, Noctuidae), *Zool. Zh.*, 56, 309, 1977.
589. **Iwaki, S., Marumo, S., Saito, T., Yamada, M., and Katagiri, K.**, Synthesis and activity of optically active disparlure, *J. Am. Chem. Soc.*, 96, 7842, 1974.
590. **Yamada, M., Saito, T., Katagiri, K., Iwaki, S., and Marumo, S.**, Electroantennogram and behavioural responses of the gypsy moth to enantiomers of disparlure and its *trans* analogues, *J. Insect Physiol.*, 22, 755, 1976.
591. **Vité, J. P., Klimetzek, D., Loskant, G., Hedden, R., and Mori, K.**, Chirality of insect pheromones: response interruption by inactive antipodes, *Naturwissenschaften*, 63, 582, 1976.
592. **Miller, J. R., Mori, K., and Roelofs, W. L.**, Gypsy moth field trapping and electroantennogram studies with pheromone enantiomers, *J. Insect Physiol.*, 23, 1447, 1977.
593. **Plimmer, J. R., Schwalbe, C. P., Paszek, E. C., Bierl, B. A., Webb, R. E., Marumo, S., and Iwaki, S.**, Contrasting effectiveness of (+) and (−) enantiomers of disparlure for trapping native populations of gypsy moth in Massachusetts, *Environ. Entomol.*, 6, 518, 1977.
594. **Bierl-Leonhardt, B. A.**, Controlled release formulations: design, application, and evaluation, presented at meeting of OILB/IOBC Working Group on Use of Pheromones in Integrated Control, Wadenswil, Switzerland, October 1 to 5, 1979.
595. **Klimetzek, D., Loskant, G., Vité, J. P., and Mori, K.**, Disparlure: differences in pheromone perception between gypsy moth and nun moth, *Naturwissenschaften*, 63, 581, 1976.
596. **Klimetzek, D. and Schonherr, J.**, Unterschiede im Anflugverhalten von *Lymantria monacha* L. and *L. dispar* L. an razemisches Disparlur, *Anz. Schädlingskd. Pflanz. Umweltschutz*, 51, 23, 1978.
597. **Miller, J. R. and Roelofs, W. L.**, Gypsy moth responses to pheromone enantiomers as evaluated in a sustained-flight tunnel, *Environ. Entomol.*, 7, 42, 1978.
598. **Silverstein, R. M., Brownlee, R. G., Bellas, T. E., Wood, D. L., and Browne, L. E.**, Brevicomin: principal sex attractant in the frass of the female western pine beetle, *Science*, 159, 889, 1968.
599. **Pitman, G. B. and Vité, J. P.**, Predator-prey response to western pine beetle attractants, *J. Econ. Entomol.*, 64, 402, 1971.
600. **Payne, T. L., Coster, J. E., Richerson, J. V., Edson, L. J., and Hart, E. R.**, Field response of the southern pine beetle to behavioral chemicals, *Environ. Entomol.*, 7, 578, 1978.
601. **Vité, J. P. and Pitman, G. B.**, Management of western pine beetle populations: use of chemical messengers, *J. Econ. Entomol.*, 63, 1132, 1970.
602. **Kinzer, G. W., Fentiman, A. F., Jr., Page, T. F., Jr., Foltz, R. L., Vité, J. P., and Pitman, G. B.**, Bark beetle attractants: identification, synthesis and field bioassay of a new compound isolated from *Dendroctonus*, *Nature (London)*, 221, 477, 1969.
603. **Pitman, G. B. and Vité, J. P.**, Field response of *Dendroctonus pseudotsugae* (Coleoptera: Scolytidae) to synthetic frontalin, *Ann. Entomol. Soc. Am.*, 63, 661, 1970.
604. **Rudinsky, J. A., Morgan, M. E., Libbey, L. M., and Putman, T. B.**, Release of frontalin by male Douglas-fir beetle, *Z. Angew. Entomol.*, 81, 267, 1976.
605. **Dyer, E. D. A.**, Spruce beetle aggregated by the synthetic pheromone frontalin, *Can. J. For. Res.*, 3, 486, 1973.
606. **Dyer, E. D. A.**, Frontalin attractant in stands infested by the spruce beetle, *Dendroctonus rufipennis* (Coleoptera: Scolytidae), *Can. Entomol.*, 107, 979, 1975.
607. **Williamson, D. L.**, Olfactory discernment of prey by *Medetera bistriata* (Diptera: Dolichopodidae), *Ann. Entomol. Soc. Am.*, 64, 586, 1971.
608. **Vité, J. P. and Williamson, D. L.**, *Thanasimus dubius:* prey perception, *J. Insect Physiol.*, 16, 233, 1970.
609. **Kennedy, B. H.**, The effect of multilure on parasites of the European elm bark beetle, *Scolytus multistriatus*, presented at Symp. on Behavior Modifying Chemicals for Elm Bark Beetles, Annu. Meet., Entomol. Soc. Am., Washington, D. C., November 27 to December 1, 1977; *Bull. Entomol. Soc. Am.*, 25, 116, 1979.

610. **Lanier, G. N., Silverstein, R. M., and Peacock, J. W.,** Attractant pheromone of the European elm bark beetle (*Scolytus multistriatus*): isolation, identification, synthesis, and utilization studies, in *Perspectives in Forest Entomology,* Academic Press, New York, 1976, 149.
611. **Elliott, W. J., Hromnak, G., Fried, J., and Lanier, G. N.,** Synthesis of multistriatin enantiomers and their action on *Scolytus multistriatus* (Coleoptera: Scolytidae), *J. Chem. Ecol.,* 5, 279, 1979.
612. **Mori, K., Sasaki, M., Tamada, S., Suguro, T., and Masuda, S.,** Synthesis of optically active 2-ethyl-1,6-dioxaspiro[4,4]nonane (chalcogran), the principal aggregation pheromone of *Pityogenes chalcographus* (L.), *Tetrahedron,* 35, 1601, 1979.
613. **MacConnell, J. G., Borden, J. H., Silverstein, R. M., and Stokkink, E.,** Isolation and tentative identification of lineatin, a pheromone from the frass of *Trypodendron lineatum* (Coleoptera: Scolytidae), *J. Chem. Ecol.,* 3, 549, 1977.
614. **Borden, J. H., Oehlschlager, A. C., Slessor, K. N., Chong, L., and Pierce, H. D., Jr.,** Field tests of isomers of lineatin, the aggregation pheromone of *Trypodendron lineatum* (Coleoptera: Scolytidae), *Can. Entomol.,* 112, 107, 1980.
615. **Henzell, R. F. and Lowe, M. D.,** Sex attractant of the grass grub beetle, *Science,* 168, 1005, 1970.
616. **Gruner, L.,** Etude de l'activite des adultes de divers Scarabeides antillais au moyen de piegeages lumineux et chimiques, *Ann. Zool. Ecol. Anim.,* 7, 399, 1975.
617. **Osborne, G. O., Chapman, R. B., and Boyd, J. F.,** Chemical attractants for males of the grass grub beetle *Costelytra zealandica:* fluorine-substituted phenols, *Environ. Entomol.,* 8, 610, 1979.
618. **Sonenshine, D. E., Silverstein, R. M., Plummer, E., West, J. R., and McCullough, T.,** 2,6-Dichlorophenol, the sex pheromone of the Rocky Mountain wood tick, *Dermacentor andersoni* Stiles and the American dog tick, *Dermacentor variabilis* (Say), *J. Chem. Ecol.,* 2, 201, 1976.
619. **Berger, R. S.,** 2,6-Dichlorophenol, sex pheromone of the lone star tick, *Science,* 177, 704, 1972.
620. **Kellum, D. and Berger, R. S.,** Relationship of the occurrence and function of 2,6-dichlorophenol in two species of *Amblyomma* (Acari: Ixodidae), *J. Med. Entomol.,* 13, 701, 1977.
621. **Leahy, M. G. and Booth, K. S.,** Perception of a sex pheromone, 2,6-dichlorophenol, in hard ticks, in *Tick-borne Diseases and Their Vectors,* Wilde, J. K. H., Ed., Edinburgh University Press, Scotland, U.K., 1978, 88.
622. **Sonenshine, D. E., Silverstein, R. M., Collins, L. A., Saunders, M., Flynt, C., and Homsher, P. J.,** Foveal glands, source of sex pheromone production in the ixodid tick *Dermacentor andersoni* Stiles, *J. Chem. Ecol.,* 3, 695, 1977.
623. **Chow, Y. S., Wang, C. B., and Lin, L. C.,** Identification of a sex pheromone of the female brown dog tick, *Rhipicephalus sanguineus, Ann. Entomol. Soc. Am.,* 68, 485, 1975.
624. **Suda, D. Y. and Cunningham, R. T.,** *Chrysopa basalis* captured in plastic traps containing methyl eugenol, *J. Econ. Entomol.,* 63, 1706, 1970.
625. **Brieze-Stegeman, R., Rice, M. J., and Hooper, G. H. S.,** Daily periodicity in attraction of male tephritid fruit flies in synthetic chemical lures, *J. Aust. Entomol. Soc.,* 17, 341, 1978.
626. **Shah, A. H. and Patel, R. C.,** Role of tulsi plant *(Ocimum sanctum)* in control of mango fruitfly, *Dacus correctus* (Tephritidae: Diptera), *Curr. Sci.,* 45, 313, 1976.
627. **Steiner, L. F.,** Methyl eugenol as an attractant for oriental fruit fly, *J. Econ. Entomol.,* 45, 241, 1952.
628. **Drew, R. A. I.,** Additions to the species of Dacini (Diptera: Tephritidae) from the South Pacific area with keys to species, *J. Aust. Entomol. Soc.,* 11,185, 1972.
629. **Qureshi, Z. A., Ashraf, M., Bughio, A. R., and Siddiqui, Q. H.,** Population fluctuation and dispersal studies of the fruit fly, *Dacus zonatus* Saunders, in *Sterility Principle for Insect Control — 1974,* Proc. Symp., Innsbruck, July 22 to 26, 1974, International Atomic Energy Agency, Vienna, 1975, 201.
630. **Wright, R. H., Chapman, J. A., and Dyer, E. D. A.,** Molecular vibration and insect attraction, *Can. For. Serv. Bi-Mon. Res. Notes,* 30(2), 10, 1974.
631. **Kostelc, J. G., Garcia, B. J., Gokel, G. W., and Hendry, L. B.,** Macrocyclic polyethers as probes into pheromone receptor mechanisms of a sciarid fly, *Lycoriella mali* Fitch, *J. Chem. Ecol.,* 5, 179, 1979.
632. **Vander Meer, R. K., Ghatak, U. R., Alam, S. K., and Chakraborti, P. C.,** (±)-des-*N*-Morphinan: a unique bridged hydrocarbon attractant for the rhinoceros beetle, *Oryctes rhinoceros;* and development of an olfactometer, *Environ. Entomol.,* 8, 6, 1979.
633. **Brossut, R., Dubois, P., and Rigaud, J.,** Le gregarisme chez *Blaberus craniifer:* isolement et identification de la pheromone, *J. Insect Physiol.,* 20, 529, 1974.
634. **Kostelc, J. G., Hendry, L. B., and Snetsinger, R. J.,** A sex pheromone complex of the mushroom-infesting sciarid fly, *Lycoriella mali* Fitch, *J. N. Y. Entomol. Soc.,* 83, 255, 1975.
635. **Roelofs, W. L. and Cardé, R. T.,** Hydrocarbon sex pheromone in tiger moths (Arctiidae), *Science,* 171, 684, 1971.
636. **Cardé, R. T., Roelofs, W. L., and Doane, C. C.,** Natural inhibitor of the gypsy moth sex attractant, *Nature,* 241, 474, 1973.

637. **Niccoli, A.**, Contributi alla conoscenza dell'azione di feromoni e di composti di sintesi sugli insetti. I. Attrattivita di tre alcheni nei confronti degli adulti del *Dacus oleae* Gmel., *Redia*, 56, 91, 1975.
638. **Jones, R. L., Lewis, W. J., Beroza, M., Bierl, B. A., and Sparks, A. N.**, Host-seeking stimulants (kairomones) for the egg parasite, *Trichogramma evanescens*, *Environ. Entomol.*, 2, 593, 1973.
639. **Linley, J. R. and Carlson, D. A.**, A contact mating pheromone in the biting midge, *Culicoides melleus*, *J. Insect Physiol.*, 24, 423, 1978.
640. **Lewis, W. J., Jones, R. L., Nordlund, D. A., and Sparks, A. N.**, Kairomones and their use for management of entomophagous insects. I. Evaluation for increasing rates of parasitization by *Trichogramma* spp. in the field. *J. Chem. Ecol.*, 1, 343, 1975.
641. **Lewis, W. J., Jones, R. L., Nordlund, D. A., and Gross, H. R., Jr.**, Kairomones and their use for management of entomaphagous insects. II. Mechanisms causing increase in rate of parasitization by *Trichogramma* spp., *J. Chem. Ecol.*, 1, 349, 1975.
642. **Carlson, D. A., Mayer, M. S., Silhacek, D. L., James, J. D., Beroza, M., and Bierl, B. A.**, Sex attractant pheromone of the house fly: isolation, identification and synthesis, *Science*, 174, 76, 1971.
643. **Richter, I., Krain, H., and Mangold, H. K.**, Long-chain (*Z*)-9-alkenes are 'psychedelics' to houseflies with regard to visually stimulated sex attraction and aggregation, *Experientia*, 32, 186, 1976.
644. **Uebel, E. C., Schwarz, M., Sonnet, P. E., Miller, R. W., and Menzer, R. E.**, Evaluation of the mating stimulant pheromones of *Fannia canicularis*, *F. pusio*, and *F. femoralis* as attractants, *Fla. Entomol.*, 61, 139, 1978.
645. **Sonnet, P. E., Uebel, E. C., and Miller, R. W.**, An unusual polyene from male stable flies, *J. Chem. Ecol.*, 3, 251, 1977.
646. **Uebel, E. C., Sonnet, P. E., Miller, R. W., and Beroza, M.**, Sex pheromone of the face fly, *Musca autumnalis* De Geer (Diptera: Muscidae), *J. Chem. Ecol.*, 1, 195, 1975.
647. **Sonnet, P. E., Uebel, E. C., Lusby, W. R., Schwarz, M., and Miller, R. W.**, Sex pheromone of the stable fly. Identification, synthesis, and evaluation of alkenes from female stable flies, *J. Chem. Ecol.*, 5, 353, 1979.
648. **Uebel, E. C., Schwarz, M., Miller, R. W., and Menzer, R. E.**, Mating stimulant pheromone and cuticular lipid constituents of *Fannia femoralis* (Stein) (Diptera: Muscidae), *J. Chem. Ecol.*, 4, 83, 1978.
649. **Uebel, E. C., Schwarz, M., Menzer, R. E., and Miller, R. W.**, Mating stimulant pheromone and cuticular lipid constituents of *Fannia pusio* (Wiedemann) (Diptera: Muscidae), *J. Chem. Ecol.*, 4, 73, 1978.
650. **Vinson, S. B., Jones, R. L., Sonnet, P. E., Bierl, B. A., and Beroza, M.**, Isolation, identification and synthesis of host-seeking stimulants for *Cardiochiles nigriceps*, a parasitoid of tobacco budworm, *Entomol. Exp. Appl.*, 18, 443, 1975.
651. **Jones, H. L., Lewis, W. J., Bowman, M. C., Beroza, M., and Bierl, B. A.**, Host-seeking stimulant for parasite of corn earworm: isolation, identification, and synthesis, *Science*, 173, 842, 1971.
652. **Sonnet, P. E., Uebel, E. C., Harris, R. L., and Miller, R. W.**, Sex pheromone of the stable fly: evaluation of methyl- and 1,5-dimethylalkanes as mating stimulants, *J. Chem. Ecol.*, 3, 245, 1977.
653. **Carlson, D. A., Langley, P. A., and Huyton, P.**, Sex pheromone of the tsetse fly: isolation, identification, and synthesis of contact aphrodisiacs, *Science*, 201, 750, 1978.
654. **Flint, H. M., Salter, S. S., and Walters, S.**, Caryophyllene: an attractant for the green lacewing, *Environ. Entomol.*, 8, 1123, 1979.
655. **Kitamura, C., Takahashi, S. Tahara, S., and Mizutani, J.**, A sex stimulant to the male American cockroach in plants, *Agric. Biol. Chem.*, 40, 1965, 1976.
656. **Knopf, J. A. E. and Pitman, G. B.**, Aggregation pheromone for manipulation of the Douglas-fir beetle, *J. Econ. Entomol.*, 65, 723, 1972.
657. **Pitman, G. B.**, Further observations on Douglure in a *Dendroctonus pseudotsugae* management system, *Environ. Entomol.*, 2, 109, 1973.
658. **Kangas, E.**, Über die in der Orientierungsphase der Borkenkäfer auftretenden Prinzipien, *Z. Angew. Entomol.*, 77, 317, 1975.
659. **Bedard, W. D., Tilden, P. E., Wood, D. L., Silverstein, R. M., Brownlee, R. G., and Rodin, J. O.**, Western pine beetle: field response to its sex pheromone and a synergistic host terpene, myrcene, *Science*, 164, 1284, 1969.
660. **Sutherland, D. W. S.**, *Common Names of Insects and Related Organisms* (1978 Revision), ESA Spec. Publ. 78-1, Entomological Society of America, College Park, Md., 1978, and revisions published in subsequent editions of *Bull. Entomol. Soc. Am.*
661. **Coppedge, J. R., Ahrens, E., Goodenough, J. L., Guillot, F. S., and Snow, J. W.**, Field comparisons of liver and a new chemical mixture as attractants for the screwworm fly, *Environ. Entomol.*, 6, 66, 1977.

The following references are for Table 1 in the contribution by Klassen et al. in Volume I, Chapter 2.

662. **McDonough, L. M.**, personal communication.
662a. **Greene, G. L., Reid, J. C., Blount, V. N., and Riddle, T. C.**, *Environ. Entomol.*, 2, 1113, 1973.
663. **Birch, M.**, Aphrodisiac pheromones in insects, in *Pheromones*, Frontiers of Biology, Vol. 32, Birch, M. C., Ed., Elsevier, New York, 1974, 115.
664. **Jacobson, M.**, *Insect Sex Pheromones*, Academic Press, New York, 1972.
665. **Du Jia Nei**, personal communication.
665a. **Brady, U. E. and Daley, R. C.**, *Ann. Entom. Soc. Am.*, 65, 1356, 1972.
666. **Nesbit, B., Beevor, P. S., Hall, D. R., Lester, R., and Williams, J. R.**, *J. Chem. Ecol.*, 6, 385, 1980.
667. **Minyailo, A. K. and Minyailo, V. A.**, *Lesn. Zh. Izv. Vyssh. Uchebn. Zaved.*, No. 3, 16, 1975.
668. **Grula, J. W., McChesney, J. D., and Taylor, O. R., Jr.**, *J. Chem. Ecol.*, 6, 241, 1980.
669. **Shumaker, E. M.**, Abstracts, Proc. U.S./U.S.S.R. Conf. on Insect Attractants, Tblisi, U.S.S.R., 1978.
670. **Roelofs, W.**, personal communication, 1980.
671. **Lewis, T. and Macaulay, E. D. M.**, *Ecol., Entomol.*, 1, 175, 1976.
672. **Wall, C. and Perry, J. N.**, *Entomol. Exp. Appl.*, 24, 155, 1978.
673. **Heikinheimo, O.**, *Norw. J. Entomol.*, 25, 102, 1978.
674. **Morizur, J. P., Bidan, G., and Kossanyi, J.**, *Tetrahedron Lett.*, 4167, 1975.
675. **Meinwald, J., Thompson, W. R., Eisner, T., and Owen, D. F.**, *Tetrahedron Lett.*, 3485, 1971.
676. **Sum, F. W. and Weiler, L.**, *J. Chem. Soc. Chem. Commun.*, 1978, 1985.
677. **Chen Te-ming, in Notes on trip to PRC, Klassen, W.**, unpublished, 1979.
678. **Langille, R. H. and Keaster, A. J.**, *Environ. Entomol.*, 2, 225, 1973.
679. **Perez Perez, R. and Hensley, S. D.**, *J. Agric. Univ. P. R.*, 57, 320, 1973.
680. **Corbet, S. A.**, *Nature (London)*, 243, 537, 1973.
681. **Hagstrum, D. W., McLaughlin, J. R., Smittle, B. J., and Coffelt, J. A.**, *Environ. Entomol.*, 7, 759, 1978.
682. **Tinbergen, N., Meeuse, J. B., Boerema, L. K., and Varossieau, W. W.**, *Z. Tierpsychologie*, 5, 182, 1942.
683. **Tinbergen, N.**, *The Study of Instinct*, Oxford University Press, New York, 1951.
684. **Chen Te-ming**, personal communication.
685. **Minyailo, V. A. and Minyailo, A. K.**, *Zool. Zh.*, 51, 1247, 1972.
686. **Minyailo, V. A. and Minyailo, A. K.**, *Khemoretseptsiya Nasekomykh, Mater. Vses. Simp., 1st*, 179, 1971.
687. **Steck, W., Chisholm, M. D., Underhill, E. W., and Peters, C. C.**, *J. Chem. Ecol.*, 6, 585, 1980.
688. **Gilbert, L. E.**, *Science*, 193, 419, 1976.
689. **Tumlinson, J.**, personal communication.
690. **Starks, K. J., McMillian, W. W., and Young, J. R.**, *J. Econ. Entomol.*, 60, 1483, 1967.
691. **Capinera, J. L.**, *J. Chem. Ecol.*, 6, 655, 1980.
692. **Roelofs, W. and Kochansky, J.**, personal communication.
693. **Wyman, J. A.**, *J. Econ. Entomol.*, 72, 865, 1979.
694. **McLaughlin, J. R.**, personal communication.
695. **Arakelian, A. O., and Nikogosian, E. Z.**, *Zashch. Rast.*, 20, 50, 1975.
696. **Aplin, R. T. and Birch, M. C.**, *Experientia*, 26, 1193, 1970.
697. **McManus, M. L. and Smith, H. R.**, *Environ. Entomol.*, 1, 793, 1972.
698. **Fitzgerald, T. D.**, *Science*, 194, 961, 1976.
699. **Fitzgerald, T. D. and Gallagher, E. M.**, *J. Chem. Ecol.*, 2, 187, 1976.
700. **Bestmann, H. J., Vostrowsky, O., and Platz, H.**, *Experientia*, 33, 874, 1977.
701. **Clearwater, J. R.**, *Comp. Biochem. Physiol.*, 50B, 77, 1975.
702. **Davis, H. G. and McDonough, L. M.**, *Environ. Entomol.*, in press.
703. **Kovalijov, B. G.**, Abstracts, Proc. U.S./U.S.S.R. Conf. on Insect Attractants, Tblisi, U.S.S.R., 1978.
704. **Brooks, T. W., Doane, C. C., and Staten, R. T.**, in *Chemical Ecology: Odour Communication in Animals*, Ritter, F. J., Ed., Elsevier/North Holland, Amsterdam, 1979, 375.
705. **Graham, H. M., Ed.**, *Pink Bollworm Control in the Western United States*, SEA ARM-W-16, U.S. Department of Agriculture, Washington, D.C., June 1980.
706. **Huber, R. T., Moore, L., and Hoffmann, M. P.**, *J. Econ. Entomol.*, 72, 222, 1979.
707. **Stern, V. M. and Sevacherian, V.**, *Calif. Agric.*, 32(7), 4, 1978.
708. **Birch, M. C., Grant, G. G., and Brady, U. E.**, *Ann. Entomol. Soc. Am.*, 69, 491, 1975.
709. **Edgar, J. A., Cockrum, P. A., and Carrodus, B. B.**, *Experientia*, 35, 861, 1979.
710. **Kuwahara, Y.**, *Appl. Entomol. Zool.*, 14, 350, 1979.
711. **Clearwater, J. R. and Sarafis, V.**, *J. Insect Physiol.*, 19, 19, 1973.

712. **Grant, G. G., Brady, W. E., and Brand, J. M.,** *Ann. Entomol. Soc. Am.*, 65, 1224, 1972.
713. **Minyailo, V. A.,** *Zashch. Rast.*, 19(1), 24, 1974.
714. **Bjostad, L. B., Taschenberg, E. F., and Roelofs, W. L.,** *J. Chem. Ecol.*, 6, 487, 1980.
715. **Redfern, R. E., Cantu, E., Jones, W. A., and Jacobson, M.,** *J. Econ. Entomol.*, 64, 1570, 1971.
716. **Jacobson, M., Redfern, R. E., Jones, W. A., and Aldrich, M. H.,** *Science*, 170, 542, 1970.
717. **Anon.,** *Agric. Res.*, 19(10), 8, 1971.
718. **Yonce, C. E. and Gentry, C. R.,** Mating disruption of the lesser peachtree borer *Synanthedon pictipes* (Grote & Robinson) and the peachtree borer, *Synanthedon exitiosa* (Say) with hollow fiber lures, *Misc. Publ. Entomol. Soc. Am.*, in press.
719. **Baryakina, I. K.,** *Borkl. Bull. V12R*, 22, 9, 1971.
720. **Baryakina, I. K.,** *Vaskhnil. M.*, 9, 1973.
721. **Culvenor, C. C. J. and Edgar, J. A.,** *Experientia*, 28, 627, 1972.
722. **Ivashchenko, I. I. and Oleshchenko, I. N.,** *Zool. Zh.*, 53, 1030, 1974.
723. **Oleshchenko, I. N., Ivashchenko, I. I., and Adamenko, E. A.,** *Skh. Biol.*, 11, 256, 1976.
724. **Ivashchenko, I. I. and Adamenko, E. A.,** *Skh. Biol.*, 6, 407, 1971.
725. **Ivashchenko, I. I. and Adamenko, E. A.,** *Zool. Zh.*, 50, 1021, 1971.
726. **Khlistovskii, E. D. and Ivashchenko, I. I.,** *Nauchn. Dokl. Vyssh. Shk. Biol. Nauko.*, No, 3, 7, 1974.
727. **Ivashchenko, I. I. and Oleshchenko, I. N.,** *Skh. Biol.*, 9, 254, 1974.
728. **Oleshchenko, I. N., Ivashchenko, I. I., and Adamenko, E. A.,** *Skh. Biol.*, 11, 256, 1976.
729. **Ivashchenko, I. I. and Oleshchenko, I. N.,** *Zool. Zh.*, 55, 1108, 1976.
730. **Pinninger, D. D.,** The use of bait traps for assessment of stored-product insect populations, *Coop. Econ. Insect Rep.*, 25, 907, 1975.
731. **Ivashchenko, I. I. and Oleshchenko, I. N.,** *Nauchn. Dokl. Vyssh. Shk. Biol. Nauki.*, 15(9), 23, 1972.
732. **Lloyd, E. P., McKibben, G. H., Knipling, E. F., Witz, J. A., Hartstack, A. W., Leggett, J. E., and Lockwood, D. F.,** *Management of Insect Pests with Semiochemicals: Concepts and Practices*, Mitchell, E. R., Ed., Plenum Press, New York, 1981, 191.
733. **Ma, H., Hummel, H. E., and Burkholder, W. E.,** *J. Chem. Ecol.*, 6, 597, 1980.
734. **Chararas, C.,** *C. R. Acad. Sci. Ser. D*, 278, 1601, 1974.
735. **Chararas, C.,** *C. R. Acad. Sci. Ser. D*, 280, 2567, 1975.
736. **Francke, W. and Heemann, V.,** *Z. Angew. Entomol.*, 82, 117, 1976.
737. **Flint, H. M., Merkle, J. R., and Sledge, M.,** *Environ. Entomol.*, 74, 301, 1981.
738. **Schwartz, P. H., Jr.,** *Control of Insects on Deciduous Fruits and Tree Nuts in the Home Orchard Without Insecticides, U.S. Dep. Agric. Home Garden Bull.*, No. 211, 1975.
739. **Coffelt, J. A., Vick, K. W., Sower, L. L., and McClellan, W. T. M.,** *Environ. Entomol.*, 7, 756, 1978.
740. **Hughes, P. R., Renwick, J. A. A., and Vité, J. P.,** *Environ. Entomol.*, 5, 1165, 1976.
741. **McNew, G. L.,** *Contrib. Boyce Thompson Inst.*, 24, 251, 1970.
742. **Vité, J. P. and Renwick, J. A. A.,** *Contrib. Boyce Thompson Inst.*, 24, 323, 1970.
743. **Fentiman, A. F. and Vité, J. P.,** U. S. Pat. Appl. Series 849,494, 1969.
744. **Vité, J. P.,** U. S. Patent Appl. Series 849,491, 1969.
745. **Bedard, W. D., Wood, D. L., Tilden, P. E., Lindahl, K. Q., Jr., Silverstein, R. M., and Rodin, J. O.,** *J. Chem. Ecol.*, 6, 625, 1980.
746. **Wood, D. L.,** in *Animals*, Noordwijkerhout, The Netherlands, 1979.
747. **Thomas, H. A. and Bradley, E. L.,** *J. Econ. Entomol.*, 68, 147, 1975.
748. **Thomas, H. A. and Hertel, G. D.,** *J. Econ. Entomol.*, 72, 342, 1979.
749. **Coffelt, J. A. and Burkholder, W. E.,** *Ann. Entomol. Soc. Am.*, 65, 447, 1972.
750. **Chuman, T., Kato, K., and Noguchi, M.,** *Agric. Biol. Chem.*, 43, 2005, 1979.
751. **Tamaki, Y.,** personal communication.
752. **Jacobson, M., Lilly, C. E., and Harding, C.,** *Science*, 159, 208, 1968.
753. **Marschall, K.,** *Proc. Int. Conf. Cocoa and Coconuts*, Kuala Lumpur, Malaysia, 1978.
754. **Alfaro, R. I., Pierce, H. D., Jr., Borden, J. H., and Oehlschlager, A. C.,** *J. Chem. Ecol.*, 5, 663, 1979.
755. **Alfaro, R. I., Pierce, H. D., Jr., Borden, J. H., and Oehlschlager, A. C.,** *Can. J. Zool.*, 58, 626, 1980.
756. **Klein, M. G.,** *Management of Insect Pests with Semiochemicals: Concepts and Practices*, Mitchell, E. R., Ed., Plenum Press, New York, 1981, 183.
757. **Burkholder, W. E.,** *Proc. Symp. on Prevention and Control of Insects in Stored Food Products*, U.S. Dep. Agric. Grain Marketing Research Laboratory, Manhattan, Kansas, 1979, 231.
758. **Burkholder, W. E.,** personal communication.
759. **Happ, G. M.,** *Nature (London)*, 222, 180, 1969.
760. **Levinson, H. Z. and Levinson, A. R.,** *Z. Angew., Entomol.*, 84, 337, 1977.
761. **Levinson, H. Z. and Levinson, A. R.,** in *Chemical Ecology: Odour Communication in Animals*, Ritter, F. J., Ed., Elsevier/North-Holland, Amsterdam, 1979, 327.

762. Keville, R. and Kannowski, P. B., *J. Insect Physiol.,* 21, 81, 1975.
763. Levinson, H. Z. and Levinson, A. R., *Entomol. Expl. Appl.,* 24, 305, 1978.
764. Francke, W., and Heemann, V., *Z. Angew. Entomol.,* 75, 67, 1974.
765. Francke, W. and Heyns, K., *Z. Naturforsch.,* 29c, 246, 1974.
766. Francke, W., Heeman, V., and Heyns, K., *Z. Naturforsch.,* 29c, 243, 1974.
767. Kerck, K., *Z. Angew. Entomol.,* 82, 119, 1976.
768. Schreck, C. E., Smith, N., Carlson, D. A., Price, G. D., Haile, D., and Godwin, D. R., *J. Econ. Entomol.,* in press.
769. Burditt, A. K. and McGovern, T. P., *Chemicals Tested as Attractants for the Caribbean Fruit Fly, 1972-1978,* SEA Agr. Rev. Manuals, ARM-S-6,U.S. Department of Agriculture, Washington, D.C., 1979.
770. Greany, P. D., Agee, H. R., Burditt, A. K., and Chambers, D. L., *Entomol. Expl. Appl.,* 23, 63, 1977.
771. Greany, P. D., Burditt, A. K., Jr., Agee, H. R., and Chambers, D. L., *Entomol. Expl. Appl.,* 21, 20, 1978.
772. Lopez, F., Steiner, L. F., and Holbrook, F. R., *J. Econ. Entomol.,* 64, 1541, 1971.
773. Nation, J. L., *Ann. Entomol. Soc. Am.,* 65, 1364, 1972.
774. Nation, J. L., *Environ. Entomol.,* 4, 27, 1975.
775. Perdomo, A. J., Nation, J. L., and Baranowski, R. M., *Environ. Entomol.,* 5, 1208, 1976.
776. Perdomo, A. J., Baranowski, R. M., and Nation, J. L., *Fla. Entomol.,* 58, 291, 1975.
777. Nation, J. L., Proc. 1977 Meeting Int. Soc. Citriculture, May 1 to 8, 1977, Orlando, Fla., 1977, Vol. 2, p. 481.
778. Prokopy, R. J., Greany, P. D., and Chambers, D. L., *Environ. Entomol.,* 6, 463, 1977.
779. Chambers, D. L., in *Chemical Control of Insect Behavior: Theory and Application,* Shorey, H. H. and McKelvey, J. J., Jr., Jr., Eds., John Wiley & Sons, New York, 1977, 327.
780. Steiner, L. F., *Proc. First Int. Citrus Symp.,* University of California at Riverside, Vol. 2, 1969, 881.
781. Steiner, L. F., Rohwer, G. G., Ayers, E. L., and Christenson, L. D., *J. Econ. Entomol.,* 54, 30, 1961.
782. Ohinata, K., Jacobson, M., and Nakagawa, S., U. S. Patent 4,152,422, 1979.
783. Coppedge, J. R., Broce, A. B., Tannahill, F. H., Goodenough, J. L., Snow, J. W., and Crystal, M. M., *J. Econ. Entomol.,* 71, 483, 1978.
784. Plimmer, J. R., Chu, H. W., and Coppedge, J. R., unpublished.
785. Peterson, R. D., Gagne, R. J., Snow, J. W., and Spencer, J. P., *Environ. Entomol.,* in press.
786. Steiner, L. F., Mitchell, W. C., Harris, E. J., Kozuma, T. T., and Fujimoto, M. S., *J. Econ. Entomol.,* 58, 961, 1965.
787. Offer, I. I., Carlson, D. A., Gadzama, N. M., and Bozimo, H. T., unpublished.
788. Bolton, H. T., Butler, J. F., and Carlson, D. A., *J. Chem. Ecol.,* 6, 951, 1980.
789. Jay, E. G., *J. Econ. Entomol.,* 55, 1011, 1962.
790. Koshelev, L. A., *Zap. Leningr. Sh. Inst.,* 212, 57, 1973.
791. Barton Browne, L., Bartell, R. J., and Shorey, H. H., *J. Insect Physiol.,* 15, 1003, 1969.
792. Kwangtung Institute of Entomology, in notes on trip to PRC, Klassen, W., unpublished.
793. Katsoyannos, B. I. and Boller, E. F., *Environ. Entomol.,* 5, 151, 1976.
794. Prokopy, R. J., Reissig, W. H., and Moericke, V., *Entomol. Expl. Appl.,* 20, 170, 1976.
795. Buriff, C. R., *Environ. Entomol.,* 2, 757, 1973.
796. Matsumura, F., Coppel, H. C., and Tai, A., *Nature (London),* 219, 963, 1968.
797. Ritter, F. J., van Oosten, A. M., and Persoons, C. J., Proc. VII Congr. IUSSI, London, 1973, 330.
798. Ritter, F. J. and Persoons, C. J., in *Integrated Control of Insect Pests in the Netherlands,* Minks, A. K. and Gruys, P., Eds., Wageningen Centre for Agriculture Publishing and Documentation, the Netherlands, 1980, chap. 6.1.
799. Ritter, F. J. and Persoons, C. J., Survey of pheromone research at the Division of Technology for Society, TNO, Delft, The Netherlands, 1979, D 79, 83.
800. Humphries, D. A., *Anim. Behav.,* 15, 82, 1967.
801. Bowers, W. S., Nault, L. R., Webb, R. E., and Dutky, S. R., *Science,* 177, 1121, 1972.
802. Blum, M. S., in *Proc. Int. Symp. on Insect Pheromones and Their Applications,* Japan Plant Protection Association, Tokyo, 1977, 49.
803. Wientjens, W. H. J. M., Lakwijk, A. C., and Marel, T. v. d., *Experientia,* 29, 658, 1973.
804. Nault, L. R., Wood, T. K., and Goff, A. M., *Nature (London),* 249, 387, 1974.
805. Levinson, H. Z. and Bar Ilan, A. R., *Experientia,* 27, 102, 1971.
806. Levinson, H. Z., Levinson, A. R., and Maschwitz, U., *Naturwissenschaften,* 61,684, 1974.
807. Youdewei, A. and Calam, D. H., *Proc. R. Entomol. Soc. London Ser. A,* 44, 38, 1969.
808. Ubik, K., Vrkoc, J., Zdarek, J., and Kontev, C., *Naturwissenschaften,* 62, 348, 1975.
809. Aldrich, J. R., Blum, M. S., and Fales, K. M., cited in Ref. 802.

810. **Park, R. J. and Sutherland, M. D.**, *Aust. J. Chem.*, 15, 172, 1962.
811. **Mitchell, W. C. and Mau, R. F. L.**, *J. Econ. Entomol.*, 64, 856, 1971.
812. **Ishiwatari, T.**, *Appl. Entomol. Zool.*, 9, 153, 1974.
813. **Lener, W.**, *Am. Zool.*, 9, 1143, 1969.
814. **Velasquez Antich, A.**, *Rev. Inst. Med. Trop. Sao Paulo*, 10, 242, 1968.
815. **Law, J. H., Wilson, E. O., and McCloskey, J. A.**, *Science*, 149, 544, 1965.
816. **Yoshida, S.**, *Entomol. Expl. Appl.*, 23, 152, 1978.
817. **Haas, A.**, *Naturwissenschaften*, 39, 484, 1952.
818. **Kitano, H.**, *Kontyu*, 43, 513, 1975.
819. **Weseloh, R. M.**, *Science*, 193, 695, 1976.
820. **Cole, L. R.**, *Anim. Behav.*, 18, 184, 1970.
821. **Rao, S. V. and DeBach, P.**, *Hilgardia*, 39, 515, 1969.
822. **Khasimuddin, S. and DeBach, P.**, *Ann. Entomol. Soc. Am.*, 68, 893, 1975.
823. **Cross, J. H., Byler, R. C., Ravid, U., Silverstein, R. M., Robinson, S. W., Baker, P. M., de Oliveira, J. S., Jutsum, A. R., and Cherrett, J. M.**, *J. Chem. Ecol.*, 5, 187, 1979.
824. **Dowell, R. J. and Horn, D. J.**, *Entomophaga*, 23, 271, 1978.
825. **Simser, D.**, M. Sci. thesis, University of Wisconsin, Madison, 1979.
826. **Kamm, J.**, personal communication.
827. **Vinson, S. B.**, *Environ. Entomol.*, 1, 409, 1972.
828. **Holldobler, B. and Maschwitz, U.**, *Z. Vgl. Physiol.*, 50, 551, 1965.
829. **Dlusskii, G. M. and Chernyshova, N. B.**, *Vestn. Mosk. Univ. Biol. Pochvoved.*, 30(6), 41, 1975; *Moscow Univ. Biol. Sci. Bull.*, 30(5/6), 90, 1976.
830. **Fink, D. E.**, *J. Agric. Res.*, 32, 1121, 1926.
831. **Hermann, L. D., Hermann, H. R., and Matthews, R. W.**, *J. Ga. Entomol. Soc.*, 9, 17, 1974.
832. **Askari, A. and Coppel, H. C.**, *Ann. Entomol. Soc. Am.*, 71, 364, 1978.
833. **Ritter, F. J., Rotgans, I. E. M., Talman, E., Verviel, P. E. J., and Stein, F.**, *Experientia*, 29, 530, 1973.
834. **Talman, E., Ritter, F. J., and Verwiel, P. E. J.**, in *Mass Spectrometry in Biochemistry and Medicine*, Friegerio, A. and Castagnoli, N., Eds., Raven Press, New York, 1974, 197.
835. **Ritter, F. J., Bruggemann-Rotgans, I. E. M., Verwiel, P. E. J., Persoons, C. J., and Talman, E.**, *Tetrahedron Lett.*, 2617, 1977.
836. **Boush, G. M. and Baerwald, R. A.**, *Ann. Entomol. Soc. Am.*, 60, 865, 1967.
837. **Hagen, K. S.**, *Proc. Hawaii. Entomol. Soc.*, 15, 115, 1953.
838. **Schlinger, E. I. and Hall, J. C.**, *Ann. Entomol. Soc. Am.*, 53, 144, 1960.
839. **vander Meer, R. K., Glancey, B. M., Lofgren, C. S., Glover, A., Tumlinson, J. M., and Rocca, J.**, *Ann. Entomol. Soc. Am.*, 73, 609, 1980.
840. **Dame, D. A. and Weidhaas, D. E.**, Proc. XVI Int. Congr. Entomol., in press.
841. **Schwartz, A. and Gerling, D.**, *Entomophaga*, 19, 483, 1974.
842. **Miller, M. C.**, *J. Econ. Entomol.*, 59, 1532, 1966.
843. **Kochetova, N. I.**, *Zool. Zh.*, 54, 1571, 1975.
844. **Viktorov, G. A.**, *Zool. J.*, 47, 1035, 1968.
845. **Wheeler, J. W., Evans, S. L., Blum, M. S., Velthuis, H. H. W., and de Camargo, J. M. F.**, *Tetrahedron Lett.*, 4029, 1976.
846. **Flint, H. M., Salter, S. S., and Walters, S.**, *Environ. Entomol.*, 8, 1123, 1979.
847. **Roth, L. M. and Willis, E. R.**, *Smithson. Misc. Collect.*, 122, 1, 1954.
848. **Roth, L. M.**, in *McGraw-Hill Yearbook of Science and Technology*, McGraw-Hill, New York, 1967, 293.
849. **Barth, R. H., Jr.**, *Psyche*, 75, 274, 1968.
850. **Engelmann, F.**, *Experientia*, 16, 69, 1960.
851. **Smyth, T.**, *Abstracts, Meet. Eastern Branch Entomol. Soc. Am.*, Entomological Society of America, College Park, Md., 1963, 22.
852. **Nolte, D. J., Eggers, S. H., and May, I. R.**, *J. Insect Physiol.*, 19, 1547, 1973.
853. **Norris, M. J.**, *Entomol. Expl. Appl.* 6, 279, 1963.
854. **Roth, L.**, *J. Morphol.*, 91, 469, 1952.
855. **Roth, L.**, *Ann. Entomol. Soc. Am.*, 62, 176, 1968.
856. **Grippo. R. S. and Huber, I.**, Abstracts, Meet. Eastern Branch Entomol. Soc. Am., *J. N.Y. Entomol. Soc.*, 86, 291, 1978.
857. **Wright, C. G.**, *J. Ga. Entomol. Soc.*, 12, 281, 1977.
858. **Cummings, B. F.**, *Proc. Zool. Soc. London*, 1914, 439.
859. **Bornemisszа, G. F.**, *Nature (London)*, 209, 732, 1966.
860. **Bornemissza, G. F.**, *Aust. J. Zool.*, 14, 371, 1966.
861. **George, J. E.**, *J. Med. Entomol.*, 18, 129, 1981.

862. **Elizarov, Y. A. and Vasiuta, A. A.**, *Parazitologiya*, 10, 136, 1976.
863. **Yamamoto, I. and Yamamoto, R.**, in *Control of Insect Behavior by Natural Products*, Wood, D. L., Ed., Academic Press, New York, 1970, 331.
864. **Kovalev, B. G.**, *Proc. Soviet-America Symp. on Integrated Control of Pests and Diseases of Agricultural Crops*, Ministry of Agriculture, Moscow, 1980, p. 10.
865. **Roelofs, W. L.**, Report on Pheromone Thrust of USSR-USA Exchange Program on Environmental Protection, Cornell University, Geneva, N.Y., 1979, unpublished.
866. **Haase, E.**, *Zool. Anz.*, 11, 475, 1888.
867. **Hathaway, D. A. and Tamaki, G.**, *J. Entomol. Soc. B. C.*, in press.
868. **Clearwater, J. R.**, *J. Insect Physiol.*, 18, 781, 1972.
869. **Ivashchenko, I. I., Oleshchenko, I. N., Adamenko, A. E., and Pleshova, V. S.**, *Zool. Zh.*, 51, 144, 1972.
870. **Ivashchenko, I. I., Oleshchenko, I. N., and Adamenko, A. E.**, *Biol. Nauki*, 59, 1976.
871. **Osgood, C. E.**, *J. Econ. Entomol.*, 64, 1038, 1971.
872. **Geynieys, P.**, *Ann. Entomol. Soc. Am.*, 18, 143, 1925.
873. **Cavill, G. W. K., Davies, N. W., and McDonald, F. J.**, *J. Chem. Ecol.*, 6, 371, 1980.
874. **Parker, H. L.**, *Macrocentrus gifuensis* Ashmead, a polyembryonic braconid parasite in the European corn borer, *U.S. Dep. Agric. Bull.*, No. 230, 1931.
875. **Cole, L. R.**, *Anim. Behav.*, 18, 184, 1970.
876. **Pasteels, J. M., Verhaeghe, J. C., Braekman, J. C., Daloze, D., and Tursch, B.**, *J. Chem. Ecol.*, 6, 467, 1980.
877. **Ikan, R., Gottlieb, R., Bergmann, E.D., and Ishay, J.**, *J. Insect Physiol.*, 15, 1709, 1969.
878. **Vinson, S. B., Frankie, G. W., Blum, M. S., and Wheeler, J. W.**, *J. Chem. Ecol.*, 4, 315, 1978.
879. **Batra, S. W. T.**, *J. Kansas Entomol. Soc.*, 53, 461, 1980.
880. **Hefetz, A., Batra, S. W. T., and Blum, M. S.**, *J. Chem. Ecol.*, 5, 753, 1979.
881. **Batra, S. W. T. and Hefetz, A.**, *Ann. Entomol. Soc. Am.*, 72, 514, 1979.
882. **Hefetz, A. and Batra, S. W. T.**, *Comp. Biochem. Physiol.*, 65B, 455, 1980.
883. **Hefetz, A. and Batra, S. W. T.**, *Experientia*, 35, 1138, 1979.
884. **Borg-Karlson, A. -K. and Tengo, J.**, *J. Chem. Ecol.*, 6, 827, 1980.
885. **Thorp, R. W.**, *Am. Midl. Nat.*, 82, 321, 1969.
886. **Moglich, M.**, *J. Chem. Ecol.*, 5, 35, 1979.
887. **Cavill, G. W. K. and Williams, P. S.**, *J. Insect Physiol.*, 13, 1097, 1967.
888. **Robertson, P. L.**, *Aust. J. Zool.*, 16, 133, 1968.
889. **Francke, W., Sauerwein, P., Vité, J. P., and Klimetzek, D.**, *Naturwissenschaften*, 67, 147, 1980.

ity to suppress, speed of suppression, and availability and distribution of attractant in the habitat, distribution pattern of the pest, dispersal characteristics of the pest, effectiveness relative to economic threshold, efficiency relative to economic threshold, and ecological selectivity.

A. Capacity of Method to Suppress a Growing Population

The dynamics of the target pest are important in determining the degree of suppressive action required to cause a downward trend in the population. For example, if conditions permit a population to increase fivefold per generation, then more than 80% of the reproductive capacity of the population must be destroyed in each generation in order to cause a downward trend in its density. Similarly, if the intrinsic rate of growth is tenfold per generation, then more than 90% of the reproductive capacity must be destroyed. As a general rule most multivoltine (more than one generation per year) insect populations can be expected to have annual growth rates of five- to tenfold, though at low densities and under favorable conditions higher rates of increase sometimes occur. Thus, for most insects, pest management systems should be designed to deal with a population with an intrinsic rate of increase of at least tenfold per generation.

B. Speed With Which Pest Population is Suppressed by the Control Method

No method of control surpasses synthetic organic insecticides in the capacity to rescue a commodity from destruction by a dense population of an insect pest. However, attractants of the specific life stage that causes damage can facilitate speedy rescue if they are deployed in poison baits, attractant toxicant devices, and traps. Also, cultural procedures such as harvesting a crop of alfalfa or sweet corn when the pest population threatens to exceed the economic threshold are highly effective if the commodity is in the proper stage of development for harvesting. Insect growth regulators and entomopathogenic preparations, on the other hand, usually act with only moderate speed, and the decision to use resistant crop cultivars must be made and implemented long before there is any indication that the insect pest will get out of hand. Pheromones and attractants that do *not* attract the damaging stage of the insect can only be employed prophylactically. For example, the sex pheromones of the pink bollworm, *Heliothis* and *Spodoptera*, must be deployed before mating has occurred because it is the larvae that do the damage. In such cases, pheromones are worthless as pest controls once the population of immature stages exceeds the economic threshold.

C. Availability and Distribution of Natural Sources of the Attractant in Habitats

As Knipling[19] noted, the overall efficiency of pheromone-baited traps (percentage of pests within the region of influence of the trap that are captured) depends upon the competition with native insects that are calling. Further, the efficiency of a trap baited with substances that are present in significant quantities in the environment (e.g., feeding and oviposition attractants produced by host plants or animals or other sources) will depend in some part on the level of emission by the trap vs. the level of emission by the competing sources. Likewise, the distribution patterns of traps and of natural sources are key factors in determining the movements of insects. However, if a powerful artificial attractant used as a bait (e.g., methyl eugenol for oriental fruit fly) is not found in appreciable amounts in the habitat of the pest, competition is not a factor. Figure 1 illustrates trap capture when there is competition between attractants in traps and attractants in nature and compares this with trap catch when there is no such competition.

D. Distribution Pattern of the Pest Population

It is unlikely that any insect population is distributed in a random fashion. Pieters

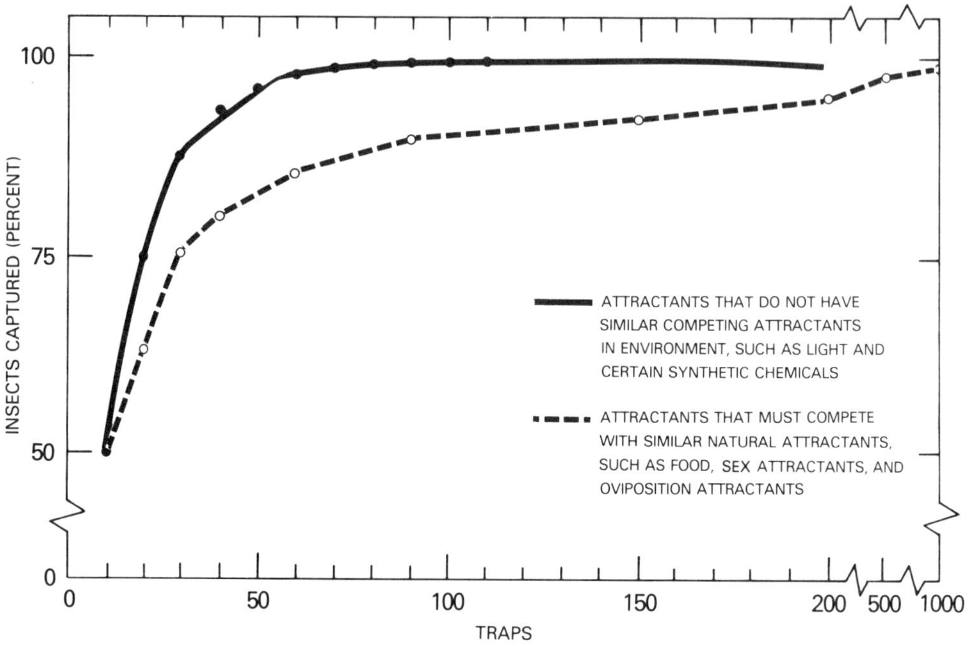

FIGURE 1. Postulated efficiency curve for attractants having different mechanisms of attraction. (Reproduced with permission from E. F. Knipling, 1979.)

and Sterling[26] found that two indexes can be used to quantify the clumping (aggregation) of cotton arthropods — the parameter K of the negative binomial distribution and Morisita's index of dispersion, I_δ.

These indexes were defined as follows:

$$I_\delta = N\frac{\Sigma x^2 - \Sigma x}{(\Sigma x)^2 - \Sigma x} \qquad (1)$$

$$N = \frac{1/\bar{x} + 1/K}{D} \qquad (2)$$

N equals the total samples and Σx equals the sum of the numbers of individual insects found in all samples. Further, $\bar{x}$ equals the mean, and D equals the level of accuracy required (i.e., normally 10%).

The values of these two indexes vary with distribution patterns as follows:

Distribution pattern	K	I_δ
Uniform	≥1	<1
Random	Approaches infinity	1
Aggregated (clumped)	0<1	>1

In fact, because of the ovipositional behavior of the boll weevil, boll weevil eggs are probably more highly aggregated than most arthropod populations. This becomes obvious when one compares the Morisita indexes a randomly distributed population (value of 1), with the Morisita index for boll weevil eggs and boll weevil adults.

	K	I_δ
Boll weevil eggs	0.16	12.83

Boll weevil adults	0.58	4.62
Heliothis eggs	0.36	8.81
Heliothis small larvae	0.88	5.10
Heliothis large larvae	0.62	3.95

Pronounced aggregation such as that for boll weevils has obvious implications for the application of attractants, entomophages, and sterile insects. A clumped distribution may be highly advantageous with entomophages that respond strongly to a kairomone or to odors emitted by the infested commodity. It can be a serious handicap to the sterile male method and to the use of pheromones, either in traps or dispersed as in the air-permeation approach to mating disruption. Thus, with a mass-trapping approach that involves pheromones, the spacing between traps and their placement is likely to be governed much more by the distribution pattern of the pest than by the rate of emission of the pheromone from the trap. On the other hand, trap placement and spacing will undoubtedly be less critical for traps baited with artificial attractants or for light traps.

E. Dispersal and Migratory Characteristics of the Pest

Dispersal, especially of gravid females, is a factor that must be addressed before any use is made of pheromones and other attractants, sexual sterility, or other genetic manipulation. For example, an experimental effort was made to use the sterile male method against the pink bollworm in a cotton field at Desert Center, Calif. an area 35 mi away from the nearest commercial cotton field that was presumed to have no indigenous population of the pest. However, in the period of one generation, the population increased about 150-fold. Such a large increase cannot be explained by excellent survival of indigenous insects because even under ideal laboratory conditions the pink bollworm cannot increase more than 70-fold per generation. Clearly, large numbers of moths had migrated at least 35 mi to the experimental field.[27]

This is why species such as the gypsy moth that have flightless females are ideally suited for management at low densities by means of pheromones or the sterile male technique; virtually all progeny are produced by females indigenous to the target area. Larval drift and transport by man of immature stages into the target area can be dealt with more easily than the influx of gravid females that can travel long distances.

Clearly, to be effective, behavioral chemicals, when used for mass trapping or mating disruption, and genetic manipulation must usually be deployed on a community-wide basis. This is often a formidable problem because it is necessary to organize a large program that requires almost total participation of all interested parties. However, the benefits can be enormous. For example, Knipling[28] calculated that more pest insects are produced when virtually complete control of the insect is achieved in 90% of an area than when only 90% control is achieved in the total area. He calculated that in a typical situation, this difference would be more than 25-fold.

F. Effectiveness Relative to the Economic Threshold

Some methods of pest control have the capacity to reverse the growth of populations or quickly decimate damaging forms only when the density of the pest population is at a certain level. For example, the sterile male method, which is ineffective against dense populations and highly effective against sparse populations is ideal for use against a pest like the screwworm for which the economic threshold on cattle is zero. Pheromones have similar strengths and limitations. On the other hand, entomophages tend to give excellent control of dense populations, but are of little value against very low populations. For example, the economic thresholds of leaf-feeding Homoptera on citrus are high and entomophages are ideal for combatting them. But entomophages tend to be of marginal value against insects that bore into fruit because the tolerance for damage to fruit is very low. Thus, no entomophages that are capable of reducing the entry of codling moth larvae into apples to less than 1% have been established in the U.S. Indeed, pests such as the codling moth which feed inside the host tend to be sheltered from unfavorable weather conditions and also from their natural enemies.

Insecticides, resistant cultivars, physical methods, and cultural methods usually are effective against all population densities, whatever the economic threshold.

G. Efficiency Relative to the Economic Threshold

Efficiency in terms of cost per insect destroyed or prevented from developing varies for different methods of pest control. Thus, an insecticide applied at a given rate will destroy approximately the same percentages of insects whatever the density. This also tends to be the case for varietal resistance, physical methods, cultural methods, and light traps. These methods are therefore most efficient when densities are high. Also, when entomophages and microbial agents cycle in the target population, they utilize the pest population as a resource to increase their numbers. Therefore, the efficiency of these agents thus tends to increase with increasing densities. Such biological agents clearly can have unsurpassed value if the economic threshold for the pest is fairly high.

Pheromones and sterile insects are of little or no value when pest densities are high but become progressively more effective and more efficient as the density declines. Therefore, with lower densities less intensive treatments with sterile insect or pheromone methods are required, which is not the case with insecticides or biological agents. Consequently, when the goal is to achieve either very low or nondetectable pest densities, the cost of pheromones and sterile males may be considerably less than the cost of insecticides or cultural or physical methods.

No single method is available that is efficient at all population densities.

H. Ecological Selectivity

In mid- and low-temperate and tropical latitudes beneficial insects tend to be the major cause of mortality of insect pest populations. In high-temperate latitudes the climate often is a more important constraint though the beneficial insects are still a factor.[29] Thus, every effort should be made to conserve indigenous entomophages, which means that the major elements in any integrated pest management system for pests that occur at damaging levels every year must possess at least a moderate degree of specificity, i.e., suppress the pest species and spare the natural enemies (ecological selectivity).

I. Guidelines for Selection and Incorporation of Control Methods Into IPM Systems

As economic entomologists become more sophisticated in analyzing pest problems and designing insect management systems in accordance with the principles of applied ecology, some general guidelines for the selection and integration of control methods into IPM systems have emerged.[30,31,19]

1. Whenever possible, ecologically selective methods should be chosen.
2. If possible, methods that are nonselective only should be used locally or as rescue treatments.
3. When the economic threshold is low or when eradication is required, methods that are efficient against high densities should be combined with methods that are efficient against low densities. (Often the action of the former will potentiate the efficiency of the latter, and in a few cases such diverse methods are mutually synergistic.)
4. When the economic threshold is moderate or fairly high, good protection can often be achieved by combining several methods each of which is efficient against moderate or high populations.

5. If the strategy of prevention is most advantageous against a pest with a low or moderate economic threshold (e.g., pink bollworm), then methods that are efficient at low densities (pheromones or sterile males) should be employed prophylactically. These measures should be supplemented if possible by cultural manipulations, use of tolerant varieties, late-season application of plant growth regulators, and spot treatments with insecticides, entomopathogens, or augmentative releases of entomophages.

The most important properties of a number of individual insect control methods and two methods used jointly are summarized in Table 2. The following comments may be useful in understanding some terms and concepts.

1. Oviposition-deterrent pheromones are important in preventing overexploitation of a host resource. Prokopy et al.[32] and Katsoyannos and Boller[33] showed that such pheromones are important in the bionomics of the apple maggot, *Rhagoletis pomonella* and the European cherry fruit fly, *Rhagoletis cerasi.* Their studies also indicated that these chemicals, when sprayed onto developing fruit, will protect it.
2. Pheromones used by social insects to recognize the brood or the queen should be useful in improving bait formulations involving entomopathogens or toxicants. Research on such pheromones of fire ants is progressing at the USDA/SEA-AR laboratories in Gainesville, Fla.[34]
3. A combination of augmentative releases of entomophages and attractants seems especially promising for dealing with a number of damaging pests. With an effective surveillance system, areas of a field needing augmentative releases could be identified and treated while the entire field was treated with either a dispersed formulation of the pheromone (to prevent reproduction of the pest) or with traps spaced sufficiently close together so they would capture a high percentage of the population. In such a scheme, insecticides or entomopathogens can be used in place of entomophages.
4. The use of pheromones in light traps was shown by Gentry and Davis[35] to increase trap catches of the cabbage looper, *Trichoplusia ni,* by roughly sixfold. Preferred color and shape were used in designing an effective trap for the apple maggot and this trap was baited with feeding attractants.[36,37] Ultraviolet-reflecting fiberglass panels are highly effective in trapping the stable fly,[38] and pyramid-shaped traps coated with a white paint containing titanium dioxide are highly attractive to face flies.[39] Thus, there appear to be many promising combinations of chemical and physical attractants.
5. Naturally occurring feeding attractants are particularly important for detecting stored product insects,[40-42] and in managing dipterans. Swormlure, a feeding attractant mixture for the screwworm, *Cochliomyia hominivorax,* is formulated with an insecticide and a feeding stimulant to produce SWASS (Screwworm Adult Suppressive System).[43,44] This bait is distributed in pellets by airplane over vast areas in Mexico as a means of reducing the screwworm populations so an effective ratio of released sterile to native males can be achieved. Swormlure in traps is also used for detecting and monitoring screwworm populations.
6. The effectiveness of feeding attractants is most influenced by the availability and distribution of competing natural food sources.[19] SWASS is effective because screwworms appear to respond to it in the same manner as to fresh cow manure. Since fresh manure is sparsely distributed on range, a few SWASS pellets per hectare compete strongly for screwworms.
7. Feeding stimulants, also known as food arrestants, have proved essential for

Table 2
PROPERTIES OF SOME INSECT CONTROL METHOD(S) THAT SHOULD BE CONSIDERED IN THE DESIGN AND USE OF INTEGRATED INSECT MANAGEMENT SYSTEMS. SOME OF THESE PROPERTIES ARE BASED ON BIOMATHEMATICAL ANALYSES THAT HAVE NOT BEEN VERIFIED IN FIELD EXPERIMENTS

Control method(s)	Degree of ecological selectivity	Comments on the effectiveness of method(s) [a]	Comments on the efficiency of method(s), etc.[b]
Naturally occurring feeding attractant in traps	Moderate or high	Effectiveness is limited by amounts of competing feeding attractant in environment. Effective at high and low pest densities.	Efficiency varies inversely with abundance of food sources. Captures same proportion at high as at low pest population densities. Also, may give estimate of density.
Naturally occurring feeding attractant combined with toxicant in bait, on trap crop, etc.	Moderate or high	Effectiveness is limited by amounts of competing feeding attractant in environment. Effective at high and low pest densities.	Efficiency varies inversely with abundance of food sources. Kills same proportion at high as at low pest densities.
Naturally occurring feeding attractant combined with entompathogen in bait, on trap crop, etc.	High (because of pathogen)	Effectiveness is limited by amounts of competing feeding attractant in environment. Effective at high and low pest densities.	If pathogen cycles, the proportion of the pest population killed will be greater at high than at low densities.
Naturally occurring feeding stimulant combined with toxicant in bait, on trap crop, etc.	Moderate or high	Effective at high and low densities. Must be well dispersed in habitat of the target pest.	Efficiency inversely proportional to amounts of competing food in environment. Kills same proportion at high as at low pest densities.
Naturally occurring feeding stimulant combined with entomopathogen in bait, on trap crop or device, etc.	High (because of pathogen)	Effective at high and low densities. Must be well dispersed in habitat of the target pest.	If pathogen cycles, the proportion of the pest population killed will be greater at high than at low densities.
Artificial[c] feeding attractant (specific) in traps.	High	No competition in attraction from natural sources. If attractancy is strong, effective at high and low densities.	Same proportion of pest population killed at all population densities. Also, may give reliable estimate of densities.

Artificial feeding attractant (specific) combined with chemosterilant in bait.	High	Effective at high and low densities. Sterile insects have additional impact.	Proportion of population prevented from reproducing is the same at high and low densities.
Artificial feeding attractant (specific) combined with toxicant in bait, on trap crop or device, etc.	High	Effective at high and low densities.	Proportion of population prevented from reproducing is the same at high and low densities.
Artificial feeding attractant (specific) combined with entomopathogen in bait, on trap crop or device, etc.	High	Effective at high and low densities.	If pathogen cycles, the proportion of the pest population killed will be greater at high than at low densities.
Sex pheromone in traps.		Effective only at low or moderate densities because of competition from wild insects. Female-attracting pheromone is more effective than male-attracting pheromone, especially if females seek to mate more than once before oviposition.	Captures a progressively higher proportion of pest population as the density declines. Fewer traps needed if female responds to pheromone than if male responds. Give reliable estimate of density only when density is low.
Sex pheromone combined with chemosterilant in bait.	High	Effective at moderate and at low densities. Effect of sexually sterilizing insects is greater than effect of killing them directly.	Proportion sterilized is greater at low than at moderate densities.
Sex pheromone combined with toxicant in bait on trap crop, etc.	High	Effective at moderate and at low densities.	Kills a progressively higher proportion of pest population as the population density declines.
Sex pheromone combined with entomopathogen in bait, on trap crop or device, etc.	High	Effective at low and moderate densities. If pathogen cycles, then effective at high densities.	If pathogen cycles, then efficiency in relation to density may be determined by pathogen.
Sex pheromone dispersed in constant release formulation (air permeation).	High	Effective only at low or moderate densities, because visual and other cues serve to assemble sexes when densities are high.	Proportion of population prevented from reproducing increases progressively as density declines.

Table 2 (continued)
PROPERTIES OF SOME INSECT CONTROL METHOD(S) THAT SHOULD BE CONSIDERED IN THE DESIGN AND USE OF INTEGRATED INSECT MANAGEMENT SYSTEMS. SOME OF THESE PROPERTIES ARE BASED ON BIOMATHEMATICAL ANALYSES THAT HAVE NOT BEEN VERIFIED IN FIELD EXPERIMENTS

Control method(s)	Degree of ecological selectivity	Comments on the effectiveness of method(s),[a]	Comments on the efficiency of method(s), etc.[b]
Artificial sex attractant (specific) in traps.	High	Effective at low and also at high densities if attractancy is strong.	Captures same proportion of high pest densities as at low densities. Also, may give reliable estimate of density.
Artificial sex attractant (specific) combined with toxicant in bait, on trap crop or device, etc.	High	Effective at all densities if attractancy is strong.	Kills same proportion at all densities.
Artificial sex attractant (specific) combined with entomopathogen in bait, on trap crop or device, etc.	High	Effective at all densities if attractancy is strong.	If pathogen cycles, then a greater proportion of population is killed at high than at low densities.
Artificial sex attractant (specific) dispersed in constant release formulation (air permeation).	High	Effective only at low densities because sex pheromone and other cues serve to assemble sexes when density is moderate and high.	Proportion of population prevented from reproducing is the same at all densities. Method is speculative.
Artificial sex attrrctant (specific and irreversible) dispersed in slow release formulation (air permeation).	High	Effective at fairly high densities if receptors of true sex pheromone are permanently complexed. Sexes assemble only by means of nonpheromonal cues.	Proportion of population prevented from reproducing is the same at all densities.
Aggregating pheromone of pest in traps.	High	Effective at low and moderate densities. Effective at higher densities than a pheromone for a single sex. (Traps destroy the sex which produces the pheromone that competes with traps.)	Captures a progressively higher proportion of pest population as the density declines. Gives reliable estimate of density only when density is low.

Aggregating pheromone combined with toxicant in bait, on trap crop or device, etc.	High	Effective at low and moderate densities. Especially effective if insect cannot produce competing pheromone until after it has fed on host.	Kills a progressively higher proportion of pest population as the density declines.
Aggregating pheromone combined with entomopathogen in bait, on trap crop, or device, etc.	High	If pathogen cycles, then effectiveness will extend to high densities.	If pathogen cycles, then efficiency in relation to density may be determined by pathogen.
Kairomone used to attract entomophages to areas with high pest densities.	High	Effective at all densities.	Proportion of population destroyed is greater at high than at moderate densities if cycling occurs.
Aggragating pheromone combined with entomopathogen in bait, on trap crop, or device, etc.	High	If pathogen cycles, then effectiveness will extend to high densities.	If pathogen cycles, then efficiency in relation to density may be determined by pathogen.
Aggregating pheromone dispersed in constant release formulation (air permeation)	High	Effective only at low or moderate densities because visual and other cues will assemble sexes present at high densities.	Proportion of population prevented from reproducing increases progressively as density declines.
Aggregating pheromone of entomophages (used to attract entomophages to areas where control is needed or to retain released ones.)	High	Effective at all densities.	Proportion of population destroyed is greater at high than at moderate densities if cycling occurs.
Kairomone used to attract entomophages to areas with high pest densities.	High	Effective at all densities.	Proportion of population destroyed is greater at high than at moderate densities if cycling
Oviposition pheromone in ovitrap.	High	Effective at low and moderate densities only. Ovitraps compete with females laying eggs on natural substrates.	Proportion of pest population destroyed is greater for low density than for moderate density. Gives reliable estimates of densities.

Table 2 (continued)
PROPERTIES OF SOME INSECT CONTROL METHOD(S) THAT SHOULD BE CONSIDERED IN THE DESIGN AND USE OF INTEGRATED INSECT MANAGEMENT SYSTEMS. SOME OF THESE PROPERTIES ARE BASED ON BIOMATHEMATICAL ANALYSES THAT HAVE NOT BEEN VERIFIED IN FIELD EXPERIMENTS

Control method(s)	Degree of ecological selectivity	Comments on the effectiveness of method(s),[a]	Comments on the efficiency of method(s), etc.[b]
Natural oviposition attractant in ovitrap.	High	Effective when natural oviposition sites are few. Ovitrap competes with natural sources of attractant.	Same proportion of eggs destroyed at all densities. Efficiency increased by reducing natural sources. Can give reliable population estimates.
Oviposition pheromone plus natural oviposition attractant	High	Effective at higher densities than either substances alone.	Reduction of natural sources would increase efficiency and effectiveness.
Artificial oviposition attractant and ovitrap.	High	Effective at moderate densities. No competition from natural resources, unless artificial attractant duplicates natural attractant.	Same proportion of eggs destroyed at all densities. Gives reliable estimates of population.
Oviposition deterrent pheromone.	High	Effective at low and high densities.	Same proportion of females deterred from oviposition at all densities.
Brood pheromone combined with toxicant or entomopathogen.	High	Effective at low and high densities.	Same proportion of colonies destroyed at all densities.
Male hair pencil calling-suppressant pheromone in constant release formulation (air permeation).	High	Not sufficiently effective to protect commodity when density is high. Some males and females may be able to assemble by means of nonolfactory cues.	Proportion of population prevented from reproducing is the same at all densities. If pheromone is a female aphrodisiac some mating may be stimulated.
Selective insecticides.	High	Effective at all densities.	Kills same proportion at all densities.

Resistant plant varieties or animal breed.	High (with exceptions)	Effective at all pest densities. (In some cases resistance trait may reduce populations of entomophages).	Kills same proportion at all densities. Usually compatible with other methods of control and may interact synergistically with methods that have greatest efficiency at low densities.
Released sterile insects.	High	Effective at low densities only.	Proportion of population prevented from reproducing increases as density declines.
Released sterile insects plus ovitraps with oviposition pheromone.	High	Effective at moderate and low densities.	Mutually synergistic methods.
Entomopathogens (specific)	High	Effective at low and high densities.	If pathogen cycles, then a greater proportion of population is killed at high than at low densities.
Parasitoids (specific) or predators (specific).	High	Little effect at low densities. If lag time between population growth of entomophages and of pest is short, then entomophages are effective against moderate and high densities. If lag time is long, then ineffective.	Proportion of pest population killed is greater at high than at moderate densities if entomophages cycle.
Parasitoids (specific) or predators (specific) released to augment indigenous entomophages.	High	Effective against moderate and high populations.	Proportion of pest population killed is greater at high than at moderate densities if entomophages cycle.

Table 2 (continued)
PROPERTIES OF SOME INSECT CONTROL METHOD(S) THAT SHOULD BE CONSIDERED IN THE DESIGN AND USE OF INTEGRATED INSECT MANAGEMENT SYSTEMS. SOME OF THESE PROPERTIES ARE BASED ON BIOMATHEMATICAL ANALYSES THAT HAVE NOT BEEN VERIFIED IN FIELD EXPERIMENTS

Control method(s)	Degree of ecological selectivity	Comments on the effectiveness of method(s),[a]	Comments on the efficiency of method(s), etc.[b]
Augmentative releases of entomophages and releases of sterile insects.	High	Effective at high and low densities.	Methods are complementary. Entomophages potentiate action of sterile insects. If entomophages cue on infested host rather than on pest, then the methods are mutually synergistic.
Light traps.	Moderate	Effective against moderate populations of a few species. Attractancy is not sufficiently strong to dramatically suppress many species.	Captures same proportion at high as at low densities. May provide reliable estimate of density.
Color traps or ultraviolet reflecting attractant.	High	Effective at moderate and low pest densities.	Kills same proportion of population at all densities. Biological or ecological correlates[d] of color are generally not understood.
Cultural measures.	Low (with exceptions)	Effective at high and low densities.	Kills same proportion at all densities.
Insecticides (broad spectrum), broadcast application.	Low	Effective at high and low densities.	Kills same proportion at all densities. If pest becomes resistant and entomophages are decimated, then pest population explodes.
Chemosterilants, broadcast application.	Low	Effective at high and low densities.	Proportion of population prevented from reproducing is the same at high and low densities.

[a] Effectiveness is the capacity to maintain or to suppress the pest populations density below the economic threshold.

[b] Efficiency is the cost of suppression per insect pest. For example, 1 kg/ha of insecticide may be required to control an insect population. The same amount is needed at all population densities. Therefore, the cost per insect killed is lowest at high densities.

[c] An artificial attractant is the creation of a chemist and consists of a chemical(s) usually not encountered by insect in its natural environment.

[d] The biological and ecological correlates of insect responses to color are best explained by an example. The stable fly, *Stomoxys calcitrans*, responds strongly to the ultraviolet light reflected from fiber glass panels. The response of the stable fly to a fiber glass panel seems to correlate closely with responses to a cow. Therefore, it can be said that the biological correlate of a fiber glass panel is a cow.

Modified from Klassen, W. and Brazzel, J. R., Insecticides as components of modern pest management practices, *Proc. FAO Conference on Ecology in Relation to Plant Pest Control*, Food and Agriculture Organization of the United Nations, Rome, 1973.

managing imported fire ants, *Solenopsis invicta* and *S. richteri.* Soybean oil is such an effective feeding stimulant for these insects that populations can be decimated by distributing bait containing only 2 to 3 g/ha of mirex so the foraging insects can encounter them during normal movement. Even with insecticides that are somewhat less effective than mirex (e.g., Amdro®*) less than 10 g/ha are required. Clearly, feeding stimulants can impart ecological selectivity when they are formulated with a broad spectrum insecticide. Also, their use may reduce the environmental load of insecticides by a factor of more than 100.

8. Artificial feeding attractants have been discovered for some tropical fruit flies. For example, methyl eugenol[45] is so highly attractive to male oriental fruit flies, *Dacus dorsalis,* that it is routinely used with an insecticide to eradicate incipient infestations from California. Also, as noted earlier, the potency of artificial feeding attractants is not offset by competition from natural sources as is that of naturally occurring feeding attractants.
9. The major virtues of sex pheromones are their high degree of ecological selectivity and their high potency. The major limitation on practical use is the competitive emission of identical material from individuals in the native population. Thus, the performance of pheromones as lures in traps or as "confusants" (to disrupt reproduction) is inseparably linked to the density and distribution pattern of the native population within the region of influence of the trap and hence to the competing quantity and location of pheromone produced by this population.
10. The use of pheromones in traps to detect the onset of emergence or of emigration is a marvelous innovation because the trap captures are so sensitive to such events and because people untrained in insect identification can make use of the data. However, quantitative interpretation of pheromone-trap data is far from simple. The following general formula for relating numbers caught in a pheromone trap to population density has been proposed by Knipling.[19]

$$P = \frac{N\,(I)\,(C)}{I - C} \qquad (3)$$

where P is the average number of the responding sex present in a unit area each day, N is the number of traps operating in the prescribed area, I is the attractancy of a trap relative to the attractant power of a calling insect, and C is the average daily rate of capture of the responding sex per trap. Knipling[19] used this formula in making a biomathematical study of trap data when mass trapping is employed to suppress reproduction in a population. He found that when the average daily rate of capture of males per trap day was 0.5 or higher for traps with attractancy equivalent to that of an unmated female, virtually all females would mate even though many males would be caught and killed. Indeed, if 85% suppression of reproduction is needed to prevent the population from reaching the economic threshold, then the average rate of capture of males per trap per day should not exceed 0.05. Similarly, for a trap with an attractancy equivalent to 5 unmated females, the average male capture rate per trap, per day should not exceed 0.25. Thus, contrary to conventional belief, high trap captures would indicate that mass trapping is failing to keep the population from growing.

11. In some instances pheromones can be used effectively to create disease epizootics in insect pest populations. Burkholder and Shapas[46] found that pheromone could be used in warehouses to lure the dermestid beetle, *Trogoderma glabrum,* into devices contaminated with the spores of *Mattesia trogodermae.* The contami-

* Amdro™ = tetrahydro-5,5-dimethyl-2(1*H*)-pyrimidinone [3-[4-(trifluoromethyl)phenyl]-1-[2-[4-(trifluoromethyl)phenyl]ethenyl]-2-propenylidene] hydrazone.

nated males returned to the population and subsequently died of the infection. Then since dermestids feed on cadavers of dead insects, the cycling of the entomopathogen was facilitated, and the population was reduced to nondetectable levels within two generations.

12. Recent studies by Simser[47] suggest that the aggregating pheromones of entomophages might be used to detect and perhaps to manage high density insect populations. Such a procedure is badly needed to increase the effectiveness of *Brachymeria* sp. against the gypsy moth. These pheromones could attract the entomophages to areas where the pest density is high and/or could prevent newly released mass-reared entomophages from leaving tree stands that need protection.
13. The possibility of permanently inactivating the olfactory receptors of insects by synthesizing analogs of attractants that bind irreversibly to receptors has been suggested by Plimmer and Klun.[48] In addition, Ferkovitch,[49] in a review of the enzymes on the antennae and body surface, stated that degradative enzymes "may be a vulnerable point of attack in the insect's communication system. Possibly inhibition of the pheromone-converting enzymes by chemical means might interfere with the perception of the pheromone." Such a procedure, if proved essentially free of hazard, would enormously expand the role of behavior-modifying chemicals in managing insects.
14. Oviposition attractants and pheromones are probably far more common and important than is generally realized since the screwworm, the black blow fly,[50] *Culex tarsalis*,[51] *Schistocerca gregaria*,[52] and the sheep blow fly, *Lucilia cuprina*,[53] all appear to possess oviposition pheromones. For example, screwworm egg masses are laid one upon another and not uniformly on the wound that attracts the flies, apparently because the first female locating the suitable oviposition site uses the oviposition pheromone to communicate to other females. Also, changes in odor of the infested host may attract individuals of the infesting species (as in the case of the screwworm) or a succession of species (as Norris[54] observed for attacked trees). Indeed, so effective is this type of chemical communication system that the date of death of decomposing vertebrate corpse can be estimated by the succession of insect species that are feeding upon it.[55]

IV. STRATEGIES FOR USE OF PHEROMONES AND OTHER ATTRACTANTS

Over the years prior to 1950 entomologists devoted much attention to the practical application of attractants and pheromone extracts and during the past 15 years chemists have made brilliant contributions. What has been lacking is adequate involvement of biologists with chemists, and this is seriously retarding progress. Biologists are needed to analyze insect behavior in order to discover the existence and function of chemical communication, to devise bioassays for each chemical component, and to analyze insect behavior in response to synthetic attractants in the field. There is also a need for greater involvement of entomologists with detailed knowledge of the life history, bionomics, and behavioral patterns of each target species. Nevertheless, a great deal of ingenuity has been demonstrated in devising ways of using pheromones and other attractants to control pest insects.

A. Uncoordinated Management of Local Populations

Management of local populations on a field-by-field basis is a desirable strategy because it allows each resource manager to act independently of neighbors and of government programs. Indeed, this is one of the strengths of synthetic organic insecticides. As effective insecticides became available, the U.S. Department of Agriculture was able to terminate community-wide programs against migratory pests such as the fall armyworm, *Spodoptera frugiperda*. Instead individual property owners were usually able to cope with the majority of insect pests.

Feeding stimulants and feeding attractants have likewise been used effectively on a field-by-field basis by preparing poison baits attractive to grasshoppers, crickets, cockroaches, silverfish, earwigs, cutworms, armyworms, white grubs, ants, house flies, fruit flies, root-maggot flies, adult oriental fruit moth, greenhouse leaf tier, corn earworm, wireworm, and tomato hornworm,[56] and against adult green June beetles, Japanese beetles, and yellow jacket wasps.[57] If baits are thinly and uniformly scattered throughout the habitat of the pest, they can provide excellent control without hazard to livestock and wild birds. However, many of these formulations are no longer widely used because they lose attractancy rapidly and are bulky and not easily applied with available equipment. More recently, as the result of greatly improved formulations, there has been a resurgence of use of baits against house flies and grasshoppers in particular.

Traps with chemical attractants can be used by individual homeowners to gain protection against the Japanese beetle and yellow jackets.

Pheromone-baited traps are widely used for detection and monitoring of local insect populations and for timing applications of insecticides. Minks[58,59] noted that monitoring with pheromone traps has three advantages: ease of handling, capture only species of interest so that no time-consuming sorting of trap catches is needed, and they are relatively inexpensive. He goes on to suggest three possible applications of pheromones in agriculture: (a) detection of presence of an insect pest, (b) determination of pest population density or of dispersal, and (c) lures for pathogen dissemination. Sensitive detection provides valuable information on presence or absence of pest species, gives early warning of insect flight, and allows determination of the number of generations, all of which can be used to identify the need for chemical control and the timing of spray applications. However, Minks points out organizational inadequacies in regard to (a) transfer of pheromone-monitoring systems to extension and advisory services and to growers and (b) the development of such pheromone modeling systems as PETE for use in integrated pest management programs over large areas.

PETE (Prediction Extension Timing Estimator), which uses pheromone trap data to obtain information on the tree fruit and selected vegetable pest complex in Michigan[60] (the redbanded leaf roller, *Argyrotaenia velutinana*; codling moth, *Cydia pomonella*; oriental fruit moth, *Grapholithia molesta*; and "tufted apple budmoth," *Platynota idaeusalis*), is just one of a number of computer-based systems used to collect data on pests, natural enemies, and crop development on a daily basis. It evaluates such data with computer-based implementation models, projects future developments, and issues advisories. Similar systems are operational in Texas and Indiana. As these systems become more numerous, the need for data from traps baited with pheromones or other specific attractants could increase, and the data probably will be used as suggested by Minks.[58,59]

Gross and Lewis,[61] Greany and Hagen,[62] and Jackson and Lewis,[63] have discussed some innovative ways of using kairomones (chemicals that benefit only the receiving species), synomones (chemicals that benefit both the sending and receiving species), and allemones (chemicals that benefit only the sending species).

Those kairomones to which entomophages respond could be used in combination with food sprays and/or artificially supplied hosts or prey to retain entomophages in fields that need protection and in sustaining populations during periods when the pest population is low. Further, if entomophages orient by means of a kairomone emitted by the pest and a synomone emitted by a noncrop host, then the joint application of the kairomone and synomone to the crop should increase retention. Moreover, the greater portion of a field could be treated with a substance that deters the pest insect (e.g., oviposition epideictic pheromone), and narrow strips could be treated with oviposition attractants and/or stimulants to create a trap crop effect. The resulting concentration of the pest could then be exploited by releases of reared entomophages and application of the appropriate kairomone; or the narrow strips could be treated with insecticides or entomopathogens.

B. Prevention

Prevention is a strategy that is widely used throughout the world.[64] It involves preventing pest populations from extending their geographic ranges through the use of quarantines and prophylactic treatments to prevent the buildup of populations that tend to explode or that cannot be readily controlled by curative treatments.

Quarantines are imposed by regulatory agencies in order to intercept emigrant pests that could otherwise migrate into uninfested areas. Thus, attractant-baited traps are placed around airports, harbors, at rest stops along highways, and along the leading edges of new infestations. In the U.S., about 200,000 traps are used in this way to prevent introduction or spread of the boll weevil, pink bollworm, gypsy moth, Japanese beetle, Comstock mealybug, California red scale, tephritid fruit flies, and other serious insect pests. Such traps serve the two functions of detection and suppression though the suppressive action is usually not adequate, and other control methods must also be used. For example, the effectiveness of traps deployed around international airports to capture the Japanese beetle is enhanced by treating the surrounding turf with insecticides, and the effect of grandlure traps is likewise enhanced by using insecticides to reduce the sources of westward-moving boll weevils in Texas. Also, parasites have been released to retard dispersal of the Comstock mealybug in California and of the gypsy moth in the northeastern U.S. Presently research is underway to develop cotton production systems for the San Joaquin Valley that may result in maturation of the crop before the diapausing generation of the pink bollworm can develop. This would strengthen the system for preventing the establishment of the pink bollworm which now consists of pheromone traps, releases of sterile males, and an enforced cotton-free period. Finally, it seems likely that pheromone-baited traps will come into greater use by regulatory officials who inspect export products to assure that they are pest free and thus in compliance with international phytosanitary conventions.

As noted, the strategy of prevention is used to keep the million acres of cotton in the San Joaquin Valley of California free from the pink bollworm. The effort has been underway for more than a decade because small numbers of feral pink bollworms are found in this valley almost every year. Since trapping studies by Stern and Sevacherian[65] indicated that these moths might be traveling from the Coachella Valley, a cotton-free period is now rigidly maintained. Also, pheromone traps are deployed at the rate of one per 40 acres and sterile moths are released. With the traps, it is possible to identify localities where feral moths are present and where additional releases may be needed.

Prevention has likewise been used against the gypsy moth since the development of disparlure for detection and suppression of low density populations has raised the possibility of retarding and perhaps halting further expansion of this pest. This insidious defoliator has been advancing out of the northeastern corridor of the U.S. at an average rate of 15 mi/year. It should be feasible to deploy a grid of traps to delimit the leading edge of the infestations, and to pinpoint outlying isolated incipient infestations. Then air permeation, mass trapping, or release of sterile males might be used to reduce low infestations to nondetectable levels. Or moderate populations could be first reduced with microbial preparations or selective insecticides during the larval stages so the pheromone could be applied against the resulting low numbers of adults. Then surviving pockets could again be delimited with traps so that suppressive measures could again be deployed. However, if this strategy is to succeed, it must be initiated soon and must be executed with thoroughness, i.e., before the pest travels much further down the Appalachian chain.

Prevention is the strategy of choice for managing the eastern subterranean termite, *Reticulitermes flavipes,* in the U.S. This insect is a valuable organism in forests and wooded area because it facilitates the recycling of dead wood and other woody debris. Yet it is so enormously destructive in wooden buildings and other wooden structures that the economic threshold in a human dwelling is zero. Losses caused by termites in the U.S. exceed $500 million, and homeowners spend vast sums for inspection and extermination by professional pest control operators.[66] Therefore, termites are presently prevented from invading wooden structures by: (1) mixing a persistent insecticide such as chlordane into the earth used to backfill around foundations, (2) using a pressure-treated plate on the foundation, (3) maintaining a distance of at least 30 cm between the soil and any wood structure, (4) trimming ornamental shrubs so that none of them contact the house, (5) or applying chlordane to outside walls of the foundation. Conceivably poisoned attractant wood blocks, the trail pheromone of the insect, or other attractant devices also could be used.

C. Area-Wide Management

Area-wide management programs for pest insects have often included attractants and feeding stimulants. The recent use of swormlure-2 against the screwworm in conjunction with the release of sterile males was mentioned earlier. It also seems likely that an oviposition attractant for the screwworm would be still more effective in suppressing high density populations and would likewise be completely compatible with the sterile male method. In fact, although SWASS is effective, it seems certain that it destroys many sterile males. Superficial analysis might suggest no merit in the joint use of pheromones and sterile males. However, Hagstrum et al.[67] demonstrated that the joint use of these methods in warehouses against the almond moth, *Ephestia cautella,* suppressed the population much more than either method alone. These authors thus feel that the joint use of these methods in advantageous against populations with heterogeneous spatial distributions.

Area-wide use of poison baits has been successful against the imported fire ant, but the strategy would be much more effective if the queen recognition pheromone could be identified, synthesized, and added to the bait.[34] Area-wide deployment of bait sprays is widely used to manage the Mediterranean fruit fly.

Since insect populations in warehouses, grain terminals, holds of ships, etc., are essentially isolated, they are excellent targets for unconventional pest management systems. Burkholder and Shapas[46] proposed that stored-grain insects may be managed by adding attractants or feeding stimulants bearing a pathogen or insecticide to small paper discs. Insects that are attracted and killed could then be removed with the bait during the cleaning process, an approach that might even replace costly and hazardous fumigation. Indeed, the introduction of such devices into the grain bulk would not cost much because colored paper discs are coming into use to code the origin or ownership of grain in order to prevent theft.

Vick et al.[68] have undertaken the development of a pest management system for stored commodities that involves the use of insect growth regulators and sex pheromones. This effort is presently focused on the Indian meal moth and almond moth on in-shell peanuts since these pests have developed resistance to malathion, and warehouse operators have had to turn to fumigants. The joint use of pheromones and insect growth regulators seems especially well-suited to warehouse situations because the buildings shelter the chemicals from the degradative actions of sunlight and rain, because the closed environment permits the maintenance of sufficiently high levels of pheromone to deter mating of adult moths, and because pheromones are effective at concentrations many orders of magnitude lower than are fumigants. They anticipate that the insect growth regulators (e.g., methoprene) will prevent the emergence of most moths and that the sex pheromone will make it difficult or impossible for those that do emerge to find mates.

Pheromone is also used in area-wide management of the pink bollworm. In fact, the sex pheromone is used in three ways to manage this pest in Arizona and in the southern valleys of California. The first involves the use of traps to detect infestations and to schedule insecticide application.[69] The second involves the use of densely spaced traps to trap out most emerging males; then supplementary spot treatments of insecticides are applied.[70] The third system involves the use of hollow fibers filled with pheromone. These fibers slowly release the pheromone so it permeates the air and thereby inhibits reproduction.[71,69] The air-permeation approach was first suggested by Beroza.[72]

A final example is the very instructive mass trapping study conducted by Lloyd et al.[73] with in-field traps against the boll weevil. The results indicated that although the boll weevil deposits its eggs in a nonrandom, highly clumped pattern, the probability of detecting the F_1 progeny of a single boll weevil per acre (0.4 ha) with two in-field traps was about 94%. Trap densities of 5.7/acre, eliminated more than 90% of emerging females. The uniform distribution of traps in cotton fields at appropriate densities should prevent overwintered or newly emerging weevils of the two sexes from aggregating because the pheromone liberated by the traps would interfere. In fact, at high trap densities the concentration of pheromone in areas where males and females emerged closely together in time and space appeared to prevent the females from detecting mates and resulted in their dispersion away from the males and to the traps. Thus, the in-field traps not only suppressed the population, but also provided a ''built-in monitoring system to serve as a guide to the type and intensity of supplemental suppressive measures that may be required. Trap capture data could identify any cotton fields that may need supplemental suppression in order to maintain subeconomic populations throughout the management area.''

In sum, pheromones are so selective in action that area-wide preventive usage seems likely to enhance the effects of natural enemies, whether populations of parasites and predators are left to build up spontaneously, conservation is practiced to facilitate build up, or augmentative releases are made. In addition, any cultural measures such as removal of alternative hosts or field sanitation to reduce overwintering pest populations should potentiate the effect of the pheromones by increasing the ratio of the synthetic material to the pheromone emitted by the native insects. Likewise, the use of moderately resistant varieties or other suppression tactics that tend to slow development of pest populations should enhance the effectiveness of the pheromone. Even when more than one major pest attacks a commodity, pheromones should be useful because several pheromones or pheromone mimics can be applied simultaneously.[74,75] Such mixtures will probably be needed to protect tree fruit orchards, sweetcorn, cotton, soybeans, and certain vegetables.

It therefore seems clear that public sector scientists should give highest priority to developing air-permeation and mass-trapping approaches against those insect species that (1) cause the greatest losses, (2) can now be managed only by methods that seriously disrupt the environment, (3) cannot be managed by any existing methods, (4) cause large losses in the aggregate but that nevertheless do not appear to justify the cost of control and managerial attention required for protection on a field-by-field basis, e.g., European cornborer and corn earworm on field corn, or (5) if introduced into the U.S. would become a major pest. A list of the ten most destructive insects and ten most destructive ticks and mites in the U.S. was prepared in 1971 in the former Entomology Research Division, Agricultural Research Service, U.S. Department of Agriculture,[76] as follows: Insects - *Heliothis* species, boll weevil, European corn borer, grasshoppers, *Lygus* species, horn fly, green peach aphid, corn rootworms, greenbug, cabbage looper; Ticks and mites — apple mites, strawberry mite, brown wheat mite, citrus mites, Rocky Mountain fever tick, American dog tick, lone star tick, west coast tick, Gulf Coast tick, poultry mite.

D. Eradication from Defined Areas

Only a few insects that have wide distribution in the U.S. have been eradicated from the U.S. These include the cattle tick, *Boophilus annulatus,* the screwworm, and the khapra beetle, The eradications were all very costly, required much labor, and entailed the inconvenience of quarantines. In addition, the bed bug, *Cimex lectularius,* which several decades ago was a problem in most low and middle income houses, is now encountered rarely. Likewise, body lice, *Pediculus humanus humanus,* are rarely encountered, and the head louse, *Pediculus humanus capitis,* now has only a foothold, and that probably because of the long hair styles of the past two decades. Cockroaches have been eradicated from most middle income single family dwellings, but they are an enormous problem in multiple family dwellings. Although there are effective baits for use against the American cockroach, *Periplaneta americana,* no effective attractant has been discovered for the German cockroach, *Blatella germanica.*

Attempts to use attractants for eradication of the oriental fruit fly, *Dacus dorsalis,* and the Mediterranean fruit fly, *Ceratitis capitata,* have had dramatic and instructive effects. The oriental fruit fly was eradicated from Rota, an island 37 mi from Guam, in 6 months by distributing cane fiber squares containing methyl eugenol, a powerful artificial attractant, and the insecticide naled at a rate of 125 squares per square mile. The total amount of naled used was less than 9 g/ha.[77-79] The Mediterranean fruit fly was eradicated from over a million acres in Florida in 1957,[80] by using delimitation and suppression iteratively. Delimitation of the infestation was accomplished with the artificial attractant, siglure, and suppression was accomplished with malathion bait sprays. However, since then the public has become reluctant to approve area-wide blanket treatments with insecticides. Therefore, in similar circumstances insecticides are supplemented with other more selective suppressive measures. That was reason that sexually sterile flies were used to eradicate the incipient infestation of the Mediterranean fruit fly population from Los Angeles in 1977.

Disparlure has likewise been used both to delimit an incipient infestation of the gypsy moth in Wisconsin and also as a mating confusant in an effort at eradication in 1979.[81]

Also grandlure, the synthesized pheromone of the boll weevil, is currently being used in a trial eradication. In fact, the boll weevil is the only widely distributed insect in the U.S. that is presently being seriously considered for eradication, and the possibility of eradicating this introduced pest is greatly enhanced by discovery that grandlure can be used effectively in in-field traps.[73] Although such an undertaking would be expensive and demanding, use of grandlure and other new techniques seem to have overcome the major technological barriers to the goal. Currently, the National Academy of Sciences is reviewing all aspects of this matter in order to advise the Secretary of Agriculture.

The most uncontroversial use of the eradication strategy is against incipient infestations of nonindigenous pest species (exotic or alien pests). In such cases, early detection is decisive. Thus, there is an urgent need to discover powerful lures for all nonindigenous species that could become serious pests if introduced into the U.S. Therefore, for a number of years, the U.S. Department of Agriculture has been engaged in a program of evaluating possible lures against these insects in areas of the world where they are indigenous.[82]

V. POTENTIAL FUTURE USE OF CHEMICAL ATTRACTANTS

The term "chemical attractant" is applied to a wide diversity of materials including natural feeding attractants, natural feeding stimulants, artificial feeding attractants, sex pheromones, artificial sex attractants, aggregating pheromones, kairomones, oviposition pheromones, natural oviposition pheromones, artificial oviposition attractants, and brood pheromones. The potential for practical use of these substances is unlimited, but many factors will affect the extent to which they will be used for insect suppression. These include: need for insect control, availability of other insect control methods, efficacy, market size, patentability, regulatory requirements, available appropriate institutional arrangements, and effective cooperation between the public and private sectors. A discussion of most of these factors appears elsewhere.[83-85] However, some understanding of the need for insect control and identification of key target pests can be obtained by examining present-day use of insecticides. For example, in 1978 the value at the user level of insecticides applied to crops was slightly over $800 million in the U.S. and over $3 billion in the world.[13] U.S. expenditures were about equal for insecticides applied to cotton, corn, and horticultural crops (slightly over $200 million); on a world basis, expenditures were greatest for cotton ($892 million), horticultural crops ($785 million), rice ($365 million), and corn ($325 million).[13] Certainly, the potential monetary advantage to be obtained from practical systems for utilizing pheromones for control of pests on these crops are considerable. However, other possible ways of controlling insects attacking forests, stored products, animals, and humans should not be overlooked.

A. Regulatory Requirements

The role of regulatory requirements in the potential use of pheromones is described in depth elsewhere.[86] However, these requirements have so far been viewed as a major barrier disincentive. Therefore, the exemption from tolerance for gossyplure on cottonseed issued by the U.S. Environmental Protection Agency[87] and the announcement that this agency proposes to issue a statement concerning the regulation of biorational pesticides are matters of great interest. The recent registration of 4 of the 5 pheromones currently approved for use appears to be related to this statement (Table 3), and specific guidelines are now being developed to clarify implementation of the new policy.[88] Thus, it does appear that the regulatory requirements for attractants, especially pheromones, will not be a significant constraint in expanding their practical use for insect suppression.

B. Cooperation Between Private and Public Sectors

The development of most insect control methods, particularly those involving considerable manipulation of chemical and biological entities, is often a lengthy and costly process. A number of issues involving the roles of private and public sectors have been reviewed previously.[84-86] The appropriate roles of the private and public sectors are always highly variable depending upon the relative value of the public good and the potential for private profit. In the case of attractants, the role of the public sector must probably be substantial because of the selectivity and limited markets, the limited patentability, and the unusually large cost associated with large-scale evaluation of efficacy. For this reason, development and use of attractants for pest control would undoubtedly be accelerated if additional resources from the public sector were pro-

Table 3
PHEROMONES REGISTERED FOR USE BY THE U.S. ENVIRONMENTAL PROTECTION AGENCY AS OF SEPTEMBER 4, 1980

Common name	Chemical name	Use pattern	Date of initial registration
Muscamone	(*Z*)-9-tricosene	Combined with toxicant for premise treatment with housefly control. Nonfood use.	Oct. 23, 1973.
Gossyplure	(*Z,Z*)-7,11-hexadecadien-1-ol acetate, and (*Z,E*)-7,11-hexadecadien-1-ol acetate	Mating disruption of the pink bollworm on cotton. Exemption from tolerance in or on cottonseed.	Feb. 9, 1978
Integralure	(*R,Z*)5-(1-Decenyl) dihydro-2 (3H)-furanone	In traps for Japanese beetle control in residential area. Nonfood use.	Mar. 26, 1978
Disparlure	*cis*-7,8-epoxy-2-methyloctadecane	Mating disruption of gypsy moth in residential and forest areas. Nonfood use.	Aug. 20, 1979
Grandlure	(1R-*cis*)-1-methyl-2-(1-methylethenyl) cyclobutaneethanol; (*Z*)-2-(3,3-dimethylcyclohexylidene)ethanol; (*Z*)-(3,3-dimethylcyclohexylidene) acetaldehyde; and (*E*)-3,3-dimethylcyclohexylidene) acetaldehyde	In traps and trap crops for boll weevil control. Nonfood use.	Aug. 20, 1979

vided for: (1) contracting with industry for research on improved methods of synthesis, thereby reducing costs; (2) evaluating formulations that have desirable controlled release and application characteristics; (3) funding research on application equipment for specialized formulations; and (4) large-scale field studies.

Also, the public sector should consider more desirable exclusive licensing arrangements.

C. Institutional Arrangements

Pheromones and other chemical attractants may continue to be used like pesticides, that is, by the grower on a field-by-field basis. This will be true for arthropod pests that feed on a wide range of host plants and are highly mobile. However, attractants are highly selective and have a different mode of action so education and technical assistance, provided through new institutional arrangements, will probably be needed for area-wide application.

Starler and Ridgway[89] reviewed a number of institutional ways augmentation of natural enemies could be used for insect control. These options, with slight modifications, are also applicable to chemical attractants. They include: (1) private producer selling directly to user, (2) private producer selling directly to user and providing pest management services, (3) private producer selling directly to user with user contracting for services of an independent pest management consultant, (4) cooperatives producing the product, providing pest management services, and applying product, (5) pest control districts purchasing product and supervising its use, and (6) product used in a state or federal program within a regulatory and enforcement framework. Thus as chemical attractants are developed, each use will have to be reviewed carefully as to determine what institutional arrangements are necessary for successful use, with the characteristics of the attractant and the pest in mind.

ACKNOWLEDGMENT

We thank Martin Jacobson and J. R. McLaughlin for providing a number of references. Also, we thank Dr. E. F. Knipling for sharing his thoughts on ways of using attractants. Dr. M. Beroza made suggestions that improved both the content and the clarity of this work.

Scientific names used in this paper were reviewed and corrected by Mrs. R. C. Lawrence and the research scientists of the Insect Identification and Beneficial Insect Introduction Institute. We are deeply grateful to these scientists for reviewing the spelling of genera, species, and authors; use of parentheses, validity of generic-specific combinations; and family placement.

REFERENCES

1. Agricultural Research Service, NRP 20620 — Insect Control in Marketing, SEA/AR, U.S. Department Agric., Beltsville, Md., 1976, 73.
2. **Klassen, W.**, Pest management: organization and resources for implementation, in *Insects Science and Society*, Pimentel, D., Ed., Academic Press, New York, 1975, 227.
3. **Pimentel, D.**, Socioeconomic and legal aspects of pest control, in *Pest Control Strategies*, Smith, E. H. and Pimentel, D., Eds., Academic Press, New York, 1978, 55.
4. U.S. Department of Agriculture, Integrated Pest Management Programs, Science and Education Administration, Beltsville, Md., unpublished report, February 2, 1979, 31.
5. U.S. Department of Agriculture, Response to the President's Environmental Message of Aug. 2, 1979, U.S. Department of Agriculture, Beltsville, Md., unpublished report, January 18, 1980, 51, plus appendixes.
6. U.S. Department of Agriculture, U.S.D.A. Policy on Management of Pest Problems, Secretary's Memorandum No. 1929, Washington, D.C., December 12, 1977, 3.
7. **Rabb, R. L.**, Principles and concepts of pest management, in *Implementating Practical Pest Management Strategies*, Purdue University, West Lafayette, Ind., 1972, 6.
8. **Ridgway, R. L., Tinney, J. C., MacGregor, J. T., and Starler, N. H.**, Pesticide use in agriculture, *Environ. Health Perspectives*, 27, 103, 1978.
9. Production and Marketing Administration, The Pesticide Situation for 1952-53, U.S. Department of Agriculture, Beltsville, Md., 1953, 36.
10. **Georghiou, G. P.**, Insecticide resistance and prospects for its management, *Residue Rev.*, 1980, in press.
11. **Stern, V. M., Smith, R. F., van den Bosch, R., and Hagen, K. S.**, The integrated control concept, *Hilgardia*, 29, 81, 1959.
12. **Ripper, W. E.**, Effect of pesticides on balance of arthropod populations, *Annu. Rev. Entomol.*, 1, 403, 1956.
13. **Anon.**, A look at world pesticide markets, *Farm Chem.*, September 65, 1979.
14. **Gilbert, C. H.**, The increasing riskiness of the pesticide business, *Farm Chem.*, April, 20, 1978.
15. **Klassen, W.**, Some overall direction in pest management research, paper given at National Pest Management Workshop, Kansas City, Missouri, February 17, 1976.
16. **Early, J. D.**, *Chemical Regulation*, Agricultural Chemicals Association, Washington, D.C., 1980, 13.
17. **Birch, M. C., Ed.**, *Pheromones*, American Elsevier, New York, 1974, 495.
18. **Jacobson, M.**, *Insect Sex Pheromones*, Academic Press, New York, 1978, 382.
19. **Knipling, E. F.**, *The Basic Principles of Insect Population Suppression and Management*, Agric. Handb. No. 512, U.S. Government Printing Office, Washington, D.C., 1979, 658.
20. **Roelofs, W. E. and Cardé, R. T.**, Responses of Lepidoptera to synthetic sex pheromone chemicals and their analogs, *Annu. Rev. Entomol.*, 22, 377, 1977.
21. **Collins, C. W. and Potts, S. F.**, Attractants for the flying gypsy moths as an aid in locating new infestations, *U.S. Dep. Agric. Tech. Bull.*, 336, 1932, 44.
22. **Butenandt, A., Beckmann, R., Stamm, D., and Hecker, E.**, Ueber den Sexual-Lockstoff des Seidenspinners "Bombyx mori." Reindarstellung und Konstitution, *Z. Naturforsch.*, B 14, 283, 1959.

23. **Inscoe, M. N.**, Insect attractants, sex attractants, pheromones and related compounds, in *Insect Suppression with Controlled Release Pheromone Systems*, Vol. 2, Kydonieus, A. F. and Beroza, M., Eds., CRC Press, Boca Raton, Fla., 1982, chap. 15.
24. **Schwartz, P. H. and Klassen, W.**, Estimate of losses caused by insects and mites to agricultural crops, in *Handbook of Pest Management in Agriculture*, Vol. 1, Pimentel, D., Ed., CRC Press, Boca Raton, Fla., 1981, 15.
25. **McGregor, R. C.**, The Emigrant Pests, Animal and Plant Health Inspection Service, U.S. Department of Agriculture, Beltsville, Md., 1973, 167.
26. **Pieters, E. P. and Sterling, W. L.**, Aggregation indices of cotton arthropods in Texas, *Environ. Entomol.*, 3, 598, 1974.
27. **Bariola, L. A., Keller, J. C., Turley, D. L., and Farris, J. R.**, Migration and population studies of the pink bollworm in the arid west, *Environ. Entomol.*, 2, 205, 1973.
28. **Knipling, E. F.**, Entomology and the management of man's environment, *J. Aust. Entomol. Soc.*, 11, 153, 1972.
29. **Le Roux, E. J.**, Ecological considerations in chemical control, *Bull. Entomol. Soc. Am.*, 10, 70, 1964.
30. **Klassen, W. and Brazzel, J. R.**, Insecticides as components of modern pest management practices, *Proc. FAO Conf. on Ecology in Relation to Plant Pest Control*, Food and Agriculture Organization of the United Nations, Rome, 1973.
31. **Knipling, E. F.**, Some basic principles of insect population suppression, *Bull. Entomol. Soc. Am.*, 12, 7, 1966.
32. **Prokopy, R. J., Reissig, W. H., and Moericke, V.**, Marking pheromones deterring repeated oviposition in *Rhagoletis* flies, *Entomol. Exp. Appl.*, 20, 170, 1976.
33. **Katsoyannos, B. I. and Boller, E. F.**, First field application of oviposition-deterring pheromone of European cherry fruit fly, *Rhagoletis cerasi*, *Environ. Entomol.*, 5, 151, 1976.
34. **Dame, D. E. and Weidhaas, D. E.**, Attractants, repellents and behavior-modifying chemicals associated with medically important insects, *Proc. Int. Congr. Entomol.*, Kyoto, Entomological Society of Japan, Tokyo, 1980.
35. **Gentry, L. R. and Davis, D. R.**, Weather: influence on catches of adult cabbage loopers in traps baited with BL only or with BL plus synthetic sex pheromone, *Environ. Entomol.*, 2, 1074, 1973.
36. **Buriff, C. R.**, Recapture of released apple maggot flies in sticky board traps, *Environ. Entomol.*, 2, 757, 1973.
37. **Kring, J. B.**, Red spheres and yellow panels combined to attract apple maggot flies, *J. Econ. Entomol.*, 63, 466, 1970.
38. **Meifert, D. W., Patterson, R. F., Whitfield, T., LaBrecque, G. G., and Weidhaas, D. E.**, Unique attractant-toxicant system to control stable fly populations, *J. Econ. Entomol.*, 71, 290, 1978.
39. **Pickens, L. G.**, Sticky panels as traps for *Musca autumnalis* populations, *J. Econ. Entomol.*, 70, 549, 1977.
40. **Levinson, H. Z. and Levinson, A. R.**, Integrated manipulation of storage insects by pheromones and food-attractants — a proposal, *Z. Angew. Entomol.*, 84, 337, 1977.
41. **Levinson, H. Z. and Levinson, A. R.**, Trapping of storage insects by sex and food attractants as a tool of integrated control, in *Chemical Ecology: Odour Communication in Animals*, Ritter, F. J., Ed., Elsevier/North-Holland, New York, 1979, 327.
42. **Pinniger, D. B.**, The use of bait traps for assessment of stored-product insect populations, *Coop. Econ. Insect Rept.*, 25, 907, 1975.
43. **Coppedge, J. R., Broce, A. R., Tannahill, F. H., Goodenough, J. L., Snow, J. W., and Crystal, M. M.**, Development of a bait system for suppression of adult screwworms, *J. Econ. Entomol.*, 71, 483, 1978.
44. **Plimmer, J. R., Chu, H. W., and Coppedge, J. R.**, A new solid formulation of the adult screwworm attractant swormlure-2, *J. Econ. Entomol.*, 1980, (in press).
45. **Beroza, M.**, *Chemicals Controlling Insect Behavior*, Academic Press, New York, 1970, 107.
46. **Burkholder, W. E. and Shapas, T. J.**, Use of entomophathogen with pheromones and attractants in pest management systems for stored-product insects, in *Microbial Control of Insect Pests; Future Strategies for Pest Management Systems*, Allen, G. E., Ignoffo, C. M., and Jaques, R. P., Eds., Selected papers from NSF-USDA-Univ. of Florida Workshop, University of Florida, Gainesville, 1978, 236.
47. **Simser, D.**, Aspects of Mating of *Brachymeria* spp., M.S. thesis, University of Wisconsin, Madison, 1979.
48. **Plimmer, J. R. and Klun, J. A.**, personal communication.
49. **Ferkovitch, S. M.**, Enzymatic alternation of insect pheromones with emphasis on *Trichoplusia ni* (Lepidoptera), in, *Perception of Behavioral Chemicals*, ASP Biological and Medical Press V. V., Amsterdam, 1980, in press.
50. **Knipling, E. F.**, personal communication.

51. **Osgood, C. E.**, An oviposition pheromone associated with the egg rafts of *Culex tarsalis*, *J. Econ. Entomol.*, 64, 1038, 1971.
52. **Norris, M. J.**, Laboratory experiments on gregarious behavior in overpositing females of the desert locust (*Schistocerca gregaria* (Forsk.)), *Entomol. Exp. Appl.*, 6, 279, 1963.
53. **Barton Browne, L., Bartell, R. J., and Shorey, H. H.**, Pheromone-mediated behavior leading to group oviposition in the blowfly *Lucilia cuprina*, *J. Insect Physiol.*, 15, 1003, 1969.
54. **Norris, D.**, personal communication.
55. **Payne, J. A.**, A summer carrion study of the baby pig, *Sus scrofa* Linnaeus, *Ecology*, 46, 592, 1965.
56. **Metcalf, C. L., Flint, W. P., and Metcalf, R. L.**, *Destructive and Useful Insects, Their Habits and Control*, McGraw-Hill, New York, 1962, 1087.
57. **Schwartz, P. H.**, Control of insects on deciduous fruits and tree nuts in the home orchard — without insecticides, *U.S. Dep. Agric. Home Gard. Bull.*, 211, 36, 1975.
58. **Minks, A. K.**, Trapping with behavior-modifying chemicals: feasibility and limitations, in *Chemical Control of Insect Behavior: Theory and Applications*, Shorey, H. H. and McKelvey, J. J., Jr., Eds., John Wiley & Sons, New York, 1977, 385.
59. **Minks, A. K.**, Monitoring of insect pests with pheromone traps: points of interest, in *Management of Insect Pests with Semiochemicals, Concepts and Practice*, 1981, Mitchell, E. R., Ed., Plenum Press, New York, 1981.
60. **Welch, S. M., Croft, B. A., Brunner, J. F., and Michels, M. F.**, PETE: an extension phenology modeling system for management of multi-species pest complex, *Environ. Entomol.*, 7, 487, 1978.
61. **Gross, H. R. and Lewis, W. J.**, Strategies for the employment of semiochemicals in pest management systems, in, *Semiochemicals: Their Role in Pest Control*, Nordlund, D. A., Jones, R. L., and Lewis, W. J., Eds., John Wiley & Sons, New York, 1980, chap. 13.
62. **Greany, P. D. and Hagen, K. S.**, Role and significance of allelochemics: prey selection, in *Semiochemicals: Their Role in Pest Control*, Nordlund, D. A., Jones, R. L., and Lewis, W. J., Eds., John Wiley & Sons, New York, 1980, chap. 2.
63. **Jackson, R. D. and Lewis, W. J.**, Summary of significance and employment strategies for semiochemicals, in *Semiochemicals: Their Role in Pest Control*, Nordlund, D. A., Jones, R. L., and Lewis, W. J., Eds., John Wiley & Sons, New York, 1980, chap. 14.
64. **Byerly, T. C.**, Preventive practices and effective pest management, in *Pest Control Strategies for the Future*, National Academy of Science, Washington, D.C., 1972, 341.
65. **Stern, V. M. and Sevacherian, V.**, Long range dispersal of the pink bollworm into the San Joaquin Valley, *Calif. Agric.*, 32, 4, 1978.
66. **Williams, L. H. and Smythe, R. V.**, Estimated losses caused by wood products insects during 1970 for single family dwellings in eleven southern states, *U.S. Dept. Agric. For. Serv. Res. Pap.*, SO-145, 10, 1978.
67. **Hagstrum, D. W., McLaughlin, J. R., Smittle, B. J., and Coffelt, J. A.**, Sterile males in sex pheromone permeated atmosphere to reduce reproduction of *Ephestria cantella*, *Environ. Entomol.*, 7, 759, 1978.
68. **Vick, K. W., Coffelt, J. A., Oberlander, H., and Silhacek, D. L.**, Pest management under post-harvest conditions: control of insect pests of in-shell peanuts with pheromones and insect growth regulators, U.S. Department of Agriculture, SEA-AR, Beltsville, Md., unpublished project document, 1978, 22.
69. **Graham, H. M., Ed.**, *Pink Bollworm Control in the Western United States*, U.S. Department of Agriculture, SEA, Beltsville, Md., ARM-W-16/June 1980, 81.
70. **Huber, R. T., Moore, L., and Huffman, M. P.**, Feasibility of area-wide pheromone trapping of male pink bollworm moths in a cotton insect pest management program, *J. Econ. Entomol.*, 72, 222, 1979.
71. **Brooks, T. W., Doane, C. C., and Staten, R. T.**, Experience with the first commercial pheromone communication disruptive for suppression of an agricultural insect pest, in *Chemical Ecology: Odour Communication in Animals*, Ritter, F. J., Ed., Elsevier/North Holland, Amsterdam, 1979, 375.
72. **Beroza, M.**, Insect attractants are taking hold, *Agric. Chem.*, 15, 37, 1960.
73. **Lloyd, E. P., Knipling, E. F., McKibben, G. H., Witz, J. A., Hartstack, A. W., Leggett, J. E., and Lockwood, D. F.**, Mass trapping for detection, suppression and integration with other suppression measures against the boll weevil, in *Management of Insect Pests with Semiochemicals, Concepts and Practice*, Mitchell, E. R., Ed., Plenum Press, New York, 1981.
74. **Mitchell, E. R.**, Disruption of pheromonal communication among coexistent pest insects with multi-chemical formulations, *BioScience*, 25, 493, 1975.
75. **Tumlinson, J. H., Mitchell, E. R., and Chambers, D. L.**, Manipulating complexes of insect pests with various combinations of behavior modifying chemicals, in *Pest Management with Insect Sex Attractants and Other Behavior-Controlling Chemicals*, Beroza, M., Ed., ACS Symp. Ser. 23, American Chemical Society, Washington, D.C., 1976, 53.
76. Memorandum of April 2, 1971, from W. Klassen to H. C. Cox.

77. **Beroza, M.**, Attractants and repellents for insect pest control, in *Pest Control Strategies for the Future,* National Academy of Sciences, Washington, D.C., 1972, 226.
78. **Steiner, L. F.**, Control and eradication of fruit flies on citrus, *Proc. 1st Int. Citrus Symposium,* University of California at Riverside, 1969, 2, 881.
79. **Steiner, L. F., Mitchell, W. C., Harris, E. J. et al.**, Oriental fruit fly eradication by male annihilation, *J. Econ. Entomol.*, 58, 961, 1965.
80. **Steiner, L. F., Rohwer, G. G., Ayers, E. L., and Christenson, L. D.**, The role of attractants in the recent Mediterranean fruit fly eradication program in Florida, *J. Econ. Entomol.*, 54, 30, 1961.
81. **Schwalbe, C. P.**, personal communication.
82. **Jacobson, M.**, personal communication.
83. **Djerassi, C., Shih-Coleman, C., and Dickman, J.**, Insect control of the future: operational and policy aspects, *Science,* 186, 596, 1974.
84. Stanford Research Institute, New, Innovative Pesticides: an Evaluation of Incentives and disincentives for Commercial Development by industry, Special Report, Menlo Park, Calif., 1977.
85. **Kydonieus, A. and Beroza, M.**, *Insect Suppression wih Controlled Release Pheromone Systems,* CRC Press, Boca Raton, Fla., 1982.
86. **Tucker, W.**, Of mites and men, *Harper's Magazine,* August 43, 1978.
87. U.S. Environmental Protection Agency, Regulation of "Biorational" Pesticides, Policy Statement and Notice of Availability of Background Document, *Fed. Regist.*, 44, 28093, 1979.
88. Environmental Protection Agency, Data Requirements for Biorational Pesticides: Preamble and Guidelines, Subpart M, Guidelines for Registering Pesticides in the United States, Draft, August 4, 1980, Washington, D.C., 406.
89. **Starler, N. H. and Ridgway, R. L.**, Economic and social considerations for the utilization of augmentation of natural enemies, in *Biological Control by Augmentation of Natural Enemies,* Ridgway, R. L. and Vinson, S. B., Eds., Plenum Press, New York, 1977.

Chapter 3

CONTROLLED RELEASE TECHNOLOGIES

Laura T. Zeoli, Agis F. Kydonieus, and Alberto R. Quisumbing

TABLE OF CONTENTS

I. INTRODUCTION

Controlled release (CR) is the permeation-moderated transfer of a biologically active material (hereafter active agent) from a reservoir to a target surface to maintain a predetermined emission or concentration level for a specified period of time. During the last decade, CR technology has received increasing attention in the face of a growing awareness that substances ranging from drugs to agricultural chemicals are often excessively toxic and sometimes ineffective when administered by conventional means.

II. ADVANTAGES AND LIMITATIONS

As shown in Table 1, the advantages of controlled release are great.[1-4] CR systems have already been used to impart important benefits in dispensing active agents, thus elevating many products to commercial success.

Although these advantages are impressive, the merits of each application have to be examined individually, and the positive and negative effects weighed carefully before large expenditures for developmental work are committed. In other words, controlled release is not a panacea, and negative effects may, at times, more than offset the benefits. Some of the disadvantages of controlled release, and other considerations that require a thorough appraisal, include: (1) cost of the CR preparation and processing may be substantially higher than that of standard formulations; (2) fate of the polymer matrix and its effect on the environment; (3) fate of polymer additives such as plasticizers, stabilizers, antioxidants, fillers, etc.; (4) environmental impact of the polymer degradation products following heat, hydrolysis, oxidation, solar radiation, and biological degradation; and (5) cost, time, and probability of success in securing government registration of the product, if such is required.

Once an application has been thoroughly investigated and found suitable, one must select (1) the CR technology that best fits the application, and (2) the basic physical form of device, and the rate-controlling polymer matrix and active agent to be used.

III. BASIC COMPONENTS OF CR DEVICES

A CR product consists of the active agent and the polymer matrix or matrices that regulate its release. Depending on the application, pharmaceutical, agricultural, and veterinary chemicals have been used as active agents in commercial and experimental CR devices.[5,6] However, the major effort in applying CR principles has been in the administration of pharmaceuticals and the application of pest control chemicals.[6]

Advances in CR technology have been rapid because polymer science has become sophisticated enough to incorporate into polymers tailor-made properties for each CR application. In selecting the polymer matrix, the following design criteria must be considered: (1) molecular weight, glass transition temperature, and chemical functionality of the polymer must allow the proper diffusion and release of the specific active agent; (2) polymer functional groups should not react chemically with the active agent; (3) polymer and its degradation products must be nontoxic to the environment, and in medical applications neither nontoxic nor antagonistic to the host; (4) polymer must not decompose in storage nor during the useful life of the device; (5) polymer must be easily manufactured or fabricated into the desired product, and should allow incorporation of large amounts of active agents without deteriorating its mechanical properties; and (6) cost of the polymer should not be excessive, thereby causing the CR device to be noncompetitive.

A list of polymers that have been used in CR formulations is shown in Table 2.[3,5,6-9]These polymers are used in coatings in microencapsulation, films in laminated structures, slabs in monolithic systems, and flakes in many erodible devices.

Table 1
SOME ADVANTAGES OF CR PROCESSES AND SYSTEMS

1. Reduce mammalian toxicity of highly toxic substances.
2. Extend duration of effectiveness of active agent at a constant level.
3. Reduce evaporative losses and flammability of liquids.
4. Reduce phytotoxicity.
5. Protect pesticides from environmental degradation.
6. Reduce leaching into the earth and transport into streams.
7. Reduce pesticide contamination of the environment.
8. Convert liquids to solids and flowable powders.
9. Control the release of active agents.
10. Separate reactive components.
11. Mask the taste of bitter materials.
12. Reduce costs because less active material is needed.
13. Promote convenience, including ease of handling.
14. Avoid patient compliance problems in drug administration.
15. Employ less total drug, thus minimizing local side effects, systemic side effects, and the drug accumulation often encountered in chronic dosing.
16. Improve drug efficiency in treatment of illness.
17. Special effects, e.g., provide sufficient controlled release aspirin to give the arthritic patient symptomatic relief on awakening despite no drug intake during the night.

Table 2
POLYMERS USED IN CR DEVICES

NATURAL POLYMERS

Carboxymethylcellulose	Zein
Cellulose acetate phthalate	Nitrocellulose
Ethylcellulose	Propylhydroxycellulose
Gelatin	Shellac
Gum Arabic	Succinylated gelatin
Starch	Waxes - paraffin
Bark	Proteins
Methylcellulose	Kraft lignin
Arabinogelactan	Natural rubber

SYNTHETIC POLYMERS

Polyvinyl alcohol	Polyvinylidene chloride
Polyethylene	Polyvinyl chloride
Polypropylene	Polyacrylate
Polystyrene	Polyacrylonitrile
Polyacrylamide	Chlorinated polyethylene
Polyether	Acetal copolymer
Polyester	Polyurethane
Polyamide	Polyvinylpyrrolidone
Polyurea	Poly (*p*-Xylylene)
Epoxy	Polymethylmethacrylate
Ethylene vinyl acetate copolymer	Polyhydroxyethyl methacrylate
Polyvinyl acetate	

SYNTHETIC ELASTOMERS

Polybutadiene	Acrylonitrile
Polyisoprene	Nitrile
Neoprene	Butyl rubber
Chloroprene	Polysiloxane
Styrene-butadiene rubber	Hydrin rubber
Silicone rubber	Ethylene-propylene-diene terpolymer

Table 3
CATEGORIZATION OF CR POLYMERIC SYSTEMS

- I. PHYSICAL SYSTEMS
  - A. Reservoir systems with rate-controlling membrane
    - 1. Microencapsulation
    - 2. Macroencapsulation
  - B. Reservoir systems without rate-controlling membrane
    - 1. Hollow fibers
    - 2. Poroplastic® and Sustrelle™ Ultramicroporous Cellulose Triacetate
    - 3. Porous polymeric substrates and foams
  - C. Monolithic systems
    - 1. Physically dissolved in nonporous, polymeric or elastomeric matrix
      - a. Non-erodible
      - b. Erodible
      - c. Environmental agent ingression
    - 2. Physically dispersed in nonporous, polymeric or elastomeric matrix
      - a. Nonerodible
      - b. Erodible
      - c. Environmental agent ingression
  - D. Laminated structures
    - 1. Reservoir layer chemically similar to outer control layers
    - 2. Reservoir layer chemically dissimilar to outer control layers
  - E. Other Physical Methods
    - 1. Osmotic pumps
    - 2. Adsorption onto ion-exchange resins
- II. CHEMICAL SYSTEMS
  - A. Chemical erosion of polymer matrix
    - 1. Heterogeneous
    - 2. Homogeneous
  - B. Biological erosion of polymer matrix
    - 1. Heterogeneous
    - 2. Homogeneous

IV. BASIC RELEASE CHARACTERISTICS OF CR PROCESSES

All CR processes (Table 3) are in one way or another controlled by the diffusion of the active agent through a polymer barrier or by an inward diffusion of an environmental fluid in the case of ''environmental agent ingression'' devices and some homogeneous ''retrograde chemical reaction'' devices.

Possibly the only systems that are not controlled by diffusion are the ''pure'' erodible devices and some heterogeneous ''retrograde chemical reaction'' devices; but even in these systems release is generally achieved through a combination of diffusion and erosion.

The active agent passes through the polymeric barrier in the absence of pores or holes by a process of absorption, solution, and diffusion down a gradient of thermodynamic activity until desorbed or removed. The transport of the active agent is governed by Fick's first law:

$$J = \frac{dM_t}{Ad_t} = \frac{-D\,dC_m}{dx} \qquad (1)$$

where J is the flux in g/cm²/sec, A is the surface area through which diffusion takes place in cm², M_t is the mass of agent released, dM_t/dt is the steady-state release rate at time t, C is the concentration of active agent in the polymeric membrane in g/cm³, dC_m/dx is the concentration gradient, and D is the diffusion coefficient of the active agent in the polymeric membrane in cm²/sec.

Equation 1 can be integrated under the proper boundary conditions for each of the systems listed in Table 3 to obtain an equation giving the amount of agent released as a function of time. In many situations, however, the mathematics become rather complicated, and no explicit equations can be derived. For more information on the mathematics of diffusion, see References 10, 11, and 12.

V. DESCRIPTION OF SPECIFIC PHYSICAL CR SYSTEMS

As shown in Table 3, the various CR systems may be categorized broadly either as *physical* or *chemical.* A description of the individual physical CR technologies follows.

A. Reservoir Systems with Rate-Controlling Membrane

These include micro- and macrocapsules, which consist of a uniformly thin polymeric coating (rate-controlling membrane) around small solid particles, droplets of liquid, or dispersions of solids in liquids, with the size of the capsules ranging dimensionally from a few tenths of a micron to several thousand microns.[1] Capsules greater than 2000 to 3000 μm are called macrocapsules. There are no real differences between micro- and macrocapsules as far as release characteristics or type of active agent that can be encapsulated are concerned. Many methods for microencapsulating active agents have been developed; these methods and the references on each are listed below:

1. Phase Separation Methods
   a. Aqueous phase separation (complex coacervation)[13-18]
   b. Organic phase separation[19]
   c. Meltable dispersion[20]
   d. Fluidized-bed spray drying[21]
   e. Spray drying[22,23]
   f. Pan coating[24]
2. Interfacial Reactions
   a. Interfacial condensation polymerization[25-28]
   b. *In situ* interfacial condensation polymerization[3,29]
   c. Interfacial addition polymerization[30]
3. Physical Methods
   a. Multiorifice centrifugation[31-33]
   b. Electrostatic encapsulation[1-3,34]

Although many of the microencapsulated products release their contents by permeation through their walls, release by other mechanisms such as by breakdown or dissolution of the walls and osmotic bursting may also be important.

Release characteristics: Fick's law predicts that a steady state will be established with the release rate being constant and independent of time, if an active agent is enclosed within an inert polymer membrane and concentration of the agent is maintained constant within the enclosure. The amount of active agent released per day is, therefore, constant for the life of the device, and

$$M_t = kt \qquad (2)$$

where k is a constant. Equation 2 applies for all geometries of the device, e.g., spheres, slabs, etc. This type of release is "zero order" and is shown as curve I in Figure 1.

Microcapsules are made with very thin polymer walls, which often are not homogeneous, allowing formation of cracks and pores observable only under magnification. As a result, microcapsules have been found to follow a "$\sqrt{t}$ order" release in some field applications.

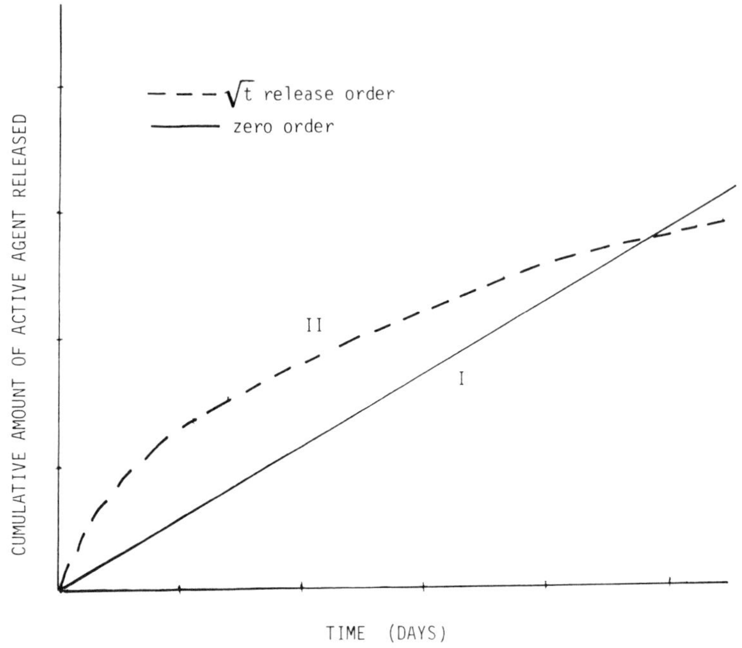

FIGURE 1. Representation of "zero order" and "$\sqrt{t}$ order" release.

In capsules with erodible membranes, the mechanism of release can be the erosion and rupture of the barrier membrane. Various delivery patterns, including essentially constant release (zero order) can be achieved by blending microcapsules of appropriate wall thicknesses.

Applications in the pheromone area: Two major types of these CR products have found a place in insect pest management. They are polyethylene vials or caps and microcapsules.

Polyethylene vials, discussed in more detail in the chapter by Daterman, have been used to dispense pheromones for monitoring purposes; for example, in trapping studies with the fall armyworm, *Spodoptera fugiperda*.[35] They have also been used for mass trapping the smaller European elm bark beetle *Scolytus multistriatus*[36,37] and the red-banded leafroller *Argyrotaenia velutinana*.[38,39]

Microcapsules, discussed at length in Volume II, Chapter 8, have been used for mating disruption of the oriental fruit moth *Grapholitha molesta*,[40] as well as for extensive work with mating disruption of the gypsy moth *Lymantria dispar*.[41-44] For example, Beroza obtained statistically valid disruption of mating with each of the population densities evaluated.[42]

B. Reservoir Systems without Rate-Controlling Membrane

These systems include hollow fibers,[45-47] impregnation in porous plastics such as MPS porous PVC sheet, millipore fibers, and Cellgard® porous polypropylene,[48,49] foams, and possibly hydrogels[50-54] and ultramicroporous cellulose triacetate.[55,56]

The simplest example is perhaps the hollow fibers which hold the active agent in a tiny tube from which it is released by diffusion through the air layer above the agent. Systems utilizing impregnated porous plastics (PVC, Cellgard®, etc.) are more complex, but in all cases, the active agent is retained by capillary action, physically imbedded in the pores. Release also occurs by diffusion through the air layer above the liquid that fills the pores. Strictly speaking, most of these systems may be considered monolithic matrix systems except that the interaction of active agent and polymer is minimal.

Release characteristics: Reservoir systems without a rate-controlling membrane generally follow a rate of release proportional to $t^{-1/2}$. The amount of agent released is then proportional to $\sqrt{t}$ ($\sqrt{t}$ order), as expressed by the equation:

$$M_t = k\sqrt{t} \tag{3}$$

Equation 3 gives a parabolic curve as shown by curve II in Figure 1. With this system a large amount of the agent is released initially, and substantially smaller, decreasing amounts are released during the last half of the life of the device.

Applications in the pheromone area: Two major types of these systems are used to dispense pheromones for insect control. These are the filters and the hollow fibers.

Controlled release filters have been used for both mass trapping and trap cropping. Grandlure, the synthetic pheromone of the cotton boll weevil, *Anthonomous grandis,* has been widely used in pest management studies because it is both a sex and an aggregation pheromone. One of the most commonly used formulations of grandlure has been a polyethylene glycol mixture impregnated in a cigarette filter[57] and enclosed in a glass vial.[58] Trap cropping of the boll weevil is discussed in more detail in Volume II, Chapter 5.

Hollow fibers, discussed at length in the Chapter 9 by Ashare et al., have been used extensively to dispense the sex lure gossyplure for suppression of the pink bollworm *Pectinophora gossypiella* by disruption of its mating communication.[59-61] These formulations were found effective. Thus, the pheromone evaporated from hollow, 104 mm long, thermoplastic fibers was comparable in effect with that achieved by use of conventional insecticides.[63]

C. Monolithic Systems

Probably the simplest and least expensive means of controlling the release of an active agent is by dispersing it uniformly in an inert polymeric matrix. Generally, the active agent is physically blended with the polymer powder and the mixture is then fused together by compression molding, injection molding, screw extrusion, calendering or casting,[64] all of which are common processes in the plastics industry.

Alternatively, the active agent is blended with elastomeric materials in the mixing step as is done with the other additives, e.g., accelerators, reinforcing pigments, stabilizers and processing aids.[65]

In either case, the active agent dissolves in the polymeric or elastomeric matrix until saturation is reached. Additional active agent, if any, remains dispersed within the polymer matrix after it is physically dispersed and molded or cured. As the agent evaporates or is otherwise removed from the surface of the monolithic device, more agent diffuses out from the interior to the surface in response to the decreased concentration gradient leading to the surface.

If the polymer used is soluble or degrades during its intended use, the monolithic device is erodible. In this case, the release of the active agent occurs by a combination of diffusion and liberation due to erosion.

Pure degradable and erodible systems, which can be defined as those that release their contents by diffusion, osmotic bursting, leaching, and any other CR mechanism, can have the very desirable property of chemically or biologically degrading after the useful life of the device has expired. Contamination of the environment with such devices is apt to be minimal.

Erodible systems can be considered a subdivision of degradable systems. They release their contents by physically "uncovering" the active agent as chemical or biological erosion of the matrix occurs. Erodible and degradable systems have been discussed more fully in various publications.[9,66-68]

If the polymer is plasticizable or swellable by an environmental agent such as water, the environmental agent invades the device, plasticizes the polymeric matrix, and thereby allows the physically bound active agent to diffuse out. This system, which includes starch xanthate[69,70] and perhaps some lignin-modified devices,[71] differs from the erodible one in that the matrix remains physically intact.

Release characteristics:

1. **Physically dissolved, nonerodible polymeric or elastomeric matrix** — Release rate in this system is proportional to $t^{-1/2}$ (same as in Equation 3) until about 60% of the active agent is released. The release rate, thereafter, is related exponentially to time, i. e.,

$$\frac{dM_t}{dt} = k_1 e^{-k_2 t} \tag{4}$$

   where k_1 and k_2 are constants. Thus, the rate of release beyond the 60% level drops exponentially. This type of release ("first order" release) is also observed in reservoir systems in which the solution of active agent within the enclosure is less than saturated.
2. **Physically dispersed, nonerodible, polymeric, or elastomeric matrix** — Release rate in this system is proportional to the $\sqrt{t}$ as long as the concentration of the active agent present (dispersed and dissolved) is higher than the solubility of the agent in the matrix. Thus, release from the dispersed systems is similar to that from the dissolved systems, except that the release rate does not decrease after 60% of the chemical has been emitted; instead, the relationship holds for almost the complete release curve.
3. **Monolithic erodible systems** — If one assumes that release of the active agent by diffusion is negligible in this system, the speed of erosion will control the release rate. Release by erosion is a surface area-dependent phenomenon, and the release will be constant (zero order) as long as the surface area does not change during the erosion process. This is true for slab-shaped devices. Eroding cylinders and spheres give delivery rates that decrease with time (owing to decreasing surface area), even though the kinetic process providing the rate-determining step is, in fact, zero order.[72]

   Applications of monolithic systems in the pheromone area: natural rubber septa (rubber sleeve stoppers with a septum) have been used for mass trapping such insects as the codling moth (*Laspeyresia pomonella*),[73-75] and are discussed in more detail in the chapter by Rothschild.

   Another monolithic system encountered in the pheromone area is the PVC "ball", which is a mixture of the active ingredient and polyvinyl chloride resin. PVC balls have been used for monitoring and mass trapping such insects as the western pineshoot borer (*Eucosma sonomana*), and are discussed at length in Chapter 7.

D. Laminated Structures

In this system, at least two and usually three polymeric films are adhered to or laminated together. The center layer of a three-layer laminate is the reservoir layer. It contains a large amount of the active agent and may be made of porous or nonporous polymeric material. The outer layers control the rate of release of the agent and are usually fabricated from a more rigid polymer than that of the reservoir. Should the reservoir layer and the outer films both be made of the same polymer, it is apparent that this system reverts to a monolithic system. The laminated structures, as used in pharmaceutical and pesticidal applications, have been discussed elsewhere.[57,58,76,77-80]

Release characteristics: Depending on the distribution coefficient between the inner and outer layers, two release characteristics are easily discernible. If the distribution coefficient of the active agent between the reservoir layer and the barrier membrane is much smaller than unity, the system approximates "zero order" release (as in a reservoir with a rate-controlling membrane), and the amount released per day is independent of time.[81] If the distribution coefficient is close to unity, the system approximates the $\sqrt{t}$ release rate of a monolithic, physically dispersed system.

Applications in the pheromone area: The three-layered polymeric laminate has proved to be particularly well suited as a dispenser of pheromones. The system is highly versatile because of the many parameters that can be varied, e.g., the size (area) of the dispenser, the thickness of the outer layers, the concentration of the pheromone, and the polymer type of both the outer and reservoir layers can be altered to give just about any desired release rate. Also, the dispensers can be fabricated in a variety of forms, including squares, ribbons, tapes, flakes and confetti, to suit a variety of application requirements.

Most of the pheromones that have been identified are long-chain, unsaturated acetates or other esters, aldehydes, alcohols, and epoxides. Many of them, aldehydes especially, are highly unstable and readily degrade when exposed to environmental conditions. The three-layer laminate is unique in its protection of these sensitive chemicals; since they are contained in the reservoir layer, they are shielded from light, oxidation, and hydrolysis. The pheromone is stored inside the dispenser until it reaches the surface where it is immediately discharged to the atmosphere. As exemplified by the typical data shown in Figure 2, the Hercon® three-layer polymeric laminate has consistently outperformed other systems in dispensing pheromones with aldehyde components, such as grandlure, the sex and aggregation phermone of the cotton boll weevil (*Anthonomous grandis*).[82] Additional applications of the Hercon® dispenser are discussed in Chapter 8 and Volume II, Chapters 1 through 8 and 10 and 11. These dispensers have been used for monitoring,[83] mass trapping,[84-86] and mating disruption.[87-89]

E. Other Physical Methods

In its simplest and most elegant design, the novel "osmotic pump" consists of a tablet containing the active agent and an osmotic "attractant" such as NaCl surrounded by a semipermeable membrane with a small orifice. When the pump is placed in an aqueous environment, the osmotic pressure of the NaCl inside the tablet draws water through the semipermeable membrane, forming a saturated aqueous solution inside the device. Because the membrane-coating is nonextensible, the saturated solution inside the device is pumped out as water is osmotically imbibed.

Also tried as a CR mechanism is the adsorption of active agents onto ion-exchange resins.[89] Thus, the adsorptive forces of these resins can decrease the release of an ionic species through an equilibrium favoring the adsorption sites of the resin, but renewal of the medium can result in very fast release. In pharmaceutical applications, the release rate depends upon pH and electrolyte concentration in the gastrointestinal tract and is therefore higher in the stomach and slows down during transit through the small intestine.

Release characteristics: With the osmotic pump, release is influenced by the osmotic pressure generated by the diffusion of water through the semipermeable membrane to form the saturated solution inside the device. The delivery rate of the solute by the pump is constant ("zero order" release) as long as excess solid is present inside the pump to form a saturated solution. The rate declines exponentially (first order) as soon as the solution inside the device drops below saturation.

Applications in the pheromone area: None of these physical methods are currently being used as pheromone dispensing systems.

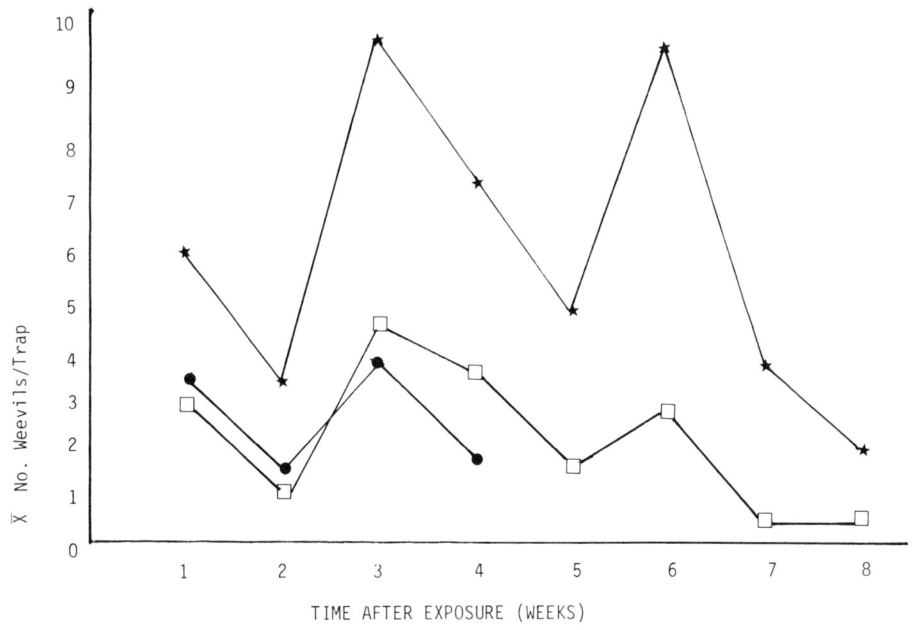

FIGURE 2. Response of native boll weevils in Mexico to various grandlure formulations. * Three layer polymeric laminate (Hercon® dispenser). ● Hollow fiber dispenser (Conrel®). □ Polyethylene glycol mixture impregnated in a cigarette filter enclosed in a glass vial.

VI. RETROGRADE CHEMICAL REACTION SYSTEMS

Controlled release may also be obtained through chemical methods. In one approach, the active agent is bonded via covalent or ionic bonds to a preformed polymer. This approach requires macromolecules with pendant functional groups capable of reacting with the active ingredients or their derivatives. The nature of the chemical bond may be varied to yield bonds with different rates of cleavage. The most common linkages are esters, anhydrides, and acetals.

The second approach involves the polymerization of difunctional monomeric pesticides. In these materials, the pesticide is released by the chemical or biological depolymerization of the system in accordance with the equation:

$$\text{Monomer} + \text{Pesticide} \underset{\text{environment}}{\overset{\text{synthesis}}{\rightleftharpoons}} \text{Polymeric Pesticide}$$

the depolymerization reaction can be purely homogeneous, purely heterogeneous (i.e., surface reactions), or some combination of the two. A heterogeneous reaction might occur, for example, because the reaction products cannot leave the reaction site in the bulk of the polymer, locally reversing the reaction, or because one degradant, such as a microorganism, is excluded from the interior of the polymer. Homogeneous and heterogeneous chemical reactions that take place in CR systems have been discussed.[90-92]

Release characteristics: Heterogeneous chemical reaction systems are similar in nature to the erodible matrix systems and "zero order" release is obtained for slabs. Owing to geometry, i.e., the changing of surface area as reaction proceeds, spherical or cylindrical systems have nonlinear release characteristics with time.

With homogenous systems, the release rate is complex and is affected by the reaction and diffusion process. For water-soluble systems,[93] release rate of pendant pesticide groups follows the conventional first order kinetics, while systems based on depolymerization reactions may be zero order if the mechanism of release comprises unzipping of polymer chains.

Applications in the pheromone area: Although none of the above mentioned techniques are being used for insect pheromones at the present time, there is the possibility that they may be found useful in insect control in the future, or perhaps in the control of other animal pests. They have been included in this section to make the subject complete.

REFERENCES

1. **Bakan, J. A.,** Microencapsulation of pesticides and other agricultural materials, in *Proc. Controlled Release Pest. Symp.,* Harris, F. W., Ed., Wright State University, Dayton, 1975, 76.
2. **Fanger, G. O.,** General background and history of controlled release, in *Proc. Controlled Release Pest. Symp.,* Cardarelli, N. F., Ed., University of Akron, Ohio, 1974, 1.18.
3. **Scher, H. B.,** Microencapsulated pesticides, in *Controlled Release Pesticides,* Scher, H. B., Ed., ACS Symp. Ser. 53, American Chemical Society, Washington, D.C., 1977, 126.
4. **Robinson, J. R.,** Controlled release pharmaceutical systems, in *Chemical Marketing and Economics Reprints,* Long, F. W., O'Neill, W. P., and Stewart, R. D., Eds., American Chemical Society, San Francisco, 1976, 212.
5. **Zweig, G.,** Environmental aspects of controlled release pesticide formulation, in *Controlled Release Pesticides,* Scher, H. B., Ed., ASC Symp. Ser. 53, American Chemical Society, Washington, D.C., 1977, 45.
6. **Bakan, J. A.,** Microcapsule drug delivery systems, in *Polymers in Medicine and Surgery,* Kronenthal, R. L., Oser, Z., and Martin E., Eds., Plenum Press, New York, 1975.
7. **Paul, D. R.,** Polymers in controlled release technology, in *Controlled Release Polymeric Formulations,* Paul, D. R., and Harris, F. W., Eds., ACS Symp. Ser. 33, American Chemical Society, Washington, D.C., 1976, 9.
8. **Cardarelli, N. F. and Kanakkanatt, S. V.,** Matrix factors affecting the controlled release of pesticides from elastomers, in *Controlled Release Pesticides,* Scher, H. B., Ed., ACS Symp. Ser. 53, American Chemical Society, Washington, D.C., 1977, 64.
9. **Neogi, A. N. and Allan, G. G.,** Controlled release pesticides: concept and realization, in *Controlled Release of Biologically Active Agents,* Tanquary, A. C. and Lacey, R. E., Eds., Plenum Press, New York, 1974, 210.
10. **Baker, R. W. and Lonsdale, H. K.,** Membrane-controlled delivery systems, in *Proceedings of Controlled Release Pesticide Symposium,* Cardarelli, N. F., Ed., University of Akron, Ohio, 1974, 40.1.
11. **Crank, J. and Park, G. S.,** Eds., *Diffusion in Polymers,* Academic Press, London, 1968.
12. **Crank, J.,** *The Mathematics of Diffusion,* Oxford University Press, London, 1956.
13. **Green, B. K.,** U. S. Patent 2,712,507, 1955.
14. **Green, B. K. and Schleicher, L.,** U. S. Patent 2,730,456, 1956.
15. **Green, B. K.,** U. S. Patent 2,800,457, 1957.
16. **Miller, R. E. and Anderson, J. L.,** U. S. Patent 3,155,590, 1964.
17. **Heistand, E. N., Wagner, J. G., and Knoechel, E. L.,** U. S. Patent 3,341,051, 1967.
18. **Brynko, C., Bakan, J. A., Miller, R. E., and Acarpelli, J. A.,** U. S. Patent 3,341,466, 1967.
19. **Reyes, Z.,** U. S. Patent 3,173,878, 1965.
20. **Herbig, J.,** U. S. Patent 3,161,602, 1964.
21. **Hall, H. and Hinkes, T.,** Air suspension encapsulation of moisture sensitive particles using aqueous systems, in *Microencapsulation: Processes and Applications,* Vandegaer, J. E., Plenum Press, New York, 1974, 145.
22. **Macaulay, N.,** U. S. Patent 3,016,308, 1962.
23. **Belcher, D. W.,** Spray drying equipment, in *The Encyclopedia of Chemical Process Equipment,* Mead, W. J., Ed., Reinhold Publishing, New York, 1964, 861.
24. **Barclay, W. H.,** Mixing equipment, in *The Encyclopedia of Chemical Process Equipment,* Mead, W. J., Ed., Reinhold Publishing, New York, 1964, 658.

25. **Vandegaer, J. E.**, U. S. Patent 3,577,515, 1971.
26. **Vandegaer, J. E.**, *Microencapsulation: Processes and Applications*, Plenum Press, New York, 1974.
27. **Ruus, H.**, U. S. Patent 3,429,827, 1969.
28. **DeSavigny, C. B.**, U.S. Patent 3,429,827, 1969.
29. **Scher, H. B.**, Great Britain Patent 1,371,179, 1975.
30. **Brynko, C.**, U. S. Patent 2,969,330, 1961.
31. **Goodwin, J. T. and Somerville, G. R.**, Physical methods for preparing microcapsules, in *Microencapsulation: Processes and Applications*, Plenum Press, New York, 1974, 155.
32. **Somerville, G. R.**, U. S. Patent 3,015,128, 1962.
33. **Somerville, G. R.**, U. S. Patent 3,389,194, 1968.
34. **Langer, G.**, et al., U. S. Patent 3,294,704, 1966.
35. **Mitchell, E. R., Copeland, W. W., Sparks, A. N., and Sekul, A. A.**, Fall armyworm: disruption of pheromone communication with synethtic acetates, *Environ. Entomol.*, 3, 778, 1974.
36. **Lanier, G. N., Silverstein, R. M., and Peacock, J. W.**, Attractant pheromone of the European elm bark beetle (*Scolytus multistriatus*): Isolation, identification; synthesis and utilization studies, in *Perspectives in Forest Entomology*, Anderson, J. F. and Kaya, H. K., Eds., Academic Press, New York, 1976, 149.
37. **Pearce, G. T., Gore, W. E., Silverstein, R. M., Peacock, J. W., Cuthbert, R. A., Lanier, G. N., and Simeone, J. B.**, Chemical attractants for the smaller European elm bark beetle *Scolytus multistriatus* (Coleoptera: Scolytidae), *J. Chem. Ecol.*, 1, 115, 1975.
38. **Roelofs, W. L., Hill, A. S., and Carde, R. T.**, Sex pheromone components of the redbanded leafroller moth, *Argyrotaenia velutinana*, *J. Chem. Entomol.*, 67, 159, 1974.
40. **Gentry, C. R., Bierl, B. A., and Blythe, J. L.**, Air Permeation Field Trials with the Oriental Fruit Moth Pheromone, presented at the 1976 Int. Controlled Release Pest. Symp., The University of Akron, September 13 to 15, 1976.
41. **Cameron, E. A.**, Disparlure: a potential tool for gypsy moth population management, *Bull. Entomol. Soc. Am.*, 19, 15, 1973.
42. **Beroza, M., Hood, C. S., Trefrey, D., Leonard, D. E., Knipling, E. F., Klassen, W., and Stevens, L. J.**, Large field trial with micro-encapsulated sex pheromone to prevent mating of the gypsy moth, *J. Econ. Entomol.*, 67, 639, 1974.
43. **Beroza, M., Hood, C. S., Trefrey, D., Leonard, D. E., Knipling, E. F., and Klassen, W.**, Field trials with disparlure in Massachusetts to suppress mating of the gypsy moth, *Environ. Entomol.*, 4, 705, 1975.
44. **Granett, J. and Doane, C. C.**, Reduction of gypsy moth mating potential in dense populations by mistblower sprays of microencapsulated disparlure, *J. Econ. Entomol.*, 68, 435, 1975.
45. **Brooks, T. W., Ashare, E., and Swenson, D. W.**, Hollow fibers as controlled release vapor devices, in *Textile and Paper Chemistry and Technology*, Arthur, J. C., Ed., ACS Symp. Ser. 49, American Chemical Society, Washington, D. C., 1977, 111.
46. **Ashare, E., Brooks, T. W., and Swenson, D. W.**, Controlled release from hollow fibers, in *Controlled Release Polymeric Formulations*, Paul, D. R. and Harris, F. W., Eds., ACS Symp. Ser. 33, American Chemical Society, Washington, D.C., 1976, 273.
47. **Coplan, M. J. and Brooks, T. W.**, U. S. Patent 4,017,030, 1977.
48. **Michaels, A. S.**, Therapeutic systems for controlled administration of drugs: a new application of membrane science, in *Permeability of Plastic Films and Coatings*, Hopfenberg, H. P., Ed., Plenum Press, New York, 1975.
49. **Chandrasekaran, S. K. and Shaw, J.**, Design of transdermal therapeutic systems, in *Contemporary Topics in Polymer Science*, Vol. 2, Pearch, E. M. and Schaefgen, J. R., Eds., Plenum Press, New York, 1977.
50. **Akkapeddi, M. K.**, et al., Cervical hydrogel dilator: a new delivery system for prostaglandins, in *Controlled Release of Biologically Active Agents*, Tanquary, A. S. and Lacey, R. E., Eds., Plenum Press, New York, 1974, 165.
51. **Anon.**, Slow-release achieved for macromolecules, *C & E News*, September 5, 1977, 26.
52. **Tollar, M.**, et al., *J. Biomed. Material Res.*, 3, 305, 1969.
53. **Drobnik, J.**, et al., *J. Biomed. Material Res.*, 8, 45, 1974.
54. **Molday, R. S.**, et al., *Nature (London)*, 249, 81, 1974.
55. **Obermayer, A. S. and Nicholas, L. D.**, Controlled release from ultramicroporous cellulose triacetate, in *Controlled Release Polymeric Formulations*, Paul, D. R., and Harris, F. W., Eds., ACS Symp. Ser. 33, American Chemical Society, Washington, D.C., 1976, 303.
56. **Nichols, L. D.**, Poroplastic and sustrelle: controlled release vehicles having broad compatibility with dissolved and precipitated pesticides, in *Proc. Controlled Release Pesticide Symp.*, Cardarelli, N. F., Ed., University of Akron, Ohio, 1974.

57. **Hardee, D. D., Graves, T. M., McKibben, G. H., Johnson, W. L., Guldner, R. C., and Olsen, C. M.,** A slow release formulation of grandlure, the pheromone of the boll weevil, *J. Econ. Entomol.,* 67, 44, 1974.
58. **Bull, D. L., Coppedge, J. R., Ridgeway, R. L., Hardee, D. D., and Graves, T. M.,** Formulations for controlling the release of the synthetic pheromone (grandlure) of the boll weevil, *Environ. Entomol.,* 2, 829, 1973.
59. **Brooks, T. W.,** et al., Experience in using a hollow fiber controlled release formulation in pheromone mediated suppression of *Pectinophora gossypiella* under humid tropical conditions, in *Proc. 6th Int. Symp. Controlled Release of Bioactive Materials,* Harris, F. W., Ed., Controlled Release Society, Wright State University, Dayton, 1979, IV-1.
60. **Gaston, Lyle, K., Kaae, R. S., Shorey, H. H., and Sellers, D.,** Controlling the pink bollworm by disrupting sex communication between adult moths, *Science,* 196, 904, 1977.
61. **Brooks, T. W. and Kitterman, R. L.,** Controlled release insect pheromone formulations based on hollow fibers and methods of application. Paper No. 77-1035, American Society of Agricultural Engineers Meeting, North Carolina State University, Raleigh, American Society of Agricultural Engineers, St. Joseph, Mich., 1977.
62. **Brooks, T. W., Doane, C. C., and Staten, R. T.,** Experience with the first commercial pheromone communication disruptive for suppression of an agricultural insect pest, in Proc. Adv. Res. Int. on Chem. Ecol. and Odour Comm. in Animals, NATO Science Committee, The Hague, The Netherlands, September 24 to 30, 1978.
63. **Gaston, L. K., Kase, R. S., Shorey, H. H., and Sellers, D.,** Controlling the pink bollworm by disrupting sex pheromone communication between adult moths, *Science,* 96, 904, 1977.
64. **Harris, F. W.,** Preparation of plastic, controlled-release pesticide formulations, in *Proc. Controlled Release Pest. Symp.,* Cardarelli, N. F., Ed., 1976, University of Akron, Ohio, 1.33.
65. **Cardarelli, N. F.,** Compounding methods for controlled release elastomers, in *Proc. Controlled Release Pest. Symp.,* Cardarelli, N. F., Ed., 1976, University of Akron, Ohio, 1.44.
66. **Yolles, S.,** et al., in *Controlled Release of Biologically Active Agents,* Tanquary, A. C. and Lacey, R. E., Eds., Plenum Press, New York, 1974.
67. **Allan, G. G.,** et al., Polymeric drugs for plants, in *Polymeric Drugs,* Donarums, L. G. and Vogi, O., Eds., Plenum Press, New York, 1974.
68. **Allan G. G., Beer, J. W., and Cousin, M. J.,** Controlled release of herbicides from biodegradable substrates, in *Controlled Release Pesticides,* Scher, H. E., Ed., ACS Symp. Ser. 53, American Chemical Society, Washington, D.C., 1977, 94.
69. **Shasha, B. S., Doane, W. M., and Russell, C. R.,** Starch encapsulated pesticides for slow release, *J. Polym. Sci.,* 14, 417, 1976.
70. **Doane, W. M., Shasha, B. S., and Russell, C. R.,** Encapsulation of pesticides with a starch matrix, in *Controlled Release Pesticides,* Scher, H. B., Ed., ACS Symp. Ser. 53, American Chemical Society, Washington, D.C., 1977.
71. **DelliColli, H. T.,** Controlled release of pesticides from Kraft lignin carriers, in *Controlled Release Pesticides,* Scher, H. B., Ed., ACS Symp. Ser. 53, American Chemical Society, Washington, D.C., 1977.
72. **Hopfenberg, H. B.,** Controlled release from erodible slabs, cylinders and spheres, in *Controlled Release Polymeric Formulations,* Paul, D. R. and Harris, F. W., Eds., ACS Symp. Ser. 33, American Chemical Society, Washington, D.C., 1976, 26.
73. **Charmillot, P. J., Baggiolini, M., Murbach, R., and Arn, H.,** Comparison de differents pieges a attractif sexuel synthetique pour le controle du vol du carpocapse (*Laspeyresia pomonella* L.) La Recherche agron. en *Suisse,* 14, 71, 1978.
74. **Proverbs, M. D., Logan, D. M., and Newton, J. R.,** A study to suppress codling moth (Lepidoptera: *Olethreutidae*) with sex pheromone traps, *Can. Entomol.* 107, 1265, 1975.
75. **MacLellan, C. R.,** Suppression of codling moth (Lepidoptera: *Olethreutidae*) by sex pheromone trapping of males, *Can. Entomol.,* 108, 1037, 1976.
76. **Kydonieus, A. F., Smith, I. K., and Hyman, S.,** A polymeric delivery system for the controlled release of pesticides, in *Proc. Controlled Release Pest. Symp.,* Harris, F. W., Ed., Wright State University, Dayton, 1975, 60.
77. **Quisumbing, A. R., Lawatsch, D. J., and Kydonieus, A. F.,** Field studies on the control of the German cockroach, *Blatella germanica* (L.), Using Hercon Roach Tape and standard sprays, in *Proc. Controlled Release Pest. Symp.,* Harris, F. W., Ed., Wright State University, Dayton, 1975, 247.
78. **Kydonieus, A. F., Smith, I. K., and Beroza, M.,** Controlled release of pheromones through multilayered polymeric dispensers, in *Controlled Release Polymeric Formulations,* Paul, D. R. and Harris, F. W., Eds., ACS Symp. Ser. 33, American Chemical Society, Washington, D.C., 1977, 283.

79. **Kydonieus, A. F., Rofheart, A., and Hyman, S.,** Marketing and economic considerations for Hercon consumer and industrial controlled release products, in Chemical Marketing and Economic preprints, ACS Symp. American Chemical Society, Washington, D.C., 140.
80. **Kydonieus, A. F. and Beroza, M.,** Insect control with the multilayered Luretape dispenser, in *Proc. Controlled Release Pest. Symp.,* Goulding, R. L., Ed., Oregon State University, Corvallis, 1977, 78.
82. **Johnson, W. L., McKibben, G. H., Rodriguez, J. V., and Davich, T. B.,** Boll weevil: increased longevity of grandlure using different formulations and dispensers, *J. Econom. Entomol.,* 69, 263, 1975.
83. **Hendricks, D. E., Hartstack, A. W., and Raulston, J. R.,** Compatibility of Virelure and Looplure dispensed from traps for cabbage looper and tobacco budworm survey, *Environ. Entomol.,* 6, 556, 1977.
84. **Tingle, F. C. and Mitchell, E. R.,** Controlled release plastic strips containing (*Z*)-9-dodecen-1-ol acetate for attracting *Spodoptera frugiperda, J. Chem. Ecol.,* 4, 41, 1978.
85. **Beroza, M., Paszek, E. C., DeVilbiss, D., Bierl, B. A., and Tardif, J. G. R.,** A 3-Layer laminated plastic dispenser of disparlure for use in traps for gypsy moths, *Environ. Entomol.,* 4, 712, 1975.
86. **Gentry, C. R., Bierl, B. A., and Blythe, J. L.,** Air permeation field trials with the Oriental fruit moth pheromone, presented at the 1976 Int. Controlled Release Pest. Symp., The University of Akron, September 13 to 15, 1976.
87. **Vick, K. W., Coffelt, J. A., and Sullivan, M. A.,** Disruption of pheromone communication in the Angoumois grain moth with synthetic female sex pheromone, *Environ. Entomol.,* 7, 528, 1978.
88. **Beroza, M., Hood, C. S., Trefrey, D., Leonard, D., Knipling, E. F., and Klassen, W.,** Field trials with disparlure in Massachusetts to suppress mating of the gypsy moth, *Environ. Entomol.,* 4, 705, 1975.
89. **Ritschel, W. A.,** Peroral solid dosage forms with prolonged action, in *Drug Design,* Ariens, E. J., Ed., Academic Press, New York, 37.
90. **Feld, W. A., Post, C. K., and Harris, F. W.,** Controlled release from polymers containing pesticides as pendent substituents, in *Proc. Controlled Release Pest. Symp.,* Harris, F. W., Ed., Wright State University, Dayton, 1975, 113.
91. **Harris, F. W. and Post, C. K.,** Synthesis of polymers containing aquatic herbicides as pendent substituents, in *Proc. Controlled Release Pest. Symp.,* Harris, F. W., Ed., Wright State University, Dayton, 1975, 113.
92. **Harris, F. W. and Post, C. K.,** Synthesis of polymers containing pendent herbicide substituents, *Am. Chem. Soc. Polym. Prepr.,* 16(2), 622, 1975.
93. **Lewis, D. H. and Cowsar, D. R.,** *Principles of Controlled Release Pesticides,* Scher, H. B., Ed., ACS Symp. Ser. 53, 1977, American Chemical Society, Washington, D.C., 152.

Chapter 4

THE SENSING, DISPERSION, AND MEASUREMENT OF PHEROMONE VAPORS IN AIR

Joseph H. Caro

TABLE OF CONTENTS

I. INTRODUCTION

It is now well-established that flying males of many species of moths locate the females for mating by sensing the airborne sex pheromone emitted by the females during their "calling" period. To use synthetic pheromones or mating disruptants (antipheromones), which are generally expensive and often unstable chemicals, in the best possible way to control or monitor moth populations, it is important that all relevant aspects of the vaporized material be fully understood and that all controllable processes be optimized. Obviously, these goals have not yet been attained, but considerable research effort has in fact been expended in their pursuit. We review here existing information in such areas as the micrometeorology governing the dispersion of vapors in field or forest atmospheres, the mechanisms by which the male moths sense and respond to the airborne odors, and the means of collecting the pheromones from the air and their subsequent qualitative or quantitative determination.

II. PHEROMONE CHARACTERISTICS AFFECTING VAPOR CONCENTRATIONS

Most known sex pheromones of the Lepidoptera are 12- to 16-carbon long-chain mono- or diunsaturated alcohols, aldehydes, and acetates. They are relatively nonpolar and many are vulnerable to oxidative decomposition. In view of the biologic necessity to volatilize the pheromones to bring about mating of the insects, reported vapor pressures are surprisingly low. The vapor pressure of gossyplure, a mixture of (*Z,Z*)- and (*Z,E*)-7,11-hexadecadien-1-ol acetates that has been identified as the sex pheromone of the pink bollworm, *Pectinophora gossypiella* (Saunders), has been reported to range from 2 to 5×10^{-5} mmHg at normal ambient temperatures,[1] which would indicate that the pheromone is only moderately volatile. Similarly, the vapor pressure of disparlure, the sex pheromone of the gypsy moth (*Lymantria dispar* (L.)), has been reported as 1 to 1.8×10^{-5} mmHg at 15 to 25°C.[2] Volatility and biodegradability are, nevertheless, high enough so that pheromones must be applied in the field in controlled release formulations to remain effective during a substantial portion of the mating period of the insect. Volatilization rate is, moreover, strongly affected by temperature. In a laboratory experiment, for example, the loss of disparlure from foil plates was twice as high at 35°C than at 20°C.[2] Rate of diffusion of pheromones from controlled release formulations, such as microcapsules, hollow fibers, or plastic laminates, will also increase as the temperature increases. Thus, a field experiment with microcapsules containing disparlure revealed a marked diurnal variation in rate of disparlure volatilization, with highest rates at midday, when temperatures were highest.[3]

Most pheromones are also sufficiently labile that they will not persist for long periods in soil and water. At 21°C, grandlure, the 4-component sex pheromone of the boll weevil, *Anthonomus grandis* Boheman, disappeared from soil within 32 hr after application and from water within 48 hr.[4] Similarly, the half-life of gossyplure at 32°C was reported as 1 day in soil and 7 days in water.[5] Contamination of soil and water by applied pheromones is, therefore, unlikely. On the other hand, the close resemblance of the long-chain hydrocarbon structures of many pheromones to the structures of plant leaf waxes might favor the rapid sorption of airborne pheromones on surfaces of plants or soil at night when the surfaces are cool and the atmosphere is stable.[6,7] Presumably, at least a fraction of the sorbed compounds would be desorbed the next day as solar radiation increases and both temperature and atmospheric turbulence increase as a consequence.

In many moth species, the sex pheromone consists of a mixture of two or more compounds or of positional or geometric isomers of the same compound. The individual compounds, being similar in structure, have similar vapor pressures and the female moth would be expected to emit the total pheromone at precise ratios of components throughout a range of temperatures.[8] Thus, measuring the aerial concentration of one component of a complex pheromone mixture should allow estimation of the complete pheromone if the precise constitution of the mixture is known.

III. SENSING OF PHEROMONE VAPORS BY INSECTS

A. Mechanism and Behavioral Response

A mechanism of insect response to odors was first formulated by J. S. Kennedy in 1939,[9] elaborated by R. H. Wright in 1958,[10] and refined, specifically with respect to the sensing of pheromone vapors, by Kennedy and co-workers in 1978[11] and Cardé and Hagaman in 1979,[12] among others. They recognized that the minuscule development of the insect brain would permit only mechanical, not purposeful, reaction to a stimulus, and that olfaction, unlike sight and sound, is a nondirectional sense. Consequently, the insect must use a subtle means of guidance to home in on an odor source. A current interpretation of pheromone perception and response, as developed largely from imaginative wind tunnel experiments in which movable floor patterns were used, is as follows:

The female moth signals sexual receptivity by emitting its sex pheromone as a vapor from a gland near the tip of its abdomen. Since air is constantly in motion, the odorous molecules are carried downwind in a plume and are dissipated by atmospheric turbulence. When a randomly flying male moth of the same species perceives the odor by impact of the vapor on its antennal receptors, it orients its body into the wind and begins upwind flight, using visual ground cues as a directional guide. This behavior is termed "optomotor anemotaxis."[13] The gradient of pheromone concentrations in the air is evidently not involved in mate-finding beyond a few centimeters from the female because the emitted vapors are rapidly attenuated by turbulent dispersion in the atmosphere.[14-16] The male zigzags as it flies toward the female, the directional reverses possibly being caused by a repetitive losing and regaining of the scent.[11] The level of pheromone in the air also affects the flight. If the concentration is high, flight is largely within the plume area, whereas if the concentration is low, the degree of lateral movement is greater.[12] When the moth loses the scent entirely, it widens its zigzags and changes its flight angle from into wind to crosswind, until it again senses the odor.[11,13] The marginal definition of the pheromone plume, which is sharper at higher pheromone concentrations, has also been postulated to influence male flight.[15,17] As the moth traverses across the plume, it repeatedly senses the edges and turns back into the plume. The steeper the concentration gradient at the edge, the easier the moth would steer up the plume. Strangely, the flying moth maintains a "preferred", relatively constant ground speed and track angle under changing wind conditions, compensating for changes in wind velocity by changing its air speed and its orientation to the wind.[11,12,18]

As the moth nears the source of the pheromone, either a female moth or a source of synthetic pheromone, it progressively reduces its air speed and turns more into the wind, presumably in response to the increasing intensity of the stimulation.[11,12] The zigzagging continues, but the distance between track reversals decreases. This behavior has been termed "reversing optomotor anemo*meno*taxis."[19] Finally, the male reaches the source, probably chiefly by short-range orientational cues provided by the pheromone diffusion gradient.[20]

B. Factors Affecting Insect Response

1. Temperature

Ambient temperature is an important cue in regulating the diurnal periodicity of the mating of Lepidoptera.[21,22] The timing of both female calling behavior and male response very often shifts as the temperature changes. In general, lower temperatures cause the response to be transferred to warmer periods of day or night:[23,25] e.g., the flight period of night-flying moths is significantly earlier on cool than on warm nights.[23]

Similarly, widespread species such as *Heliothis zea* (Boddie), the corn earworm, exhibit peak mating earlier in the evening in cooler climates.[7] Many species of moths and beetles seem to have an optimum temperature for flight, above and below which flight, and thereby catch in pheromone-baited traps, is reduced.[24,26]

2. Wind Speed

One direct effect of variation in wind speed on the behavior of moths is the apparently adaptive decrease in the period of pheromone emission by female moths with increase in winds. This was shown clearly in experiments with the cabbage looper, *Trichoplusia ni* (Hübner), in which emission periods were 20, 12, 7, and 5 min at wind speeds of < 10, 30, 100, and 300 cm/sec, respectively.[16]

With both moths and beetles, wind velocity affects communication distance, with direct effects on trap catches. The direction and intensity of the effect apparently differ with insect species and probably with location of traps in the field. Thus, the catch of both the western corn rootworm beetle, *Diabrotica virgifera* Le Conte, and the northern corn rootworm beetle, *D. longicornis* (Say), increased with velocity of winds ranging from 0 to 7 m/sec.;[26] The catch of the red bollworm moth, *Diparopsis castanea* Hampson, was highest at winds of 1 to 2.5 m/sec, with both greater and lesser winds depressing activity;[27] and the attraction of soybean looper moths, *Pseudoplusia includens* (Walker), to a pheromone source was little affected by the velocity of winds ranging from 0 to 2 m/sec.[28] The general influence of wind velocity on pheromone-mediated communication distance is discussed more fully below in the section on the latter subject.

3. Light Intensity

Studies of light intensity as an independent variable showed that it distinctly affects insect responses to pheromones. Solar radiation intensity was the dominant factor in trap catches of the northern corn rootworm moth, with catches decreasing very sharply as the intensity increased from zero. By contrast, trap catches of the western corn rootworm moth were highest at a light intensity of 0.19 cal cm^{-2} sec^{-1}, a level that is typical of the condition about 2 hr before sunset.[26] In an earlier study,[29] cabbage looper moths were most responsive at a light intensity near that of moonlight.

4. Time of Day

Mating of practically all lepidopteran species takes place during distinct periods of the day. Environmental cues setting the rhythm include sunrise, sunset, and changes in ambient temperature.[21] For example, the mating of gypsy moths peaks from 1100 to 1500 hours,[30] and that of the pea moth, *Cydia nigricana* (Stephens), peaks at 1700 hours.[31] Activity, as indicated by moth trap catch, peaked at both 1000 and 2000 hours for the western corn rootworm and around midnight for the northern corn rootworm,[26] and median catches of the codling moth, *Cydia pomonella* (L.), occurred over the entire season within 1 hr after sunset.[24]

5. Atmospheric Concentration of Pheromone

We have already seen that the level of pheromone in the atmosphere can affect the flight path and air speed of the mate-seeking male moth. The concentration can also influence the behavior of the insect in other ways. For one, the higher the concentration, the higher the percentage of male moths in a population that respond to the pheromone after a given period of exposure,[32] and the faster the moths respond to the stimulus. For example, in a study with the cabbage looper, male response per unit time was markedly faster at greater pheromone concentrations.[33] In this connection, Aylor and co-workers[32,34] have proposed that, since male moths respond quickly to high concentrations of pheromones and since peak concentrations may be as much as 25 times greater than would be calculated by average diffusion equations, the important value for predicting moth response is the instantaneous peak concentration, rather than some time-averaged concentration.

An important feature involving the atmospheric concentration of pheromone is the saturation of the antennal receptor organs of the male moth, producing habituation and loss of response, that occurs on contact of the insect with high levels of the pheromone. This effect provides a basis for the atmospheric permeation technique of insect control. Actually, the level of habituation achieved is probably related to the duration, as well as the intensity, of the stimulus,[21,34] but most work in this area has focused on concentration effects. It was estimated that concentrations about 10^5 times the threshold of response would be required to saturate the male receptors,[35] but later results suggested that concentrations only 10^2 to 10^3 times the threshold are sufficient to disrupt long-distance response.[36,37] Total abolition of response may require impractical concentrations for complete mating disruption of some species, but lesser concentrations would still have value, in that the chances of mate finding would be reduced.[21] Tests with several moth species showed that concentrations only 3 to 100 times that which is optimum for male attraction clearly reduced trap catches of males.[38]

6. Composition of Stimulus

The response of male moths to an olfactory stimulus is affected in certain cases by the composition of the vapors. Racemic disparlure, for example, has been utilized to control the gypsy moth by atmospheric permeation, but it now appears that the male moth is responsive only to the (+)-enantiomer. The (−)-enantiomer is, in fact, inhibitory. In wind tunnel tests, the (−)-enantiomer has been shown to reduce the percentage of moths initiating anemotactic flight, the duration of flight, and the ground speed of the flying moths.[39] For those species that utilize mixtures of two or more compounds as the sex pheromone, mixtures having about the natural ratio of components emitted by the female will elicit male response at lower concentrations than mixtures having other ratios.[8]

C. Response Threshold Concentrations

The threshold of male response is one of the three parameters that are fundamental to communication between moths in the field, the other two being the rate of pheromone emission by the "calling female" and the rate of dispersion of vapors in a moving air mass.[40] Reported concentrations for the threshold of male response to a pheromone stimulus are shown in Table 1. In 1958, Wright[10] used an arbitrary threshold value of 10^{-11} $\mu g/cm^3$, equivalent to about 10^5 molecules per cm^3, in his analysis of olfactory guidance of insects. Hirooka and Suwanai[44] developed a predictive model in 1978 that supported Wright's value. However, measurements on individual moth species have shown that threshold concentrations can differ by as much as three orders of magnitude among various species, and that many moths respond to concentrations as low as 10^3 to 10^4 molecules per cm^3 (Table 1).

Attempts have been made to quantitate the relationship between the pheromone emission rate of the female moth and the male response threshold. The original work was done by Bossert and Wilson,[14] who proposed the equation:

$$(W)_O = 0.1256\ UX^{7/4} \qquad (1)$$

Table 1
RESPONSE THRESHOLDS OF MALE MOTHS TO AIRBORNE SEX PHEROMONES

Insect	Response threshold		
	μg/cm³	molecules/cm³	Ref.
Trichoplusia ni	3×10^{-12}	8×10^{3}	16,40
Bombyx mori (L.)		10^{3}	41
Diparopsis castanea		9.2×10^{3}	36
Pectinophora gossypiella (Saunders)	5×10^{-8} (a)		42
Trogoderma glabrum Herbst	10^{-9}		43
Unspecified (arbitrary value)	10^{-11}	10^{5}	10
Model insect, similar to *Hyphantria cunea* (Drury)	$10^{-11} - 10^{-10}$	$10^{5} - 10^{6}$	44

[a] Expressed as weight of pheromone at source, μg.

where $(W)_o$ is the ratio of pheromone release rate to the threshold concentration in the air; U is the average wind velocity, cm/sec; and X is the distance from the source, cm. Hirooka and Suwanai[44] later suggested a supposedly more precise relationship:

$$(W)_O = \pi U G^2 X_M^2 \quad (2)$$

where G is gustiness ($= \sqrt{y^2}/X$, where y^2 is the variance of wind variation, cm²) and X_M is the maximum effective distance downwind from a virgin female. Regardless of the precise quantitation, it is evident that, as the wind velocity increases, an increased emission rate is required to achieve the threshold concentration at any given distance from a pheromone source.

In 1978, Roelofs[8] presented a hypothesis, based on threshold concentrations, for male response in those moth species that utilize a binary mixture of isomers as the sex pheromone. Plots of atmospheric concentration against isomer ratios in the mixtures yielded an "attraction area" for a particular insect species, bounded on the lower edge by the threshold curve for flight activation and on the upper edge by the curve for insect disorientation. The shape of the attraction area helps to explain why high concentrations of isomer ratios other than the natural ratio emitted by the female moth will attract male moths.

D. Communication Distance

Because of the many interrelating climatic, topographic, and biologic factors involved, an inherent high degree of variability exists in any measure of maximum communication distance between male and female moths in the field. Nevertheless, efforts have been made to quantify distances, largely by developing predictive models. Bossert and Wilson[14] formulated a relationship for the mean maximum distance of pheromone communication in moving air:

$$X_M = \left(\frac{8Q}{UK}\right)^{4/7} \quad (3)$$

where Q is the pheromone emission rate, μg/sec, and K is the threshold of response, μg/cm³. They then estimated that gypsy moths could communicate over distances in excess of 4000 m, basing their calculations on results of earlier trapping studies. Gypsy moths had, in fact, been observed to respond to a pheromone stimulus in the forest at distances of up to 900 m.[45] In contrast, a model predicting communication among cabbage loopers indicated a mean maximum distance of only 200 m,[16] and a model for *Trogoderma glabrum* communication gave a maximum distance of only several meters.[43]

Equation 3 shows that the maximum communication distance is inversely related to the wind velocity. The boundaries of the biologically effective space leading downwind from a pheromone source are shaped as an elongated half-ellipsoid,[40] and as the wind velocity increases, the length of the ellipsoid decreases because the increased turbulence in the atmosphere produces more rapid dispersion of the emitted vapors. This inverse relationship does not apply, however, at extremes of both high and low wind velocities. Because communication is limited both by the distance over which the pheromone can be carried by the wind before the female ceases emission and by the time required for the male to fly to the female,[16] an optimum wind speed must exist for moth mating. Sower and co-workers[16] developed a relationship that allows for the limitations:

$$X_M = \frac{Ut\,(r-U)}{r} \qquad (4)$$

where t is the time, in seconds, that the female moth continuously emits pheromone and r is the air speed of the male moth, cm/sec. These authors contended[40] that communication between cabbage looper moths would be highly unlikely beyond 1250 m at wind speeds of 100 cm/sec or more, and suggested that the optimum wind speed for communication was 25 cm/sec. At wind speeds lower than 25 cm/sec, communication distance decreases rapidly, possibly a result of pheromone deposition on ground or leaf surfaces[40,43] or because of the limiting duration of pheromone emission by the female moth.[16] An optimum wind speed, 1.0 m/sec, was found in field tests with the "tobacco cutworm", *Spodoptera litura* (F), below which the maximum distance of male response decreased with decreasing wind.[46]

IV. DISPERSION OF PHEROMONE VAPORS IN THE ATMOSPHERE

When a pheromone is vaporized into the atmosphere from a source in field or forest, the vapor disperses in a manner dictated by the combined effect of several meteorological and topographical factors. The nature of the dispersion is highly important to the success of the control technique being applied, whether it be trap catches of male moths or mating disruption by atmospheric permeation. In the latter case, dispersion characteristics are particularly important when the pheromone is released from large, spaced evaporators, rather than from closely spaced small sources, because layers and pockets of disruptant-free air may occur if the spacing between evaporators is large.[21,47]

In general, as a pheromone is carried downwind from a source, it forms an elongated plume consisting of numerous separate filaments or lamellae. As they move downwind, the filaments writhe, twist, change shape, break up, and radiate out from the longitudinal axis of the plume.[10,30] Despite this action, the general structure of the plume can persist at large distances from the source under relatively stable environmental conditions. Because of the filamentous nature, gradients of pheromone concentration are substantial even well inside the plume, with the degree of variability depending on the height of the source, the distance downwind, and the distance from the axis of the plume, as well as on the characteristics of the atmosphere and the terrain.[48]

The shape of the plumes generated by pheromone-containing traps of various designs is apparently critical to success in catching male moths. Up to 30-fold differences in catches of a single moth species have been obtained with traps of different designs containing the same doses of pheromone.[17] The more elongated the plume, the greater the trap catch, presumably because the insects can trace a narrow, concentrated scent more easily than a more diffuse one.[17,49] This supposition has been supported in numerous experiments in which pheromone traps were placed at several heights above ground in infested areas. Generally, the male moths were not strongly attracted to the traps located at heights other than those encompassing the normal flying patterns of the insects.

A. Wind Velocity and Turbulent Transfer

The extent and rate of dispersion of pheromone vapors in the atmosphere depend largely on the wind velocity, since winds cause atmospheric turbulence and attendant transfer of the vapors. The wind-generated turbulent dispersion overwhelms the diffusion properties of the pheromone, so that molecular diffusion is relatively unimportant.[14,32] A puff of pheromone emitted from a source is dispersed by the action of turbulence through being caught up in eddies of various sizes ranging from a few millimeters to many meters, all acting simultaneously to transport and fragment it.[34] As noted earlier, as the wind velocity increases from zero, the pheromone will be dispersed over greater distances until a maximum is reached; then, higher winds produce greater turbulence and a more rapid dispersion, resulting in a decrease in effective range of the pheromone.[40,46] The higher turbulence also affects the character of the filaments within the pheromone plume; they increase in number, become thinner and more changeable, and the space between them decreases.[11]

Winds must, of course, be sufficient to disperse the pheromone enough to have it perform its function. In the trapping of male moths, for example, even the most efficiently designed trap will not function well in an area of restricted air movement, such as in a hollow or when surrounded by dense foliage.[50] In the forest, turbulence is, in fact, often very low. Because of both vegetative aerodynamic drag and a daytime temperature inversion condition produced by warming of the canopy of trees, the winds under the canopy are often too light to measure. As a consequence, vapor plumes emanating from pheromone sources meander widely but exhibit little turbulent transfer until they reach an area under an opening in the canopy. There, turbulent energy is considerably greater, mixing occurs, and vapors are rapidly dispersed. Beyond the canopy opening, turbulence decays as the air is convected under the canopy.[32,34]

B. Aerial Pheromone Concentrations

The dispersion of pheromone vapors in the atmosphere may be represented directly by measurement of concentrations of the chemical at various points in the field or forest. Mathematically, the average concentration, C (g/cm^3), of pheromone downwind in the X direction from a source emitting pheromone at a steady rate, Q (g/sec), may be expressed by a Gaussian equation:

$$C_{(x,y,z)} = \frac{Q}{2\pi S_y S_z U} \exp\left[-\frac{y^2}{2 S_y{}^2} + \frac{z^2}{2 S_z{}^2}\right] \qquad (5)$$

where S_y and S_z are the standard deviations of plume dimensions in the crosswind, y, and vertical, z, directions, and U is the average horizontal wind speed, cm/sec.[51] At a specific distance from the source, the Hirooka and Suwanai model[44] yielded the relationship:

$$C(x) = Q/\pi G^2 x^2 U \qquad (6)$$

where C(x) is the concentration of pheromone, g/cm^3, at distance x, cm, from the source, and G is gustiness, as defined in Equation 2.

The mechanism of the decrease in pheromone concentration after emission into the turbulent atmosphere has been described by Aylor and co-workers.[34] The most dramatic effect is produced when a puff of pheromone encounters an atmospheric eddy of about the same size as the puff. The puff is disintegrated by the turbulence and the concentration decreases sharply. Eddies of other sizes, which constantly coexist with the puff-sized eddies, produce only slight decreases in concentration.

As noted earlier, Aylor[32] has taken the view that the common use of time-averaged concentrations to evaluate response to pheromones is faulty because, as the time of averaging increases, the average concentration in the composite sample decreases. This he claims is a result of wide meandering of plumes, so that a fixed sampler would frequently experience zero concentration. However, the disparity between average and Aylor's preferred peak concentrations would be negligible when the atmosphere is permeated by broadcasting many small pheromone sources to control insect mating. Measurements have shown that, in this type of treatment, the plumes overlap to provide a relatively uniform aerial concentration of pheromone throughout the treated area.

Equation 6 indicates that average concentrations are strongly affected by wind velocity, decreasing as the winds increase. A contributing factor in this relationship is the dilution of the initial concentration at the source.[40] In a recent field experiment conducted by the author and co-workers with a mating disruptant of the corn earworm, *Heliothis zea*,[7] we found the inverse relationship of wind speed with aerial pheromone concentration applied during daylight hours, but generally not at night. The results suggested that the plants and soil may act as sinks for the chemical vapor at night. In another of our experiments, conducted with the gypsy moth pheromone in woodlands,[52] ambient temperature was the dominant factor affecting aerial pheromone concentrations. Measured concentrations increased directly with temperature, a result of increased rate of emission from the microencapsulated source. In this experiment, plots of aerial concentration of pheromone vs. time showed that the pheromone dissipated at a near constant rate for 2 weeks and thereafter at a much slower rate, suggesting the existence of two mechanisms of release from the source.

V. DETERMINATION OF PHEROMONES IN AIR

Development of accurate, specific, and highly sensitive analytical methods for pheromones and related compounds in air is a prerequisite to successful direct measurement of the dispersion of these vapors in the atmosphere. Such methods would also be applicable to measurement of emission rates from formulations in the laboratory. Nevertheless, only scattered work has been done in the area. Most past analytical work on pheromones, which has been summarized in a comprehensive review by Inscoe and Beroza,[50] has centered on the isolation and identification of individual pheromones or components of pheromone mixtures. Here, we review information on trace analysis, particularly as it applies to determinations in air.

In general, analysis of pheromones for identification or for measurement of emission rates has been by gas chromatography with a flame-ionization detector[50,53] or by high-performance liquid chromatography with silver-loaded supports.[54,55] However, greater sensitivity than that which is possible with these methods is necessary for optimum trace analysis; this can best be achieved by appropriate derivatization of the subject pheromone and subsequent gas chromatographic analysis with capillary columns and electron-capture, flame photometric, or thermionic detectors.[50] An effective cleanup procedure is necessary with these techniques to remove possible interferences arising from volatile nonpheromonal ingredients of the formulation being used or from other sources in the ambient environment. Other approaches that seem feasible for trace analysis of pheromones include the use of radiolabeled pheromones[56,57] or an enzyme bioluminescent assay.[58] The latter which is still in the early stages of development, is specific for aliphatic long-chain aldehydes and is claimed to be extremely sensitive and much faster than capillary-column gas chromatography.

A sensitive method utilizing electron-capture detection was reported in 1973 for analyzing (*Z*),(*E*)-9,12-tetradecadien-1-ol acetate, a sex pheromone of several species of moths.[59] The isolated acetate was hydrolyzed to the alcohol, then conversion was made to the trichloroacetyl derivative. Subsequent gas chromatography was quantitative down to 1 ng of the derivative. In our laboratory, we have developed electron-capture gas chromatographic methods for minute quantities in air of both disparlure, the gypsy moth sex pheromone,[60] and (*Z*)-9-tetradecen-1-ol formate (*Z*-9-tdf), a synthetic mating disruptant of *Heliothis* spp.[61] In the former, a *vic*-dibromide derivative, to which the electron-capture detector is sensitive, was formed by reaction of the disparlure with diphenylphosphine dibromide. In the latter, a similar derivative was prepared by direct bromination of the formate. Quantitative limits in the two methods are about 0.2 ng disparlure and 2 to 3 ng *Z*-9-tdf per cubic meter of air.

A. Sample Collection

Most past work on the collection of pheromones from air has involved the stripping of an air stream that had been passed over live insects, with the purpose being to isolate and identify the volatile biologically active chemicals. Several different procedures were used, including collection of the volatiles in cold traps, absorption in lanolin, triglycerides, or gas chromatographic liquid phases, and adsorption on charcoal or powdered cellulose.[62] However, none of these methods is acceptable for quantitative trace analysis because of inherent limitations on airflow rates or inefficiencies in trapping or recovery.

To overcome these difficulties, the method of choice has been the trapping of the volatile pheromones on a bed of solid, porous adsorbent such as Tenax®, Porapak®, XAD resin, Chromosorb®, molecular sieves, and the like. This technique is highly efficient, permits long-term collection at relatively high airflow rates, and produces samples that are sufficiently stable that they can be analyzed at a convenient time after collection. Potential disadvantages of the method include the possibility of breakthrough of very volatile compounds and the gradual degradation of oxygen-sensitive materials as air is passed through the bed.[62]

The earliest use of solid adsorbents, although liquid-coated, for trapping airborne organics related to pheromones was in 1966.[63] The adsorbent was Chromosorb® A, coated with a high-molecular-weight liquid substrate and with an antioxidant added to reduce degradation of sensitive compounds. Most more recent studies with an uncoated adsorptive solid have used Porapak Q as adsorbent.[56,62,64,65] A bed of this material gave acceptable results when it was systematically evaluated as trapping agent for a series of pheromone-like model compounds.[62] Recoveries of long-chain alcohols and acetates ranged from 77 to 89%, and alcohols of six carbons or more did not break through the 20 × 2.2-cm bed even after 27 m^3 of air had been drawn through at a flow rate of 0.16 m^3/hr. Compounds sensitive to air oxidation were also recovered in good yield.

In our laboratory, considerations of achievable air flow rate, bed capacity, and cost of adsorbent led us to select type 4A molecular sieves for trapping disparlure[60] and Chromosorb® 102 for trapping (*Z*)-9-tetradecen-1-ol formate.[61] Actually, any of the solid adsorbents cited above are probably acceptable for trace analysis if the surface area of the bed is high enough that breakthrough will not occur during the course of the sample collection.

B. Removal From Samples

Once collected on an appropriate adsorbent, the pheromones must be isolated for further analysis. Perhaps the simplest procedure is thermal desorption, in which the adsorbate is swept directly onto the analytical chromatographic column by sudden heating. However, excessively contaminated samples are often a problem with this approach, and frequently a derivative of the pheromone to which the detector being used is sensitive must be prepared before analysis can be conducted at trace levels. Consequently, extraction with an organic solvent, which provides more flexibility, is commonly used instead of thermal desorption.

Sex pheromones and other behavior-controlling chemicals are generally nonpolar, so they can be extracted with solvents of low or moderate polarity, such as hexane, benzene, or dichloromethane.[66] The latter is the most commonly used solvent for direct extraction of pheromones from insect tissues.[50] Quantitative extraction from the porous polymeric adsorbents, now generally used as trapping agents, is quite simple, requiring only agitation with the solvent or brief Soxhlet extraction. For efficient extraction from molecular sieves, a somewhat polar solvent is necessary. We have employed either 1:1 hexane-diethyl ether[60] or 9:1 hexane-acetone, the latter to avoid the possible problem from peroxide formation in ether.

Because the extractability of pheromones from adsorbents may decrease as the samples age in storage, care must be taken, when analyzing large numbers of field samples, to measure and, if necessary, to correct for the effect. In replicated aging tests using our published methods,[60,61] we found that the average recovery of disparlure decreased regularly from 92% on immediate extraction to 75% after 50 days of storage. By contrast the recovery of (*Z*)-9-tetradecen-1-ol formate was quantitative after 40 days. The appropriate correction was made in all reported disparlure analyses.

C. Cleanup of Extracts

Air samples taken in field or forest may contain a variety of substances that can interfere with the quantitative determination of pheromones, so that a cleanup step is generally included in the analytical scheme. The interferences may arise from natural emissions such as plant volatiles, from nearby urban or industrial sources, or from components of the pheromone formulation such as stickers, thickeners, or antioxidants. In measurements in the author's laboratory, air taken from different agricultural locations in the eastern U.S. was found to vary widely with respect to degree of contamination.[61]

Cleanup techniques that have been used with pheromone-containing extracts include liquid-solid column chromatography with Florisil[60] or silica gel,[61] gel permeation chromatography,[66] and ultraviolet photolysis.[61] Solvent partitioning and high-performance liquid chromatography (HPLC) column cleanup offer other options. Analytical method development for specific pheromones should include testing of the several cleanup possibilities to optimize the procedures.

D. Confirmation of Identity

Several means are available to confirm the identity of pheromone peaks occurring in gas chromatograms. These include (1) measurement of retention times by gas-liquid chromatography (GLC) on two or more columns, both polar and nonpolar; (2) HPLC; (3) thin-layer chromatography (TLC), including argentation TLC; (4) mass spectrometry (MS); (5) proton magnetic resonance spectrophotometry; (6) GLC-MS; (7) infrared spectrophotometry, and (8) GLC of derivatives.[50] Published examples of some of these techniques include GLC of pheromone derivative on three columns at two column temperatures,[59] high-resolution GLC coupled with quadrupole mass fragmentography,[67] and conversion of a derivative of a mating disruptant to a second derivative with a different GLC retention time.[61]

REFERENCES

1. **Iyer, P. P., Yates, W. E., Akesson, N. B., and Horgan, P. M., Jr.**, Cotton Pink Bollworm Control Using the Sex Pheromone, Gossyplure-I.: Controlled Release of Gossyplure, Paper 78-1502, presented at Winter Meeting, American Society of Agricultural Engineers, Chicago, Ill., Dec. 18 to 20, 1978.
2. **Weintraub, F. P., Vylegzhanina, G. F., Chernichuk, L. L., and Konyukhova, I. A.**, Disparlure and Some Juvenile Hormone Analogs' Persistence in the Environment, U.S.E.P.A., Off. Res. Devel. Rep. EPA-600/9-78-003, Cincinnati, 1978.
3. **Caro, J. H., Bierl, B. A., Freeman, H. P., Glotfelty, D. E., and Turner, B. C.**, Disparlure: volatilization rates of two microencapsulated formulations from a grass field, *Environ. Entomol.*, 6, 877, 1977.
4. **Henson, R. D., Bull, D. L., Ridgway, R. L., and Ivie, G. W.**, Identification of the oxidative decomposition products of the boll weevil pheromone, grandlure, and the determination of the fate of grandlure in soil and water, *J. Agric. Food Chem.*, 24, 228, 1976.
5. **Henson, R. D.**, Environmental fate of gossyplure, *Environ. Entomol.*, 6, 821, 1977.
6. **Plimmer, J. R., Bierl, B. A., Webb, R. E., and Schwalbe, C. P.**, Controlled release of pheromone in the gypsy moth program, in *Controlled Release Pesticides*, Scher, H. B., Ed., ACS Symp. Ser. 53, American Chemical Society, Washington, D.C., 1977, chap. 15.
7. **Caro, J. H., Glotfelty, D. E., and Freeman, H. P.**, (*Z*)-9-Tetradecen-1-ol formate: distribution and dissipation in the air within a corn crop after emission from a controlled-release formulation, *J. Chem. Ecol.*, in press, 1980.
8. **Roelofs, W. L.**, Threshold hypothesis for pheromone perception, *J. Chem. Ecol.*, 4, 685, 1978.
9. **Kennedy, J. S.**, The visual responses of flying mosquitoes, *Proc. Zool. Soc. London*, Ser. A, 109, 221, 1939.
10. **Wright, R. H.**, The olfactory guidance of flying insects, *Can. Entomol.*, 90, 81, 1958.
11. **Marsh, D., Kennedy, J. S., and Ludlow, A. R.**, An analysis of anemotactic zigzagging flight in male moths stimulated by pheromone, *Physiol. Entomol.*, 3, 221, 1978.
12. **Cardé, R. T. and Hagaman, T. E.**, Behavioral responses of the gypsy moth in a wind tunnel to air-borne enantiomers of disparlure, *Environ. Entomol.*, 8, 475, 1979.
13. **Kennedy, J. S. and Marsh, D.**, Pheromone-regulated anemotaxis in flying moths, *Science*, 184, 999, 1974.
14. **Bossert, W. H. and Wilson, E. O.**, The analysis of olfactory communication among animals, *J. Theoret. Biol.*, 5, 443, 1963.
15. **Farkas, S. R. and Shorey, H. H.**, Chemical trail-following by flying insects: a mechanism for orientation to a distant odor source, *Science*, 178, 67, 1972.
16. **Sower, L. L., Kaae, R. S., and Shorey, H. H.**, Sex pheromones of lepidoptera. XLI. Factors limiting potential distance of sex pheromone communication in *Trichoplusia ni*, *Ann. Entomol. Soc. Am.*, 66, 1121, 1973.
17. **Lewis, T. and Macaulay, E. D. M.**, Design and elevation of sex-attractant traps for pea moth, *Cydia nigricana* (Steph.) and the effect of plume shape on catches, *Ecol. Entomol.*, 1, 175, 1976.
18. **Marsh, D.**, Behaviour of flying male moths approaching a source of the female sex pheromone, *Pestic. Sci.*, 7, 633, 1976.
19. **Kennedy, J. S.**, Olfactory responses to distant plants and other odor sources, in *Chemical Control of Insect Behavior*, Shorey, H. H. and McKelvey, J. J., Eds., Academic Press, New York, 1977, 67.
20. **Smith, R. L., Flint, H. M., and Forey, D. E.**, Air permeation with gossyplure: feasibility studies on chemical confusion for control of the pink bollworm, *J. Econ. Entomol.*, 71, 257, 1978.
21. **Cardé, R. T.**, Utilization of pheromones in the population management of moth pests, *Environ. Health Perspect.*, 14, 133, 1976.
22. **Cardé, R. T., Comeau, A., Baker, T. C., and Roelofs, W. L.**, Moth mating periodicity: temperature regulates the circadian gate, *Experientia*, 31, 46, 1975.
23. **Webb, J. W. and Berisford, C. W.**, Temperature modification of flight and response to pheromones in *Rhyacionia frustrana*, *Environ. Entomol.*, 7, 278, 1978.
24. **Batiste, W. C., Olson, W. H., and Berlowitz, A.**, Codling moth: diel periodicity of catch in synthetic sex attractant vs. female-baited traps, *Environ. Entomol.*, 2, 673, 1973.
25. **Comeau, A., Cardé, R. T., and Roelofs, W. L.**, Relationship of ambient temperatures to diel periodicities of sex attraction in six species of Lepidoptera, *Can. Entomol.*, 108, 415, 1976.
26. **Bartelt, R. J. and Chiang, H. C.**, Field studies involving the sex-attractant pheromones of the western and northern corn rootworm beetles, *Environ. Entomol.*, 6, 853, 1977.

27. **Marks, R. J.**, The influence of climatic factors on catches of the red bollworm, *Diparopsis castanea* (Lepidoptera: Noctuidae) in sex pheromone traps, *Bull. Entomol. Res.*, 67, 243, 1977.
28. **Turner, W. K., Hamilton, E. W., and Lee, F. L.**, Effect of wind speed and direction on the approach of soybean loopers to a pheromone source in the field, *Fla. Entomol.*, 61, 19, 1978.
29. **Shorey, H. H. and Gaston, L. K.**, Sex pheromones of noctuid moths. III. Inhibition of male responses to the sex pheromone in *Trichoplusia ni* (Lepidoptera: Noctuidae), *Ann. Entomol Soc. Am.*, 57, 775, 1964.
30. **Doane, C. C.**, Flight and mating behavior of the gypsy moth, in *Perspectives in Forest Entomology,* Anderson, J. F. and Kaya, H. K., Eds., Academic Press, New York, 1976, chap. 10.
31. **Lewis, T., Wall, C., Macaulay, E. D. M., and Greenway, A. R.**, The behavioral basis of a pheromone monitoring system for pea moth, *Cydia nigricana, Ann. Appl. Biol.*, 80, 257, 1975.
32. **Aylor, D. E.**, Estimating peak concentrations of pheromones in the forest, in *Perspectives in Forest Entomology,* Anderson, J. F. and Kaya, H. K., Eds., Academic Press, New York, 1976, chap. 13.
33. **Mayer, M. S.**, Attraction studies of male *Trichoplusia ni* (Lepidoptera: Noctuidae) with new combination of olfactometer and pheromone dispenser, *Ann. Entomol. Soc. Am.*, 66, 1191, 1973.
34. **Aylor, D. E., Parlange, J.- Y., and Granett, J.**, Turbulent dispersion of disparlure in the forest and male gypsy moth response, *Environ. Entomol.*, 5, 1026, 1976.
35. **Wright, R. H.**, Metarchons: insect control through recognition signals, *Bull. Atomic Sci.*, 21, 1, 1965.
36. **Marks, R. J.**, Laboratory evaluation of the sex pheromone and mating inhibitor of the red bollworm *Diparopsis castanea* Hampson (Lepidoptera, Noctuidae), *Bull. Entomol. Res.*, 66, 427, 1976.
37. **Shorey, H. H., Gaston, L. K., and Saario, C. A.**, Sex pheromones of noctuid moths. XIV. Feasibility of behavioral control by disrupting pheromone communication in cabbage loopers, *J. Econ. Entomol.*, 60, 1541, 1967.
38. **Roelofs, W. L. and Cardé, R. T.**, Responses of lepidoptera to synthetic sex pheromone chemicals and their analogues, *Ann. Rev. Entomol.*, 22, 377, 1977.
39. **Miller, J. R. and Roelofs, W. L.**, Gypsy moth responses to pheromone enantiomers as evaluated in a sustained flight tunnel, *Environ. Entomol.*, 7, 42, 1978.
40. **Sower, L. L., Gaston, L. K., and Shorey, H. H.**, Sex pheromones of noctuid moths. XXVI. Female release rate, male response threshold, and communication distance for *Trichoplusia ni, Ann. Entomol. Soc. Am.*, 64, 1448, 1971.
41. **Campion, D. G.**, Sex pheromones for the control of lepidopterous pests using microencapsulation and dispenser techniques, *Pestic. Sci.*, 7, 636, 1976.
42. **Cook, B. J., Shelton, W. D., and Staton, R. T.**, Antennal responses of the pink bollworm to gossyplure, *Southwest. Entomol.*, 3, 141, 1978.
43. **Shapas, T. J. and Burkholder, W. E.**, Patterns of sex pheromone release from adult females, and effects of air velocity and pheromone release rates on theoretical communication distances in *Trogoderma glabrum, J. Chem. Ecol.*, 4, 395, 1978.
44. **Hirooka, Y. and Suwanai, M.**, Role of insect sex pheromone in mating behavior. I. Theoretical consideration on release and diffusion of sex pheromone in the air, *Appl. Entomol. Zool.*, 11, 126, 1976.
45. **Schwink, I.**, A study of olfactory stimuli in the orientation of moths, in *Proc. 10th Int. Congr. Entomol., Montreal,* Vol. 2, International Congress of Entomology, College Park, Md., 1958, 577.
46. **Nakamura, K.**, The effect of wind velocity on the diffusion of Spodoptera litura (F.) sex pheromone, *Appl. Entomol. Zool.*, 11, 312, 1976.
47. **Farkas, S. R., Shorey, H. H., and Gaston, L. K.**, Sex pheromone of lepidoptera. The use of widely separated evaporators of looplure for the disruption of pheromone communication in *Trichoplusia ni, Environ. Entomol.*, 3, 876, 1974.
48. **Jones, C. D.**, Ion concentration variations at short distances downwind of continuous and quasicontinuous point sources, *Pestic. Sci.*, 8, 84, 1977.
49. **Lewis, T. and Macaulay, E. D. M.**, Towards rational design and elevation of pheromone traps, *Pestic. Sci.*, 7, 634, 1976.
50. **Inscoe, M. N. and Beroza, M.**, Analysis of pheromones and other compounds controlling insect behavior, in *Analytical Methods for Pesticides and Plant Growth Regulators,* Vol. 8, Zweig, G. and Sherma, J., Eds., Academic Press, New York, 1976, chap. 2.
51. **Gifford, F. A., Jr.**, An outline of theories of diffusion in the lower layers of the atmosphere, in *Meteorology and Atomic Energy,* Slade, D. H., Ed., U. S. Atomic Energy Commission, Oak Ridge, Tenn., 1968, 65.
52. **Plimmer, J. R., Caro, J. H., and Freeman, H. P.**, Distribution and dissipation of aerially-applied disparlure under a woodland canopy, *J. Econ. Entomol.*, 71, 155, 1978.

53. **Bierl, B. A. and DeVilbiss, D.,** Insect sex attractants in controlled release formulations: measurements and applications, in *Proc. Int. Controlled Release Pestic. Symp.,* Harris, F. W., Ed., Wright State University, Dayton, Ohio, 1975, 230.
54. **Houx, N. W. H. and Voerman, S.,** High-performance liquid chromatography of potential insect sex attractants and other geometrical isomers on a silver-loaded ion exchanger, *J. Chromatogr.,* 129, 456, 1976.
55. **Lam, S. and Grushka, E.,** Silver loaded aluminosilicate as a stationary phase for the liquid chromatographic separation of unsaturated compounds, *J. Chromatogr. Sci.,* 15, 234, 1977.
56. **Look, M.,** Determining release rates of 3-methyl-2-cyclohexen-1-one antiaggregation pheromone of *Dendructonus pseudotsugae* (Coleoptera: Scolytidae), *J. Chem. Ecol.,* 2, 481, 1976.
57. **Schneider, D., Kasang, G., and Kaissling, K. E.,** Determination of the olfactory threshold of *Bombyx mori* with tritium-labeled bombykol (Ger.), *Naturwissenschaften,* 55, 395, 1968.
58. **Meighen, E. A., Slessor, K. N., and Grant, G. G.,** Development of a Bioluminescent Assay for Low Levels of Spruce Budworm Pheromone, Canada/U.S. Spruce Budworm Program, unpublished manuscript, 1980.
59. **Kuwahara, Y. and Casida, J. E.,** Quantitative analysis of the sex pheromone of several phycitid moths by electron-capture gas chromatography, *Agric. Biol. Chem.,* 37, 681, 1973.
60. **Caro, J. H., Bierl, B. A., Freeman, H. P., and Sonnet, P. E.,** A method for trapping disparlure from air and its determination by electron-capture gas chromatography, *J. Agric. Food Chem.,* 26, 461, 1978.
61. **Caro, J. H., Freeman, H. P., and Bierl-Leonhardt, B. A.,** Determination of (*Z*)-9-tetradecen-1-ol formate, a *Heliothis* spp. mating disruptant, in air by electron-capture gas chromatography following photolytic cleanup, *J. Agric. Food Chem.,* 27, 1211, 1979.
62. **Byrne, K. J., Gore, W. E., Pearce, G. T., and Silverstein, R. M.,** Porapak-Q collection of airborne organic compounds serving as models for insect pheromones, *J. Chem. Ecol.,* 1, 1, 1975.
63. **Silverstein, R. M. and Rodin, J. O.,** Insect pheromone collection with absorption columns. I. Studies on model organic compounds, *J. Econ. Entomol.,* 59, 1152, 1966.
64. **Baker, J. L., Hill, A. S., and Roelofs, W. L.,** Seasonal variations in pheromone trap catches of male omniverous leafroller moths, *Platynota stultana, Environ. Entomol.,* 7, 399, 1978.
65. **Peacock, J. W., Cuthbert, R. A., Gore, W. E., Lanier, G. N., Pearce, G. T., and Silverstein, R. M.,** Collection on Porapak-Q of the aggregation pheromone of *Scolytus multistriatus* (Coleoptera: Scolytidae), *J. Chem. Ecol.,* 1, 149, 1975.
66. **Beroza, M.,** Microanalytical methodology relating to the identification of insect sex pheromones and related behavior-control chemicals, *J. Chromatogr. Sci.,* 13, 314, 1975.
67. **Buser, H.-R. and Arn, H.,** Analysis of insect pheromones by quadrupole mass fragmentography and high-resolution gas chromatography, *J. Chromatogr.,* 106, 83, 1975.

Chapter 5

EPA REGISTRATION REQUIREMENTS FOR BIOCHEMICAL PESTICIDES, WITH SPECIAL EMPHASIS ON PHEROMONES

Gunter Zweig, Stuart Z. Cohen, and Frederick S. Betz*

TABLE OF CONTENTS

* Although the authors have endeavored to be as accurate as possible in the description of protocols and policies, the opinions expressed by them are, nevertheless, their own and do not necessarily represent the official policy of the U.S. Environmental Protection Agency.

I. INTRODUCTION

Pheromones used in agriculture to control insect pests are registered by the U.S. Environmental Protection Agency (EPA) under the provisions of the FIFRA (Federal Insecticide Fungicide and Rodenticide Act) and its several amendments. When the latest major pesticide legislation was enacted in 1972 (and amended in 1975 and 1978), the authors of the law undoubtedly intended to regulate the use of pesticides such as the synthetic, manmade chemicals and some naturally occurring chemicals, like pyrethrins and rotenone. No specific provisions or exemptions from the normal regulatory process were provided for biological pesticides such as insect and plant growth regulating hormones, and pheromones.

Although a number of microbial pesticides, e.g., *Bacillus thuringiensis* and pheromones like gossyplure, disparlure, and grandlure, have received registration or experimental use permits, no definitive guidelines governing the registration requirements for these "atypical" pesticides have been written until now.

The EPA has recognized that microbial pesticides, insect and plant hormones, and pheromones may be more desirable environmentally than synthetic pesticide chemicals, and, therefore, should receive encouragement for development by lessening stringent registration requirements deemed necessary for the conventional pesticides. For example, Steven D. Jellinek, former head of the Office of Pesticides and Toxic Substances, recently stated that " . . . biorational agents can augment — in some cases supplant — the use of toxic chemicals in controlling target pests . . . without affecting beneficial insects, wildlife, and humans. This makes them [biorational agents] extremely attractive from an environmental as well as crop protection point of view."[1]

Thus, the EPA has recently undertaken to draft a set of guidelines which describe the data requirements for registering "biorational pesticides," a term which encompasses microbials (viruses, bacteria, protozoa and fungi) and biochemicals (insect and plant growth regulators and semiochemicals, especially pheromones).

A draft of these guidelines,[4] Subpart M, Data Requirements for Biorational Pesticides, has been presented recently to the EPA FIFRA Scientific Advisory Panel for comments. Although it is not expected that drastic changes will be made in the draft guidelines, this chapter is based on this draft (September 29, 1980) and must, therefore, be considered a preliminary assessment.

In this chapter we shall confine our discussion to the sections of the guidelines covering "biochemicals" and more specifically, pheromones. The topics that will be covered are the following: product analysis, toxicology, residue analysis, nontarget organism hazard (including terrestrial animals, aquatic animals, plants, and nontarget insects), and environmental fate. The reader is advised to obtain a copy of the Guidelines, Subpart M, once they appear in the Federal Register, so that he may refer to specific details of particular interest to him.

II. BACKGROUND AND PHILOSOPHY

In 1978, the EPA published revised, proposed guidelines for the registration of pesticides[2] which stated that "the data requirements for those living organisms or viruses which are pesticides will be determined on a case-by-case basis . . . " Since data requirements were not defined in this document, EPA issued a policy statement on May 14, 1979, on the regulation of biorational pesticides.[3] Biorational pesticides were defined to include biological pest control agents and certain naturally occurring biochemicals that exhibit an inherently different mode of action than that of most synthetic organic and inorganic pesticides currently registered by the EPA. The policy statement further noted that EPA intended to develop specific guidelines for the registration of

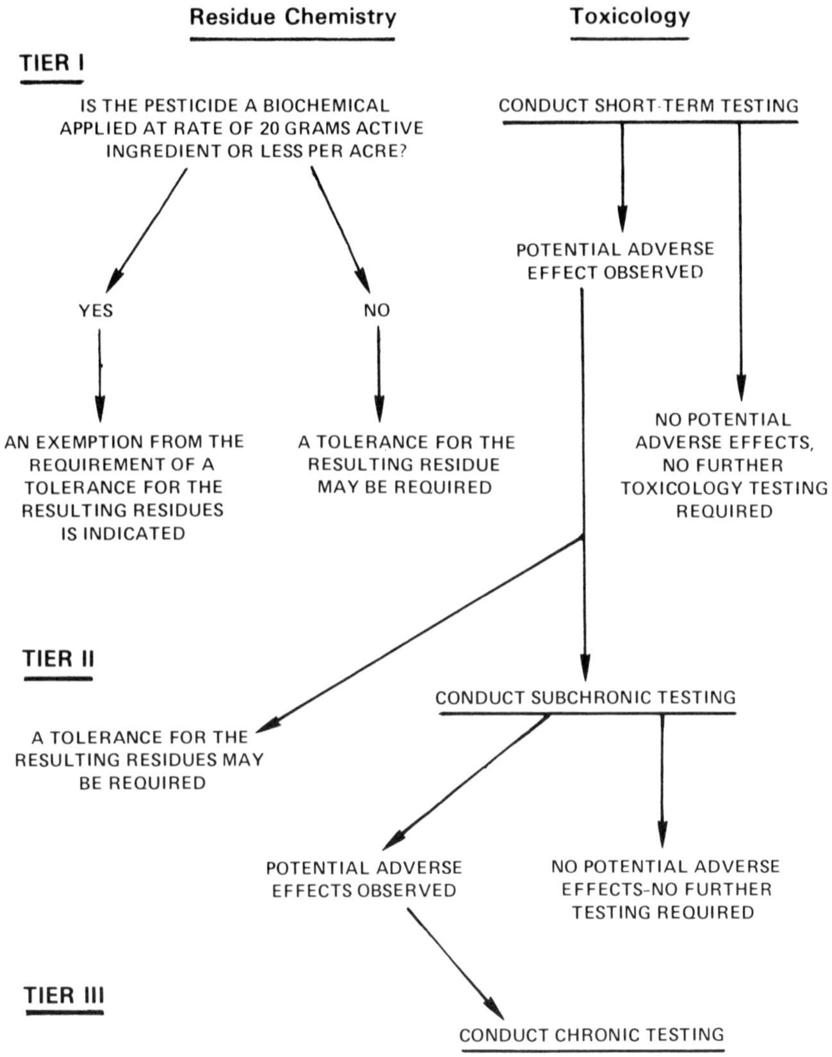

FIGURE 1. Tier testing scheme for residue chemistry/toxicology.

biorational pesticides consisting of test protocols for human and environmental safety which would be unique for biorationals and could simplify the registration process for these products. Although the registrant will not be relieved of the burden to prove safety, the EPA, nevertheless, recognizes the fundamentally different mode of action of biorationals and also the inherently lower risk posed by them.

Pheromones act by altering behavior of specific target insects rather than by killing target insects and, sometimes, nontarget insects. It is the personal opinion of the authors that pheromones may not necessarily be considered "pesticides" in accord with Congressional intent although they do meet the strict definition of the law.

III. APPROACHES TO TESTING REQUIREMENTS FOR PHEROMONES

The Agency proposes a tier testing scheme for biochemicals, including pheromones with tests of progressive complexity and expense. This is a departure from the approach taken for the registration requirements for conventional pesticides. One notable excep-

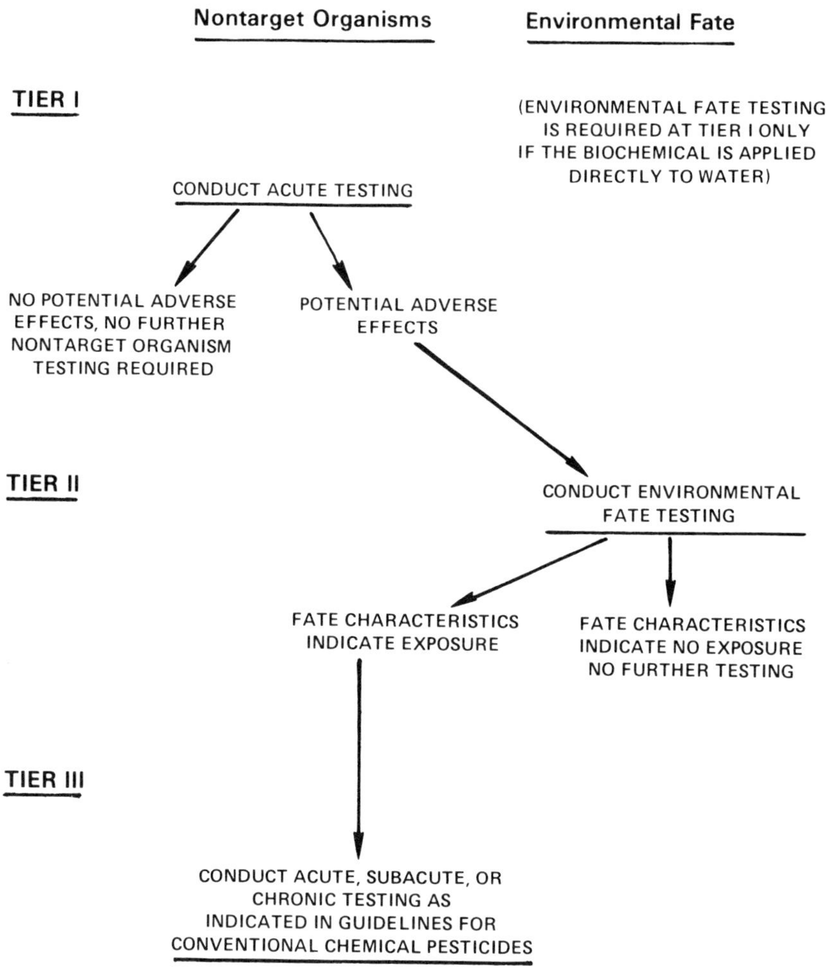

FIGURE 2. Tier testing scheme for nontarget organisms/environmental fate.

tion is the ecological effects testing for the conventional pesticides scheme, of tests, progressing from simple, inexpensive, to complex, more costly tests. Figures 1 and 2 depict the proposed tier system. As may be seen in these figures, the decision to enter a higher stage or tier of testing is determined by comparing results of lower tier testing with predetermined triggers or decision criteria. The purpose of such schemes is to require the most extensive testing of only those compounds that are likely to be most hazardous and to ensure that only limited testing be performed on less hazardous compounds. Thus the economic burden on registrants of less hazardous compounds would be minimized, a situation especially appropriate to pheromones. With the exception of ecological effects testing, current registration testing guidelines for conventional pesticides do not allow this approach as noted above. They are based on the presumption of high risk and require extensive testing for conventional pesticides. The selection of the appropriate test battery in the registration guidelines for conventional pesticides is often determined by exposure patterns (e.g., greenhouse use, repeated dermal contact), rather than by results of Tier I tests.

Two major issues are involved in designing tier testing schemes: the design of the decision criteria or triggers for progression within and between tiers, and the availability of valid, short term Tier I tests that can serve as the initial screen for risk evaluation.

Decision criteria are judgmental standards used to decide whether to proceed from lower tiers to higher tiers of testing. An example of such a decision criterion might be the following: if a toxicologically significant amount of chemical is likely to leach to the aquatic environment, then further testing would be required.

Triggers are specific and predetermined semiquantitative cutoff points. An example of a trigger might be the following: If it is likely that, as a result of leaching, the estimated environmental concentration of the biochemical pesticide is greater than one tenth the LC_{50} for the most sensitive aquatic species tested, then further testing would be required.

Volatility from water is another trigger used in the proposed guidelines. Henry's Law Constant is an estimate of the potential for hydrophobic substances to escape from water, i.e., volatility from water. If Henry's Law Constant (the ratio of the vapor pressure of a compound in atmospheres (atm) and water solubility expressed as mole compound/m^3 water) of a pheromone exceeds or equals 10^{-4} atm-m^3/mole, then reduced ecological effects testing is warranted. The rationale for this approach is that volatile pheromones will dissipate readily, resulting in low exposure to aquatic and terrestrial organisms. Most pheromones would probably meet this criterion. A more detailed technical discussion of this criterion appears in the preamble to the environmental fate testing guidelines for biochemicals.[4] Also a good discussion of this process was written by Mackay and Yuen.[5]

The advantage of triggers and decision criteria in testing schemes is that specific guidance is given on the extent of testing. Triggers and criteria provide the basis for the risk analysis process used to decide whether to register a pesticide. The disadvantage of decision criteria is that they require subjective decisions and might, therefore, become controversial. It is expected that the trigger/criteria issue will elicit much public comment.

If a tier scheme is to perform properly, the lower tier tests must be reliable indicators for further testing. For example, a lack of absorption in the solar spectrum might or might not be an indicator for further photolysis testing. In the affirmative, this lower tier, a simple absorption spectrum, is used as an indicator for direct photolysis. If however, this lower tier is used as an indicator for indirect or sensitized photolysis, the absorption spectra may not be a valid screening measurement for direct photolysis. Thus, the tier testing scheme for pheromones may be limited in part by the availability of unequivocal lower tier tests.

A relevant question is whether it is necessary to enter the testing tiers, i.e., should any testing be required at all? For example, environmental chemistry tests are used to generate exposure data; but if toxicological concerns are not raised, should these tests be required? It was decided that in order to delete the environmental chemistry tests, a base tier of toxicological and environmental effects tests must be followed. If these results should show no cause for concern, further tests are not needed.

A new concept in the proposed guidelines for biorational pesticides is the term "maximum hazard testing" which involves the design of Tier I tests for toxicology and hazard to nontarget organisms that impose conditions of maximum challenge (in terms of pesticide dose, route of exposure, and age or life stage of the test organism). Negative findings (i.e., no adverse effect) in these tests would eliminate further testing.

Another innovative concept concerns the degree of environmental exposure resulting from pesticide use. For example, if exposure is limited by the nature of the formulation, rate of application, use site or volatility, this would be the basis for reduced data requirements. Most pheromones developed thus far produce limited exposure to humans and nontarget organisms and are, therefore, of limited hazard potential.

It is recognized that routine procedures may not be suitable for the evaluation of all pesticides so that additional tests are being developed and existing methods and protocols modified. Testing requirements should be flexible enough to take into account slight differences in analytical instrumentation and biological materials among laboratories. In spite of these differences, test protocols must be standardized to allow interlaboratory comparisons.

Due to the unique nature of biorational pesticides, EPA may issue waivers of tests deemed inappropriate for a particular substance or permit a modification of testing procedures to accommodate the particular properties of an unusual product. The decision to accept an altered or modified test protocol will be made by EPA on a case-by-case basis. Specific provisions for data requirements for the registration of biochemical pesticides, including pheromones, are described in the following sections.

IV. PRODUCT ANALYSIS

Descriptive data and analytical tests for naturally occurring or synthetic biochemicals parallel those required for conventional pesticides. This information includes product identity (product name, trade name, etc.); confidential statement of formula and chemical data (e.g., structural formula, Chemical Abstracts name, molecular weight). Other descriptive material includes the percentage of active ingredient, expressed in suitable units and identification of organism(s) from which the biochemical substance was isolated or associated. Physical-chemical properties of the active ingredient and product must be stated. Water solubility, vapor pressure, and octanol/water partition coefficients are critical properties of pheromones and will therefore be required. Other ingredients, so-called inerts, in the product must be listed quantitatively and chemically identified.

Another requirement for product chemistry is the description of the manufacturing process, including possible formation of byproducts, which themselves may pose hazard to man and the environment.

Even if the registrant were not required to obtain these data, he undoubtedly has this information identifying his product, in addition to some of the basic physical-chemical property data, in order to meet his own quality standards. Some difficulty may be encountered in quantifying percentages of all possible isomers, including optical isomers, but quality control of production requires that a uniform product enter the market in order to achieve standardized response and efficacy. Therefore it is not anticipated that product analysis requirements will pose a hardship on the registrant, but more experience with the registration process may result in a modification of present requirements.

V. TOXICOLOGY

When proposing toxicological testing requirements for pheromones, the Agency recognizes that these chemicals are efficacious in very small amounts when used in the field. They are active on target insects by altering their mating behavior, and yet it must be ascertained whether these chemicals will pose hazards to man. The potential hazards of other biological agents, such as viruses and bacteria, are much greater than those for the pheromones. The problem to be resolved is how to design minimum toxicological testing for pheromones to enable EPA scientists to make scientific judgments on the relative safety of these compounds.

Therefore, the minimum tests that should be performed include acute laboratory animal toxicity determinations by the dermal, oral, and inhalation routes. These acute toxicity tests, together with tests on irritancy, hypersensitivity, cellular immune responses, and a battery of tests to detect potential for DNA, chromosomal, and genetic damage, provide the initial basis to assess the safety of the material. These tests constitute Tier I testing, and if all tests indicate no potential for significant adverse effects, no further testing is required.

With low rates of exposure and absence of human toxicity, negative results from the short term tests for carcinogenicity and mutagenicity are sufficient to make a judgment on low expectation of risk to humans from cancer, and no further tests in this area of concern will be required when no potential for significant adverse effects is indicated.

The toxicology scheme provides for further testing if any of the Tier I tests, including in vitro mutagenicity have been found to be positive. These additional tests, Tiers II and III, consist of subchronic and/or chronic studies of general toxicity, teratogenicity, cellular immune response, and oncogenicity. However, there is little expectation that such further testing will be necessary for the pheromones.

VI. RESIDUE ANALYSIS

Biochemical pesticides applied at rates greater than 20 g active ingredient per acre per application, whose use may result in residues on food or feed, may not be used on food or feed unless a tolerance or an exemption from the requirement for a tolerance has been established by the Agency. Residue studies involve time-course dissipation analyses, metabolism studies, and development of specific methods of residue analysis. However, it is anticipated that most pheromones will be exempted from the establishment of food or feed tolerance, based on the belief that the low exposure aspects of the delivery system, the exceedingly small application rate, and the high volatility of the compounds make it implausible that any measurable residues on food or feed will result. Biochemicals that require testing in toxicology Tier II or beyond will be subject to the full range of residue chemistry testing requirements for the establishment of a tolerance.

VII. NONTARGET ORGANISM HAZARD

The draft of proposed requirements for ecological effects testing consists of tier testing schemes for terrestrial animals, aquatic animals, plants, and nontarget insects. The ecological effects testing schemes have been integrated with the environmental fate testing scheme (discussed later in this chapter) in the following manner: generally, Tier I short term tests for ecological effects would be required for most biochemicals. If the test results indicate the potential for significant adverse effects, the biochemical would be tested in the environmental fate tier. If significant exposure, e.g., a concentration in the environment exceeding one tenth of the LC_{50} or EC_{50} in the most sensitive species tested, is indicated by the environmental fate tests, further ecological effects tests would be required.

In various parts of the draft proposed guidelines, reductions in data requirements will be allowed based on high volatility method of dispensing, low exposure, etc.

The individual testing schemes for nontarget organism hazard assessment are discussed below.

Terrestrial Animals — If the Henry's Law constant for a pheromone exceeds 10^{-4} atm-m^3/mole, Tier I of the testing scheme would require that one avian single-dose oral study be performed.* If this criterion is not met, then one additional avian test (avian dietary toxicity) would be required. This testing is less than required by the registration guidelines for conventional pesticides. Thus, Tier I tests may trigger environmental fate tests, which in turn may trigger additional terrestrial organisms tests.

* See Section III above for a brief discussion of the Henry's Law constant.

The additional (Tier III) tests would be as specified in Subpart E[6] of the registration guidelines for conventional pesticides ("Hazard Evaluation: Wildlife and Aquatic Organisms").

Aquatic Animals — Pheromones would qualify for reduced testing in most cases. However, if the biochemical is applied directly to water, testing as specified in Subpart E is required and Tier II fate testing is automatically required. Tier I would consist of one freshwater fish acute bioassay and one aquatic invertebrate acute bioassay. Tier progression follows the pattern described above, although additional tests specified in the environmental fate section of the biorational pesticide guidelines may be required. Those test results could then trigger additional testing as described in Subpart E.[6]

Plants — The plant testing scheme in the draft biorational pesticide guidelines will be nearly identical to the tier testing required for conventional pesticides, Subpart J[6] ("Hazard Evaluation: Nontarget Plants and Microorganisms"). Tier I of the biorational pesticide guidelines for plants would measure seed germination and vegetative vigor for terrestrial plants and general growth for aquatic plants. After Tier I, tier progression would be the same as stated in Subpart J[6] for Tiers II, III, and IV. Exemptions from this scheme would be granted based on the natural occurrence pattern of the biochemical, the ambient concentrations, and use patterns.

Beneficial Insects — Beneficial insect testing requirements would be divided into requirements for aquatic insects and terrestrial insects. With regard to aquatic insect testing, biochemicals applied directly to water would be tested as described in Subpart L[6] of the conventional pesticide guidelines (Hazard Evaluation: Nontarget Insects) while all others would be tested as described previously under aquatic animals. For terrestrial insects, if adverse effects are noted during efficacy testing and/or other data indicate potential for adverse effects on nontarget insects, then environmental fate testing would be required.

IX. ENVIRONMENTAL FATE

Environmental fate tests would usually only be triggered if the results of Tier I ecological effects tests indicate the potential for significant adverse effects. Further, persistence studies would be triggered when the results of the transport studies (e.g., adsorption, volatility) indicate the potential for significant transport of pheromone active ingredients to the nontarget population of concern (e.g., terrestrial animals). Even if persistence studies are required by the results of previous testing, the requirements are less extensive than those for conventional pesticides. For example, rotational crop studies are not required for pheromones, primarily due to analytical limitations. Because of the presumption of low risk of pheromones, it is proposed that some estimation procedures would be allowed. This would introduce a slightly greater margin of error in the environmental exposure assessment at a great cost saving to the registrant. One such procedure might involve the measurement of ultraviolet-visible absorption spectrum to estimate rates of photolysis. This estimation procedure is based on absorption cross sections (calculated directly from the spectra), solar or actinic flux (available in tables), and assuming quantum yields (usually assumed to be 1.0 for screening purposes). Other structure-activity relationships can be used to predict vapor phase reactivity with hydroxyl radicals, ozone, and an approximation of the resulting transformation products.

Typically, the initial level of environmental fate testing would include measurement of volatility and leaching from pheromone dispensers, sorption to soil, vapor pressure, water solubility, and the octanol-water partition coefficient. The latter three measurements are already required in the product analysis guidelines (see Section IV). If the potential for significant exposure to sensitive nontarget species were indicated (see above), persistence studies to be required would be hydrolysis, soil metabolism, soil photolysis, and possibly aquatic metabolism and photolysis. (Results similar to those described above [a certain fraction of the LC_{50} or EC_{50} of a sensitive species] would indicate the need for further ecological effects testing.)

An example of how this testing process works follows. If there is concern for safety of aquatic organisms (indicated by Tier I ecological effects tests), but the use site is terrestrial, the registrant would first be required to assess the pheromone transport potential with measurements of volatility, water solubility, etc. With these data, one could estimate the potential for chemicals to partition between media. If a simple mass balance calculation indicates that significant quantities of pheromone could be transported to the aquatic environment, persistence studies such as hydrolysis would be required. If, in addition, the compound absorbs light in the 295 to 700 nm range, aquatic and soil photolysis studies would also be required.

IX. SUMMARY

Tests to meet the EPA requirements for the registration of pheromones have been reduced and/or simplified by a proposed tier testing scheme. Negative results in simple first tier tests to assess possible hazard to man (toxicology tests) and nontarget organisms eliminate further environmental fate studies. Conversely, positive results from first-order tier tests will trigger more detailed studies which are very similar to those required for conventional pesticides.

ACKNOWLEDGMENTS

Development of these guidelines has been a multidisciplinary task. Thus efforts of several people should be acknowledged — W. Preston, D. Urban, R. Holst, A. Vaughn, I. Mauer, W. Woodrow, B. Roessler, A. Keller, G. Beusch, E. Britton, and R. Carsel. In addition, an expert panel of the American Institute for Biological Sciences has provided much of the substantive material for the toxicology section of these guidelines. Further, Messrs. Urban, Holst, and Vaughn provided several helpful comments on this chapter.

REFERENCES

1. **Jellinek, S. D.,** Building on Clues from Nature, in a speech at Sandoz Corporation at Wasco, Calif., June 15, 1979.
2. Environmental Protection Agency; Registration of pesticides in the United States proposed guidelines, *Fed. Regist.*, 43 (132), 29707, 1978.
3. Environmental Protection Agency; Regulation of "Biorational" Pesticides, *Fed. Regist.*, 44 (94), 28093, 1978.
4. Environmental Protection Agency; Subpart M, Data Requirements for Biorational Pesticides, September 29, 1980 draft.
5. **Mackay, D. and Yuen, T. K.,** Volatilization rates of organic contaminants from rivers, *Water Pollut. Res., Can.*, 15(2), 83, 1980.
6. Environmental Protection Agency; 1978. op cit.

Section II
Formulations, Methods of Application, and Insect Monitoring

Chapter 6

INSECT MONITORING TECHNOLOGY FOR EVALUATION OF SUPPRESSION VIA PHEROMONE SYSTEMS

P. D. Lingren, J. R. Raulston, A. N. Sparks, and W. W. Wolf*

TABLE OF CONTENTS

* This paper reports the results of research only. Mention of a pesticide in this paper does not constitute a recommendation for use by the USDA nor does it imply registration under FIFRA as amended. Also mention of a commercial product in this paper does not constitute an endorsement of the product by the USDA.

I. INTRODUCTION

Two decades ago Beroza[1] proposed an interesting new approach to insect control through confusion of males by sex attractants sprayed over large areas. A short time thereafter, Wright[2] proposed the "Metarchon" method of insect control whereby one would use nontoxic behavior-modifying stimuli including sex attractants (pheromones). Wright[3] also noted that the potential of sex attractants in practical pest control was very great but that rapid movement toward practical usage would require the highest type of creative thinking at all levels. He realized that lack of information concerning insect behavior was a major limiting factor in the development of such a procedure and stated, "what we need to know more about now are the external agencies that tell a particular species of insect to do a particular thing at a particular time. Then we can give it the signal at the wrong time or place, or repeat the signal so often and so strongly that it becomes too fatigued to respond anymore."

The implications of Wrights[3] statements concerning insect behavior truly challenged all concerned with pest control, especially scientists involved in investigations of insect behavior. This is especially true when one considers that many of our pests are nocturnal and that, in fact, little is known about their specific behaviors or the time and places that these specific behaviors occur, let alone the external factors to which the insect acts or reacts. There are many reasons for this lack of behavioral information, among them weak human night vision, lack of information about the way catches in pheromone traps relate to density or mating behavior, and the fact that pest insects can be controlled with chemical insecticides though the relief is only temporary. Among other considerations, the factors discussed above probably influenced top federal officials to recently categorize behavior of insects as being one of the primary and "toughest research problems not being adequately addressed" by scientists.[4]

Although insect behavior has, as a field, been seriously neglected, techniques have been developed over the past several years that allow us to study the behavior of nocturnal insects. As a result, a beginning has been made in gathering information on the behavior of certain serious lepidopteran pests. We report here on nocturnal lepidopteran pests that attack field crops and whose nocturnal behavior has been studied to some extent by one or more of the authors of this chapter. The insect pests include: *Heliothis* spp. [bollworm,[5-7] *H. zea* (Boddie), and tobacco budworm,[8-17] *H. virescens* (F.)]; pink bollworm,[18-29] *Pectinophora gossypiella* (Saunders); cabbage looper,[8,30-35] *Trichoplusia ni* (Hübner); soybean looper,[8,30] *Pseudoplusia includens* Walker; fall armyworm,[36] *Spodoptra frugiperda* (J. E. Smith); cotton leafperforator,[37] *Bucculatrix thurberiella* Busck; and tobacco hornworm,[8,38-39] *Manduca sexta* (L.). Pheromones that are at least as attractive as the virgin females have been identified and synthesized for all of these species except for the cotton leafperforator and the tobacco hornworm. Attention will be directed primarily toward the tobacco budworm and pink bollworm because our experience with these species, as it relates to behavior and pheromones has been more extensive.

A complete literature review of the subject will not be presented; instead, only selected references are cited. Our objectives are (1) to better acquaint the reader with some of the techniques available for studying nocturnal behavior; (2) show how the behavioral information can be used to more rapidly and accurately assess the impact of pheromones on mating in a population; and (3) encourage those involved in the development of pheromonal or other behavior-altering systems to utilize available technology to their best advantage. We recognize that efficacy in any successful control program is related directly to the reduction in infestations and damage. However, we suggest that at an early stage direct field observations of effects of pheromone on insect behavior that culminate in mating are essential to the rapid development of a sound insect control program. This is especially true when one must depend on indirect information such as captures in pheromone traps, and one does not know how these captures relate to mating or to population dynamics and densities.

II. TOOLS FOR OBSERVATION OF NOCTURNAL INSECT BEHAVIOR

Today, we have several tools that can be used to gather the behavioral information needed to assess the impact of a pheromone on a nocturnal insect population. Those that allow direct observations of behavior in the natural habitat include the use of visible light (lanterns, flashlights, or headlamps) or night-viewing devices such as night vision goggles. Other tools such as traps, mating tables, tethered females, radar, and genetic markers can be used to make indirect observations of behavior. Unfortunately information collected by indirect methods ultimately requires interpretation as it relates to the true behavior of the population under study. Therefore, the use of indirect methods such as pheromone traps or mating tables to ascertain the degree of mating disruption by pheromone treatments without establishing the relationship of these methods to population density and actual mating is in our opinion a risky business.

III. EVALUATION OF TOBACCO BUDWORM MATING DISRUPTION WITH PHEROMONES VIA DIRECT NOCTURNAL OBSERVATIONS

Sound evaluation of a proposed mating suppression system requires a thorough knowledge of the basic nocturnal activity patterns exhibited by the insect and the way these activities normally vary within a population. Diel behavioral patterns of feeding, oviposition, flight activity, pheromone secretion, and mating have been described for the tobacco budworm.[8,9,11] These studies coupled with mating studies[12-14] and pheromone trap response behavior studies[10,15-16,22] have led to the development of techniques and use of equipment that are broadly applicable to the nocturnal study of many Lepidoptera. They have also resulted in a relatively good understanding of the nocturnal activity of the tobacco budworm and so are important to the assessment of pheromones as mating disruptants for the species. Based on our studies, a practical method of assessing the effects of pheromones on the nocturnal behavior of the tobacco budworm follows.

A. Population Assessment

The performance of pheromones or pheromone analogues as mating disruptants should be assessed relative to the sexual status and size of the population at the time of investigation. It is therefore important to obtain qualitative and quantitative information on the adult population. This can be done by observing and collecting adults during their feeding, oviposition, and resting periods (sundown to midnight). With data collected in this manner, Lingren et al.[11] produced the Diel patterns shown in Figure 1.

Data of this type can be obtained by observers who use visible light but many nocturnal insects will respond to such light either negatively (escape response) or positively (attraction). As a result, most early workers used red filters to dampen the beam of white light because they felt it would interfere with the normal behavior of the species under observation.[5-6] This greatly restricted their field of view and resulted in few observations of field activity. However, we have since determined that most of these responses can be avoided by locating the insect with a bright beam of light and then covering the beam with a red filter or by averting it so the insect is located in the outer fringes of the beam. For example, feeding and ovipositing adult noctuids that we have observed orient upwind and move slowly from plant to plant in a hovering fashion

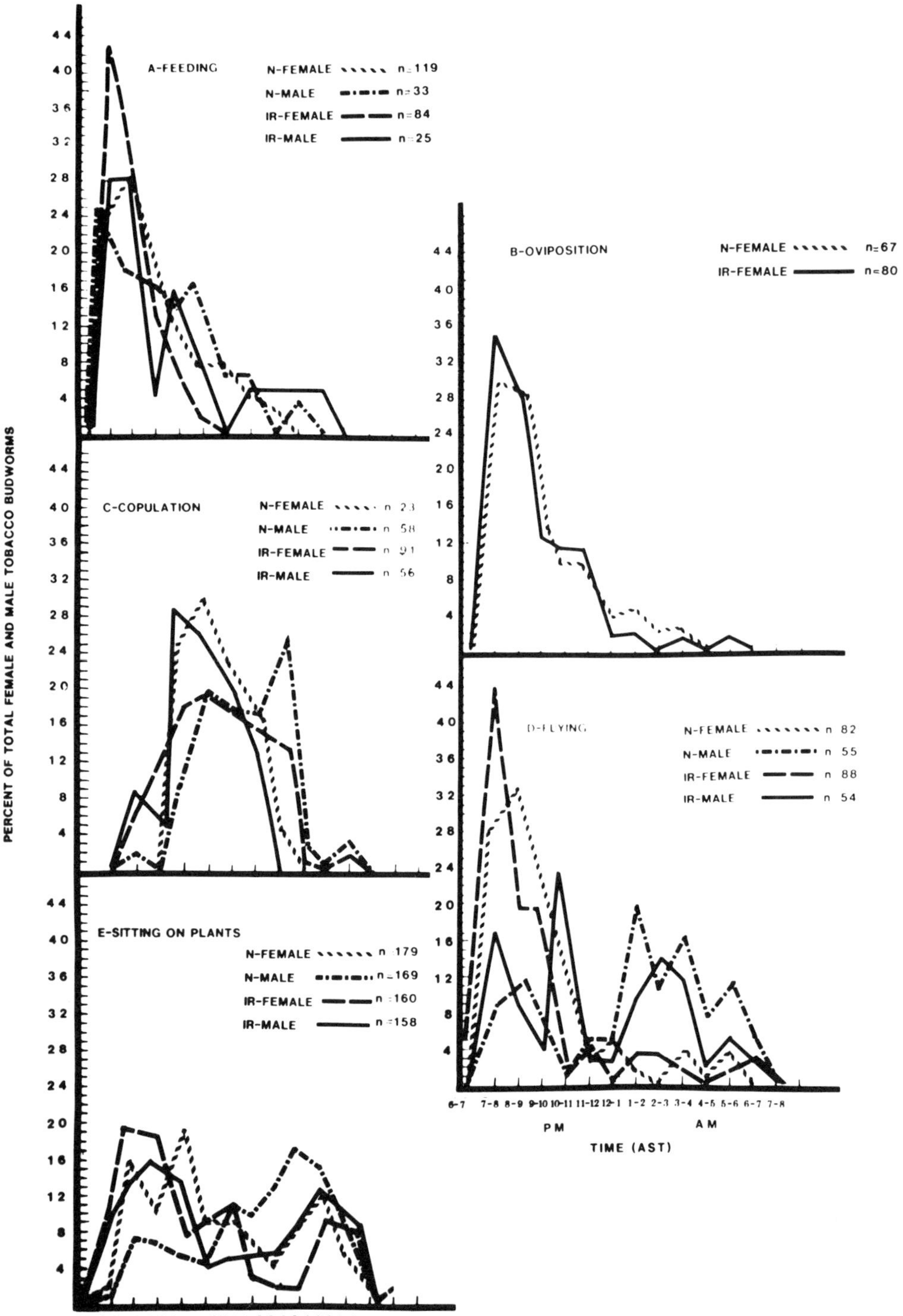

FIGURE 1. Diel patterns of nocturnal behavior of native (N), laboratory-reared, irradiated and released (IR) and light-adapted (L) tobacco budworm adults in pigeon pea and *Bastardia* sp. Data reflect one night of observation at Isabela, P.R., and 25 nights of observation on St. Croix, U.S.V.I. All observations were made between April 11 and May 28, 1974. Behavioral traits observed were as follows: A = feeding, B = oviposition, C = copulating, D = flying, and E = sitting on plants.[11]

inspecting the plants for feeding and oviposition sites.[10] Also, when these insects are in various other behavioral modes such as resting, calling, copulating, or recently emerged, they are very passive in terms of flight activity and respond to the bright beam minimally. Therefore, a great deal of information on the site and type of nocturnal activity can be obtained by using visible light. We found that a 6-V headlamp with a 6.35-cm polished reflector was the most satisfactory for locating the adults, either by general reflection from the body or by the brilliant green reflection of their dark-adapted eyes. Species such as the tobacco budworm,[11] pink bollworm,[10] and even the cotton leafperforator[37] are easy to locate in the field at distances from 3 to 8 m by using this technique. When located, the adults can normally be approached cautiously and captured in a plastic cup or with an insect net.

Adults collected by these means can be used to characterize the population in the study area. For example, the populations within pheromone-treated and untreated fields can be compared quantitatively either on a capture per unit time (such as capture per man hour) or unit area basis. The collections also allow one to establish the comparative sex ratios. However, in this case, the observer must search the plants for nonmoving adults since the activity patterns (flying and sitting) of males and females do not always coincide.[11] Also, mating pairs, calling female and newly emerged adults occupy similar niches of the plant during the mating activity period and except for the calling females they are primarily immobile in terms of flight activity. During this same period sexually active males exhibit an oblique-to-the-wind mate-searching fast-flight movement that makes their capture very difficult.[16] Therefore, collections of single adults during the mating activity period for the purpose of establishing population density and sex ratio will bias results. On the other hand, quantification of the male's movement during its sexually active period can give valuable insight concerning the way pheromone permeation of the air affects searching behavior. Such data can best be obtained by using night vision goggles and infrared light to observe movement per unit time in predetermined areas and comparing data from treated and untreated areas.[40]

Indeed, the sex ratio and mating status of females are very important for population assessment because they yield information on the age structure and relative sexual attractiveness of the populations. We now know that tobacco budworm adults emerge at night, usually between 10 p.m. and 2 a.m.[41] Eclosion of females is more extensive than eclosion of males early in a generation cycle, and generally, neither sex mates during the first night of emergence; therefore, the ratio of virgin female to male is high in field collections early in a generation cycle. Then, as emergence of a population continues, the sex ratio at the emergence site begins to reverse. Later, as the population ages and begins to disperse, sex ratios tend to equalize, and there is an increase in multiple matings and a decrease in female virginity.

Dissections of field-collected females too yield important information about the age structure and relative sexual attractiveness of the population. A preponderance of multiple-mated females indicates an old population with many sexually nonreceptive females[14] (tobacco budworm females normally do not attract males for a period of 1 to 3 days after a successful mating).[12] Thus, in an older population, the number of sexually active females is reduced relative to the number of active males that can mate every night. The result is an excess male potential and intensified competition between males for mates.[14] On the other hand, a preponderance of virgin females suggests a newly emerging population comprised primarily of sexually attractive females. In this situation competition between males for mates would be relatively less intense, and competition between females would be more intense. Also, competition between females and pheromone traps would be more intense and catches in traps should be reduced.

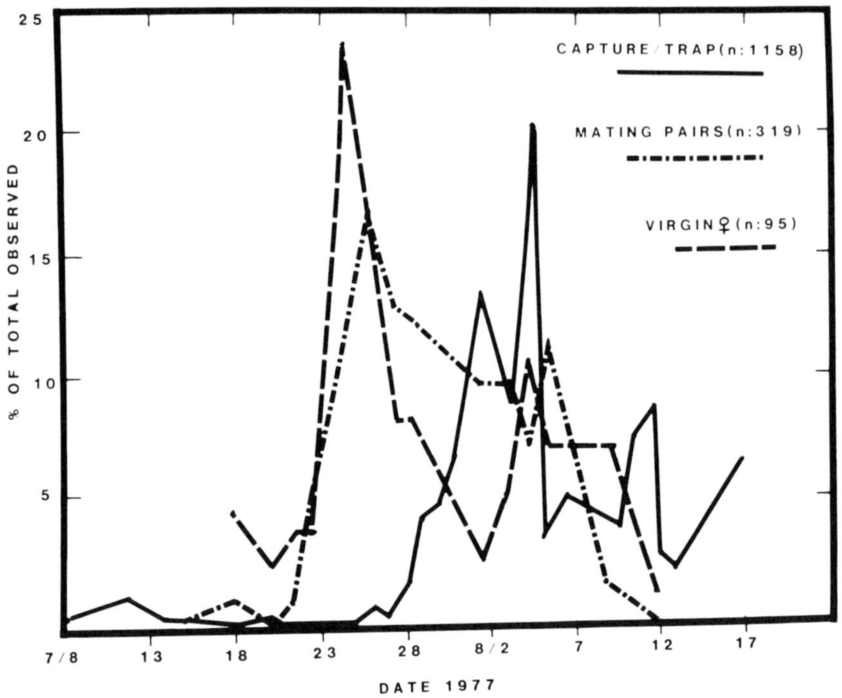

FIGURE 2. Comparison of emergence (no. of virgin females) and mating of tobacco budworms to capture of males in directional traps baited with five virgin females. (From Raulston, J. R., Lingren, P. D., Sparks, A. N., and Martin, D. F., *Environ. Entomol.*, 8, 349, 1979. With permission.)

Plainly then, the sexual status of a population will affect the performance of applied mating disruptants and such evaluation methods as male capture in pheromone traps. However, collections of adults and dissections of females over time makes it possible to map population cycles (Figure 2) and obtain the background data necessary for making comparisons between pheromone-treated and untreated areas. The data presented in Figure 2 show that peak tobacco budworm female virginity (used as an emergence indicator) preceded the mating peak by 1 day and preceded peak male capture in traps by 9 days. Thus the emergence of the population was rapid, the virgin population mated rapidly, and peak male capture in traps occurred well after peak mating. Obviously, information of this type is necessary to determine whether data collected from pheromone-treated and untreated areas represent the same type of population, that is, that both have an equal degree of susceptibility to mating disruptants.

B. Mating Comparisons

Virgin female tobacco budworms generally obtain the first mates during the night and peak mating of tobacco budworms normally occurs between 1 and 4 a.m.[8-9,11,12-14] Mating pairs are normally located in the upper ⅓ of the crop canopy, usually on leaves at the outer periphery of the plant where they are positioned in a characteristic V configuration. Pairs can be detected easily with a headlamp, and they are relatively immobile, which allows one to capture them with ease. Mating pairs are normally collected hourly from a specified area or for a specified time so direct comparisons between treated and untreated areas can be made. The females can also be dissected to determine their prior mating status. Such information coupled with data obtained from single adult collections (sundown to midnight) can be used to assess treatment

Table 1
COMPARISON OF DENSITY, SEX RATIOS, EMERGENCE, MATING, AND AGE STRUCTURE OF ADULT TOBACCO BUDWORM POPULATIONS IN ADJACENT COTTON FIELDS NEAR PHOENIX, AZ, 1979. FIELDS TREATED WITH A COMBINATION OF INSECTICIDES BY AIR AT CA. 2 A.M. ON NIGHT AFTER PRETREATMENT COUNTS WERE TAKEN

Field	Status of adults captured	Total no. captured	Ratio ♀ ♂	% Virgin ♀	Average no. spermatophore/♀	No. captured/ man-hour
			1 night before treatment (Sept. 5)			
1	Singles	30	1:1	13.3	1.8	5
	Mating pairs	3	—	0	2.3	—
2	Singles	41	1:1.9	92.9	0.1	8.2
	Mating pairs	5	—	20	1.8	—
			1 night after treatment (Sept. 7)			
1	Singles	7	1:2.5	100	0	1.8
	Mating pairs	0	—	—	—	—
2	Singles	76	1:7.4	100	0	15.2
	Mating pairs	0	—	—	—	—
			3 nights after treatment (Sept. 9)			
1	Singles	3	1:1	100	0	1
	Mating pairs	5	—	60	0.4	—
2	Singles	10	1:1	60	0.4	3.3
	Mating pairs	7	—	100	0	—

effects on mating. Then, if one considers possible differences in densities and age structure (relative susceptibility to mating by determining sex ratios, level of virginity and mating status from collections of single adults) of the adult populations, the data can be compared with that obtained from captures in pheromone traps or other indirect methods to obtain a valid and rapid assessment of the effects of the treatments on mating in the population.

C. Population Comparisons

Data in Table 1 show the results obtained when populations of adult tobacco budworms in two nearby fields of cotton were compared in respect to density and age structure. On September 5, 1979 three observers collected single adults and mating pairs. Treatment with a combination of insecticides was applied by air at ca. 2 a.m. the night of September 6 and collections were made again at one and three nights after treatment. Previous seasonal history of the fields showed very light infestations in field one and a medium infestation in Field 2 which had been treated with insecticide in mid-August. Both fields were fruiting well, and no major insect damage was visible.

The pretreatment counts showed that the two fields contained two completely different adult populations: that in Field 2 consisted of twice as many males as females, and 92.9% of the females were virgin; that in Field 1 consisted of equal numbers of males and females, and only 13.3% of the females were virgin. Then, since the females begin their emergence cycle 1 to 2 days before the males and newly emerged adults are docile and easily collected, as in Field 2, the pretreatment population there was indicative of an emerging population 1 to 2 days past peak emergence (high level of virginity, sex ratio in favor of the males). The population in Field 1 was old as indicated by equal sex ratio, low level of virginity and a preponderance of single females that were multiple mated (±1.8 spermatophores/female ranging from zero to five). In view of the low densities that occurred in this field in the previous generation, these adults probably had moved into the field from other areas. There were also quantitative differences in the two populations. A larger population was present in Field 2, as indicated by the capture per man hour data. However, three and five mating pairs were collected, respectively, in Fields 1 and 2. Thus there were no major differences in numbers mating in the fields before treatment.

Data collected the first night after the insecticide application reinforced the assessment of the population based on pretreatment counts. The old population that had been present in Field 1 appeared to have been annihilated, and the few adults collected were newly emerged (100% virgin females and no mating — females do not mate on the night of emergence). The emerging population in Field 2 showed a 7.4:1 male-female ratio and no mating, evidence that the population had passed peak emergence and that the previously emerged population had been annihilated. The fact that no mating or mated females were found in either field supported the conclusion that sexually mature adults had been killed by the insecticide application. Then, in view of the large differences in density (1.8 and 15.2 adults captured per man hour for Fields 1 and 2, respectively) and previous record of infestation, it is clear that Field 2 had received much more infestation pressure the previous generation than Field 1.

Data collected the third night after treatment indicated that "recovery" of the population had begun as a result of continued emergence in Field 2. Meanwhile, young, sexually mature adults had apparently begun to move into Field 1 from Field 2 since there was a high percentage of virgin females and increased mating. The mating pattern was similar in Field 2 but the reduction in virgin females there compared with the number present previously indicated that most emergence had occurred. The adults were now dispersing (less dense), and the population was beginning the "aging" process. This conclusion is further demonstrated by the fact that the sex ratio in both fields had become equal.

Our conclusion concerning the populations observed in the two fields may be somewhat speculative, but the data do indicate striking differences in nearby fields. Such differences may occur in the evaluation of programs such as pheromone treatments, which deal with a mobile adult insect, and they demonstrate the need for a complete and thorough assessment of infield native populations in terms of their sex ratio, age structure, and density as a prerequisite to program evaluation. For instance, if our conclusions about the emergence and dispersal of the population from Field 2 to Field 1 are correct and if the population in Field 2 had been large, a large infestation would have occurred in Field 1 regardless of where mating had taken place. Under these circumstances, immature infestation records would have yielded little information about the area where mating had taken place and conclusions could be drawn that were totally false. The result would be disastrous if mating disruption was the method being evaluated.

Characterization of the adult population as described can also provide important information relative to pheromonal control or the need, timing, and type of other necessary control. For instance, nocturnal collection of adults makes it possible to immediately separate species such as the tobacco budworm and bollworm, which coinhabit field crops such as cotton (this is important because it is very difficult to separate immature forms of the two species). This determination plus information on adult density and age structure would allow one to determine the target pest and possibly project levels of infestation at an early stage in a given generation. For instance, if our observations are correct concerning the short time involved in the emergence of a generation of tobacco budworms (a large majority in 5 days) (Figure 2), then the establishment of adult-immature density relationships should allow for a better projection of subsequent densities of immature from adult collections. In addition, if pheromonal control procedures have failed, knowledge concerning the time and type of adult and larval activity would still allow adequate time to attack the population with insecticides at a vulnerable point. For example, we know that adult tobacco budworms emerge and mate during specific times of the night, and we know that they are located near the top of the plants and are quite immobile at this time. The data presented in Table 1 demonstrate good kill of adults by insecticides applied at night during the emergence and mating cycle, probably a result of direct contact of the insecticides on the adults. Then three applications might be required to contain an adult population during the emergence cycle of a given generation. Also, larvae have been observed moving from fruiting forms to the upper leaves of cotton plants to molt during the night and after they molt they eat their cast skins. The exact timing and extent of this activity is not fully known; however, if it is extensive and our information on adult kill is accurate, then a shift in timing of insecticide applications (to nighttime) based on the behavior and dynamics of the population could result in more effective and efficient control. Indeed, nighttime applications of insecticide on cotton in the western U.S. are becoming common.

Finally, if a majority of each generation of a tobacco budworm population emerges over a short period and ages as shown in Figure 2, there are several days between peak adult emergence and peak egg hatch because of the time required for adult sexual maturity, preoviposition, egg deposition, and embryonic development. Such a time lag should allow adequate time for evaluation of the impact of pheromonal control (using direct nocturnal observations) and decisions concerning the need for other types of control. Moreover, if relatively accurate relationships can be established between densities of the adults and subsequent immature infestations, one may have much greater latitude for making decisions concerning the future need for control in a generation cycle.

D. Application of Night Vision Equipment

Night vision goggles developed by the Department of Defense[42] have, in recent years, been used in the study of nocturnal insect behavior.[10] The goggles utilize natural light but are sensitive in the near infrared. Therefore, even on very dark nights, one can use auxillary infrared lighting to obtain relatively clear images of insects that are less than 1 mm in size at distances of several meters. The goggles provide a 40° field of view and allow one to observe nocturnal insects in their natural environment. Techniques are available that allow these observations to be put on film.[10,43]

The goggles have been used to study such aspects of the nocturnal behavior of tobacco budworms as trivial population movement,[10] male searching behavior,[10] size of pheromone plume,[10] pheromone trap efficiency[10,15,16] and response to pheromone components.[16] As a result of these studies, and of others, we know that the male mate searching behavior is a circadian activity and is not triggered by the presence of pheromone. Males can move relatively long distances in search of mates[11] but the pheromone plume from a female or a trap is relatively short.[10] Females have been observed marking plants with pheromone and moving in what appears to be an active attempt to locate mates and they have also been observed rejecting mates after attracting them.

More recently, the senior author has adapted a monocular night viewing device that is attached to a DC portable TV system for direct or delayed remote viewing of insect behavior in the field (Figure 3). The TV system includes a Javelin LL-900 series portable video camera, tape recorder, and adapter (Javelin Electronics, Los Angeles, Calif. 90045). The camera is adapted to a monocular Noctron IV night vision viewer as a single reflex camera and is equipped with various lenses and adapters for 35 mm and movie cameras (Texas Division, Varo Inc., Garland, Tex. 75040). With this system,

FIGURE 3. Monocular night vision TV set-up: (A) Javelin LL-900 series portable video camera (B) Motion picture/video camera adapter, Model 9842, Varo, Inc. (D) f/1.8 telephoto lens with C mount adapter, Model 9819, Varo, Inc. (E) Javelin VT-900 portable video tape recorder (F) TV monitor.

we have obtained excellent video pictures in the field at night. Undoubtedly, other systems are likely available and adaptable to the monocular night viewing device. Indeed, there are other video cameras available commercially that allow viewing in the near-infrared. With proper infrared lighting, such systems can provide relatively good images of nocturnal insects for study of their behavior.[44] Unfortunately, these systems are usually AC and not portable, so they are generally restricted to laboratory usage.

IV. USE OF INDIRECT METHODS OF EVALUATION

A. Trapping

In most studies made to establish the efficacy of pheromones or pheromone analogues in mating disruption of tobacco budworms, little or no attention has been paid to the behavior of the insect.[45-46] The index of disruption of pheromonal communication on mating has been based on reduction of trap capture. Indeed, the state-of-the-art of indexing pheromonal disruption is clearly indicated by Roelofs and Cardé[47] who defined pheromone components in terms of the effects on trap capture. Nevertheless, such trapping studies can be quite useful as an aid in identifying pheromone components or analogues that affect insect behavior.

The fact that virgin females and synthetic pheromones in traps will attract and capture males is indisputable. Various segments of the behavior of *Heliothis* spp. and pink bollworm responding to pheromone traps has been described by one or more of the authors of this chapter in five relatively recent papers.[7,10,15-16,22] Results of these studies make it clear that many variables such as population mating status, trap design and placement, weather, attractant, concentration of attractant, and pheromone plume affect captures. Indeed, all these factors also affect trap efficiency which varies greatly among designs. The question then is how to decipher the meaning of trap capture as it relates to population density and actual mating in a mobile population.

Studies of the response of male tobacco budworm to a pheromone trap show that the insects move upward and from areas of low light levels to higher light levels when they enter a trap;[15] pink bollworms respond similarly but orient downward.[22] As a result of this information, more efficient pheromone traps with greater trap capture potential have been designed for both species. The use of such standardized traps that have adequate capture potential and known response-capture efficiency will help in deciphering the meaning of trap capture data obtained from fields where pheromonal control is being attempted. However, if we are to measure populations of nocturnal noctuids with pheromone traps, we must be capable of deciphering the meaning of a catch of 5 males per trap per night or 25 males per trap per night. If either catch was presented without additional information, all one could reliably say is that male adults are in the area: 5 males per trap could represent the catch of males from a light population in which most of the females are virgin (Figure 2) and 25 males per trap could represent a population of the same density that was older (heavily mated). However, the difference in the two catches could also reflect the efficiency of the two traps (providing they were different designs) in catching males.[16,48] There is no doubt that one can influence trap capture of the target species by pheromone treatments but, until we relate this information in a positive manner to the actual mating in pheromone-treated areas, trap capture data will continue to be questionable.

In a recent review, light traps are compared with other physical and mechanical methods of trapping adult lepidopteran insects.[49] Most of these devices capture at least some females, and these can be used to gain added information about the mating status and thus the age structure of a population. Such captures are important to the assessment of the impact of pheromonal control agents on a population. However, so few adult insects of species like the tobacco budworm are captured in light traps that it is difficult to relate captures to densities or mating status of the population because we really don't know if a capture represents the true status of the population. Further, passive malaise traps used to assess the mating status of pink bollworm populations have shown reductions in multiple matings as a result of treatment with hexalure.[24] This is encouraging, and further investigation of malaise traps as a tool for monitoring the impact of pheromones on a population is certainly warranted. However, back up information that can only be obtained by direct nocturnal observation is needed so we can determine the relationship between the true population and that portion of the population that is captured. Only then will one be able to decipher the true meaning of information obtained from such captures.

B. Mating Tables

Mating tables are another device that can be used to obtain indirect information on mating in a population. Clipped-wing females can be placed on mating tables[50] or on plants surrounded with screen cages with slits large enough to allow entry of responding males.[52] The females can then be observed for mating. Previously tables were used to assess such aspects of the nocturnal behavior of the tobacco budworm as calling activity of females, mating interaction of laboratory-reared and native populations, mating barriers, and searching behavior of males.[9,11] More recently the mating tables described by Lingren et al.[11] have been modified for a study of mating activity of pink bollworms in a pheromonal control program.[52] Smaller tables (mini-tables)[53] are now being used to investigate the impact of pheromones on mating of tobacco budworms and pink bollworms in the field. These mini-tables consist of a 1.9-ℓ ice cream carton attached to a stake. A cotton plant terminal is placed in the carton, which is subsequently baited with a single-clipped-wing virgin female. The cartons are then checked hourly for mating or all females are collected at dawn and taken to the laboratory for dissection to determine the degree of mating. Some females escape from the cartons but minimum loss of females (<10% and <25% for tobacco budworm and pink bollworm, respectively) can be obtained by placing them in the cartons about one hour before in-field mating begins. Also, this technique has the advantage over the larger table in that the females can be spaced throughout a test area and mating levels of 90% have been obtained in untreated fields containing high density populations.

A major problem with mating tables as well as pheromone traps is that each unit emits one pheromone plume and that the plume is associated with objects which may affect the behavior of the male when he is responding to the females. However, the greatest disadvantage is the inability to monitor mating in the population of native females, the most important consideration in a pheromone suppression system. Certainly, our experiences tell us that tobacco budworm and pink bollworm females are not passive in seeking mates: they mark plants with pheromone(s) and move to new locations when unable to attract mates at a given site; also, they have been observed to reject mates after attracting them. The extent of this type of behavior has not been fully quantified, but any active participation (movement) by native females to find mates would diminish the accuracy of data obtained from such fixed measuring devices as traps and mating tables.

C. Radar

The tobacco budworm[54-55] and pink bollworm[56-57] are highly mobile insects and they may even be migratory. Thus the time and extent of the movement of each sex and mating status at the time they move is important to pheromonal control. Equipment such as night vision goggles,[10] nets attached to airplanes,[58] and traps on towers[59-60] have all been used by researchers attempting to obtain information on night-flying insects. However, equipment of this type can sample only a small portion of the total air space occupied by airborne insects. The need to increase sampling volume and explain insect airborne behavior has led to the use of special radar equipment.

Events leading to the use of radar in entomological studies, results of various field trials, and current capabilities of radar were recently discussed by Riley.[61] Radar equipment modified specifically for entomological research was used first in 1968.[62] Since then studies have shown that radar can provide relatively good information about insect size, numbers per unit volume, altitude, flight speed and direction, wing beat frequency, time of flight, rate of climb, flight ceiling, layering during dispersal, orientation during flight, and the concentrating effects of convergent wind fields.[63]

The most extensive use of radar for studying dispersal of one species of insect was conducted in Canada.[64] This multidisciplinary study produced new information about the dispersal of spruce budworm moths, *Choristoneura fumiferana* (Clemens), and the way meteorological events influenced airborne insect densities and displacement, par-

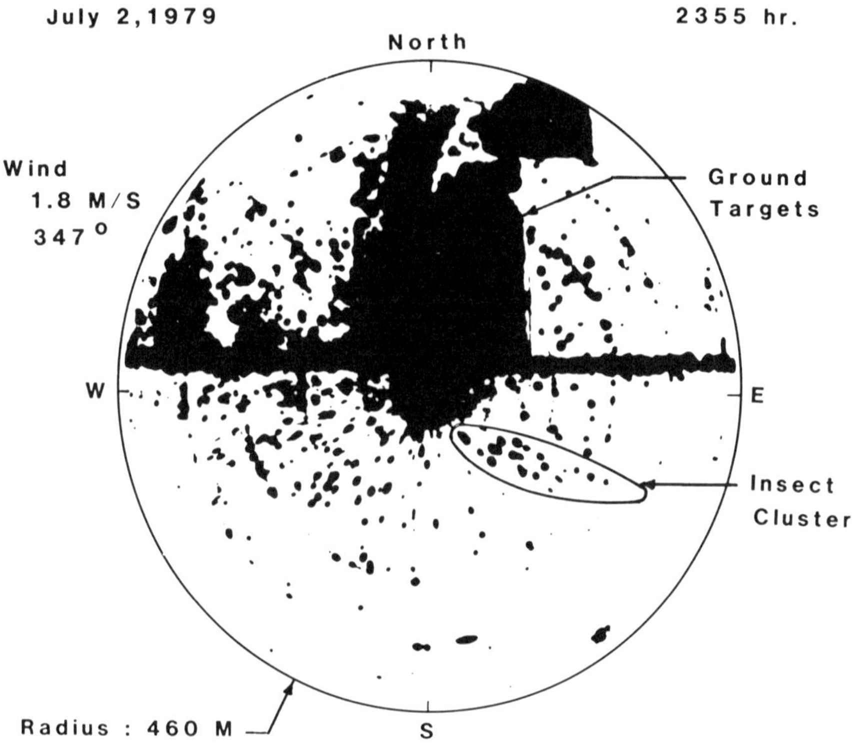

FIGURE 4. Radar display during one antenna revolution (3 sec) showing individual insects (dots) 3 to 9 m above a cotton field. Dark areas north of the radar were ground targets such as trees and fence rows. A cluster of insects southeast of the radar was moving toward the south-southwest at 6.4 m/sec. Antenna elevation angle was 1.0°.

ticularly long distance dispersal. This work showed the importance of behavior that occurs near the ground prior to, during, and at the termination of dispersal. Radar is capable of providing information about this activity but special locations and modifications may be necessary to minimize interference from plants, ground, and other obstacles.

With night vision goggles, we have observed tobacco budworm males moving just above the canopy of cotton fields in clusters of 50 or more during the active mating period.[16] Orientation appeared to be diagonal to the wind, and flight appeared to be more rapid than during other periods of the night. With radar, moths were detected during the mate searching period between 3 and 9 m above the ground over a cotton field (Figure 4). The velocity and direction of individual insect displacement was recorded by photographing the radar display every 3 sec and wind direction and velocity were similarly recorded by tracking small helium-filled balloons. Insect air speed and heading were determined from the wind and insect displacement vectors. Thus, displacement of the insect cluster shown in Figure 4 was 6.4 m/sec on an azimuth of 209°. A balloon released 5 min earlier indicated the wind was blowing at 1.8 m/sec from an azimuth of 347°. The difference between the wind vector and insect displacement yielded an insect air speed and heading of 5.2 m/sec and 22°, respectively. To account for the insects being displaced over the ground toward the 209° azimuth, they had to crab 35° with respect to the wind. In fact, in the study, the radar operator could determine when an insect approached an observer in the field and relayed information via radio to the observer. The field observer then used night vision goggles to confirm the location, height, and flight direction of the insect. In this way, radar was

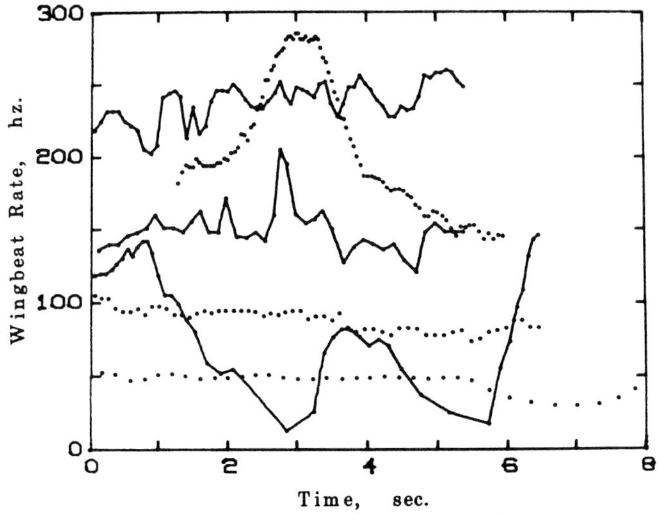

FIGURE 5. Variation of wingbeat rate of random insects detected within 10 to 20 m above cotton field by radar October 4, 1979. Each point is average frequency of ten consecutive wingbeats.

used to detect individual insects and follow their trajectories over much greater distances than an observer with only night vision goggles; however, if the insect changed altitude it flew out of the radar beam. Also, the radar beam rotation rate (20 rev/min) was too slow to resolve fine details of trajectories. Better trajectory resolution could be attained if the scanning rate could be increased or if individual insects could be tracked automatically. In addition special radar circuits that detect only moving targets might enable a radar to detect insects even closer to the ground. Certainly, studies of this type should lead to a better understanding of the airborne movement of insects such as the tobacco budworm near the crop canopy, and this information would be extremely important to the evolution of population suppression procedures such as pheromones.

Insect wing beat frequencies too may be determined by sampling the radar signal returned from individual insects. As the insect beats its wings, the reflected radar signal is modulated. Schaefer[63] reported that wing-beat rates of desert locusts, *Schistocerca gregaria,* (Forskal), were very constant during periods of 10 sec when these insects were in a definite migratory behavioral mode. Measurements of the wing beat frequencies of unknown insect targets near Phoenix, Arizona in 1979 also revealed relatively constant wing beating for some of the insects; however, others exhibited fluctuating wing beat rates (Figure 5) that may reflect some sort of behavioral maneuvering. Indeed, even though the wing beat rates of some insects varied by factors of two or more, wing beat rates provided information about the diversity of species that were airborne at a particular time: a frequency distribution of wing beat rates of 287 insects indicated that the highest proportion of insects had average wing beat rates near 50, 130, and 250 Hz (Figure 6). Insects such as cabbage loopers and honey bees, *Apis mellifera* L., have wing beats up to 60 and 200 Hz, respectively.[65]

We believe that radar can be a useful tool for studying various phases of activity of airborne insects, but the experiments must be carefully planned and executed so interferences from obstacles on the ground or unwanted targets such as birds, bats, and nontarget insects do not bias the results. Close cooperation between entomologists, meteorologists, and radar personnel is necessary to realize the full potential of this approach. For example, current entomological radars are generally modified commercial marine radars, and there is a relatively long wait before data collected even during

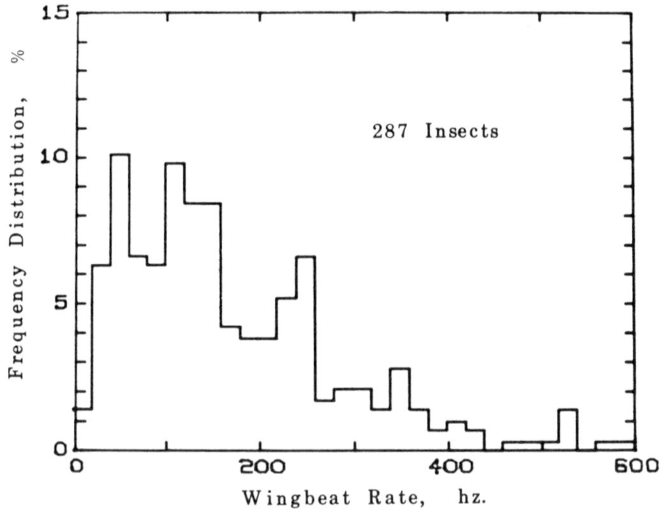

FIGURE 6. Frequency distribution of insect wingbeat rates detected within 10 to 20 m above cotton field by radar during October 1979 (8 nights). Class interval = 20 Hz.

marine radars, and there is a relatively long wait before data collected even during short periods of real time can be processed. More sophisticated radar technology which already exists could provide improved classification of targets, trajectory information, and quantification of density estimates. Modern data processing equipment could also be used to convert the data to a more useful and timely format. Radar is certainly the best tool available today to study the flight of insects in the air mass. Coupled with careful biological and meteorological observations and well-planned experimentation, it will undoubtedly provide a better understanding of insect flight behavior. Therefore, radar should definitely be considered an important tool for studies of insect behavior and of the impact of pheromones on insect populations.

D. Laboratory Evaluations

The effectiveness of pheromones as an insect control agent cannot be fully evaluated in a laboratory. Olfactometers, electroantennographs, and other laboratory measurements are beneficial in preliminary investigations, but they tell little about the usefulness of pheromones placed in the actual ecological niche of the insects. In some cases, laboratory and even field cage evaluations have actually led to incomplete[66] or erroneous results.[67,68]

E. Field Cage Evaluations

Field cage evaluations are somewhat more useful than laboratory evaluations once the pheromone has been identified. However, these evaluations should not be taken as conclusive evidence that pheromones will or will not perform satisfactorily as a control technique. On the other hand, they do provide a less expensive means than field tests to gain insight on the effectiveness of given concentrations against a given population of caged insects that were reared in the laboratory. There are two major reasons why field cage data are limited for pheromone control tests aside from the fact that plants grown under field cages are generally inferior and sometimes quite dissimilar to plants grown under regular cropping conditions. The first is the use of laboratory-reared insects in a seminatural ecological niche as a substitute for the feral insect. Laboratory-reared tobacco budworms released into the environment have proven inferior in ability to compete with the feral population for mates and space,[13-14] and even though improvements in laboratory rearing are continuing, there remain significant differences in behavior of the laboratory reared vs. the feral nocturnal insects with which we have worked.[11] The second major limitation involves the size of the plot or cage. The larger the cage, the more likely the insect is to display behavior patterns normally associated with those of feral populations in host plants under natural cropping conditions. Nevertheless, some valid observations of nocturnal behavior that were later confirmed by open field observations have been reported for several nocturnal species studied in 0.4-ha cages.[8]

Field cage tests were recently conducted at Tifton, Ga., to examine the possibility of utilizing any one or a combination of the chemicals identified as components of the pheromones of tobacco budworm and bollworm to disrupt mating.[69] The cages were 9.1 × 30.5 × 3.5 m and contained preferred host plants of *Heliothis* spp. The chemicals identified by Klun et al.[70] and field tested by Sparks et al.[7,16] were evaporated individually and in combination from cigarette filters (1,853 to 12,335/ha) at rates of 185 mg, 18.5 g, and 185 g/ha. Then a minimum of 50 virgin adult pairs of each species were separated by sex and the sexes were released on opposite ends of a cage. Each component or combination of components was tested for one to five nights by making direct nocturnal observations of behavior and mating in treated and untreated cages throughout the nocturnal mating activity period. In some cases, initiation of mating in treated cages was delayed by 1 to 2 hr, but, by the end of the period, there were insignificant differences in the occurrence of mating. Possibly the resumption of mating after the initial delay reflected a degradation in some of the pheromone components because changes in tobacco budworm male behavior occurred in earlier tests when the pheromone was dispensed from cigarette filters in traps.[16] Also, the reduction in mating was for only one night in field tests against very low density populations when virelure was applied in Hercon® dispensers and there was no overall reduction between applications.[71] These test results therefore suggest that the currently identified pheromones in the formulations tested have limited use as disruptants for *Heliothis* spp. though they do not prove that these materials will not disrupt mating when they are applied in adequate formulations over large acreages of a host crop containing a low density population of *Heliothis* spp. However, the real point is that data obtained from field-cage tests are of limited use in determining the potential of pheromones to manipulate populations of insects. The degree of validity of such tests is probably proportional to the size of the cage and the insect species being evaluated.

V. PINK BOLLWORM

A. Behavioral Information

Literature documenting observations of the nocturnal field behavior of the pink bollworm is fragmented and limited.[18,20] Nevertheless, those observations that have been made plus those that have been made in the laboratory and data on activity sequences that have been obtained from trap catches suggest a sequence of nocturnal activities such as feeding, oviposition, and mating similar to the sequence described for the tobacco budworm.[19,21-29] Peak mating activity seems to occur at ca. 2:30 a.m. during warm summer nights;[27,72] cool temperatures (<17.7°C) result in earlier mating activity (as early as 6 p.m. and usually over by midnight); and cool temperatures and wind speeds in excess of 8 m/sec influence flight activity and the site of activities such as mating.[20] On warm and calm nights, the mating activity takes place in the upper canopy; but when winds are excessive or temperatures are cool, many adults move down into the canopy and mate on the ground or on ground trash.[20,72] This behavior coupled with small size, color, and sporadic movement on the plants makes direct field observations of behavior much more difficult with these moths than with the tobacco budworm.

Pink bollworm adults usually emerge during the morning (8 to 10 a.m.).[56] Females emerge first during a generation and field studies made using genetic markers indicated a short emergence period for a given generation (over 80% in 1 week).[73] After emergence the adults move up into the plants to spread and dry their wings and remain, very passive, for ca. 12 hr.[74] Few females or males mate during the first night after emergence.[74] Females generally mate every 2 to 3 days but only 1 mating is required to lay a full complement of fertile eggs. With this information about biology and behavior, one can collect single adults and mating pairs in the field at night and use them to characterize the age structure of the population. Specific studies of the mating status of the population as it relates to capture in pheromone traps have not been conducted with the pink bollworm, but similarities in the biology and behavior with that of the tobacco budworm suggest that competition between females and traps would be comparable for the two species.

B. Pheromone Impact Assessment

For the most part, reductions in capture of pink bollworm males in pheromone traps has been a primary tool used in evaluating the impact of pheromone on the populations. Nevertheless, reduced larval infestations have been demonstrated in several cases following treatments with hexalure and gossyplure,[24,52,75-79] and in these test, techniques such as mating tables[52] and passive malaise traps[24,71] have indicated a reduction in mating. However, trap capture has never been reduced to zero nor has mating been completely inhibited between applications, regardless of the indirect methods used in making such determination. Also, population suppression was quite variable in some tests and was lost later in the season when densities normally become higher. Many physical factors such as the pheromone release formulation, pheromone purity, rate of application, timing of application, and canopy coverage could have influenced the results. However, a more adequate knowledge of behavior such as the number, range, and degree of movement of gravid females, and the degree of mating inhibition regardless of the reason(s) would have provided a more valid means of assessing the sources of variability. In fact, we feel that a rapid assessment of actual mating inhibition can be accomplished through the use of direct nocturnal observation techniques. Then after the relationship between direct and indirect techniques of measuring mating suppression are established, it may be possible to use indirect techniques such as pheromone traps to obtain a rapid and valid assessment that would only require verification by limited direct observations of the population age structure.

Projections of the age structure of pink bollworm populations appear possible by using heat unit accumulations for development of the insect[80-82] and its primary host plant (cotton),[83-86] since pink bollworms overwinter in a diapause state and emerge in the spring when the larvae can feed on the cotton fruit, and the immature forms develop at a rate that depends primarily on temperature. As a result, an F_1 generation will build up depending on the availability of fruit and it synchrony with the emergence of the overwintering parents. Thus, integration of information on planting dates and time of overwintering emergence with temperatures should allow relatively good predictions of population development and generation cycling. Of course, these predictions would have to be validated through extensive testing and population cycling will not be the same at different latitudes because of temperature differences during the late winter and spring.

Table 2
COMPARISON OF TRAP CAPTURE OF MALE TOBACCO BUDWORMS WITH MATING OF CLIPPED-WING FEMALES ON MINI-MATING TABLES AND MATING OF NATIVE FEMALES FOLLOWING EVAPORATION OF PHEROMONE COMPONENTS FROM CIGARETTE FILTERS. TREATED AND UNTREATED AREAS WERE 0.4-HA PLOTS OF FRUITING COTTON CONTAINED IN A 4.9 HA FIELD OF COTTON.

Treatments	Concentration of pheromone components (mg/ha)	No. native mating pairs/ha[a]		% Reduction in		
		Treated	Untreated	Response to pheromone[b]	Mating on mini-tables[c]	Male capture in traps[d]
7-Component pheromone	1236	2.5	7.4	100	100	100
Virelure (*Z*-11-hexadecenal + *Z*-9-tetradecenal-16:1 ratio)	371	9.9	0	100	87	100
Z-11-hexadecenal	1236	69.2	0	71	29	86
Z-9-tetradecenal	1236	9.9	0	77	54	99

Note: Pheromone components dispersed from 500 cigarette filters attached to upper leaves of cotton plants and spaced equidistance apart; treated and untreated plots separated by 200 m. Test conducted near Brownsville, Tex., 1979, for 3 nights with all treatments.

[a] Mating pairs collected from 12M to 4 a.m. from ⅓ of each plot.
[b] Based on number of male responses to single cigarette filters inoculated with 150 μg of 7 component pheromone per 5 min period in treated and untreated plots between 12M and 1 a.m. Night vision goggles were used to observe responses.
[c] Matings from 49 mini-tables spaced equidistance apart in each plot and each baited with single virgin clipped-wing female.
[d] Based on capture in wind-oriented live trap, each baited with 5 virgin females and placed in center of treated and untreated plots.

VI. RELATIONSHIPS OF DIRECT AND INDIRECT METHODS OF EVALUATION

As noted previously, little attention has been given to the establishment of numerical relationships between direct and indirect methods of measuring mating disruption of tobacco budworm and pink bollworm by pheromone applications. However, Raulston[87] conducted a preliminary field study in which comparison between in-field mating, male responses to a point source consisting of the seven-component pheromone, mating or mini-tables, and capture in traps baited with virgin females were made after dispensing various components of the tobacco budworm pheromone from cigarette filters (Table 2). All components tested show drastically reduced trap capture. Mating on mini-tables was affected less than reductions of capture in traps in all treatments except the seven-component pheromone. As expected, reductions in capture in traps was similar to reductions in response to a point source of the pheromone. More matings were observed in treated plots than in untreated plots except when the seven-component pheromone was applied. In this treatment, mating was reduced by 66%.

Trap captures and in-field mating of pink bollworm after treatment with microencapsulated gossyplure are compared in Table 3.[88] The data show an average reduction of 94% in trap capture in the treated plots over an 8-day period after treatment but only a 66% reduction in mating. However, at 45 m from the treated area, capture in traps was reduced only 76% and there was no concomitant reduction in mating. At 136 m, in-field mating increased by 50%, and capture in traps was reduced 16%. On

Table 3
COMPARISON OF CAPTURE OF MALE PINK BOLLWORMS IN PHEROMONE TRAPS AND MATING OF NATIVE FEMALES FOLLOWING APPLICATION OF MICROENCAPSULATED GOSSYPLURE AT A RATE OF 7.4 GRAMS AI PER HA TO TWO 0.4 HA PLOTS WITHIN A 6 HA FIELD OF FRUITING COTTON

Plot	Distance from treatments (m)	% Reduction or increase over 8 days after treatment in[a]	
		Mating	Trap capture
Treated		66 (2.5)	94 (203)
Untreated 1	45	0 (1.0)	76 (128)
Untreated 2	136	+50 (1.0)	16 (114)

Note: Single adult collections showed a relatively high density, heavily mated (2.2 spermatophores/female; range 0-7) population with less than 4% virgin females; test conducted near Phoenix, Ariz. from Oct. 17-26, 1979. Mating pairs collected from each plot by 2 to 3 observers for 3 hr during mating activity period which, because of relatively cool temperatures, occurred before midnight.

[a] Numbers in parenthesis indicate numbers found mating/man hour of search and capture/trap/night at 1 night before treatment; percentages indicate the average reduction or increase in mating or trap capture that occurred over 8 nights following treatment.

the night of treatment, though mating at this location was not affected, capture in traps was reduced by 100%.

In-field mating of both the tobacco budworm and pink bollworm was very limited in the test plots. However, captures of males in traps were above 100 per trap per night in untreated plots in the tobacco budworm test and in the pink bollworm test captures were above 100 per trap per night on the night before treatment. Captures of this magnitude suggest a population density that should have allowed for greater collections of mating pairs from the untreated areas. However, collections and dissections of single females in both tests showed old populations (mostly multiple mated with few virgins). Therefore, as previously discussed, few matings would have occurred during the test period even though relatively large numbers of males were present that could respond to pheromone traps.

The evidence presented concerning the relationships between the direct and indirect means of measuring mating reduction for tobacco budworm and pink bollworm is preliminary; it does, however, indicate different levels of response that are probably related to the fact that females are not passive in the mating process. For example, we have observed females of both species calling, moving, and then calling again. Pheromone traps and mating tables do not measure this type of activity. In particular, this movement means that the chance of a native female locating a mate should be greater than that of a clipped-wing female on a table. Pheromone traps should be even less effective than clipped-wing females because they are more passive, less than 50% efficient, and normally are distributed over only a small portion of the treated area.

In the case of the pink bollworm, a 28% difference between actual in-field mating and trap capture reduction was observed following treatment with gossyplure (Table 3). Differences in larval infestations and trap capture reductions of this magnitude have been reported in several tests where gossyplure has been applied for pink bollworm control. Also similar relationships have been shown between reductions in mating of tethered females (about 60%) and larval infestations of the cabbage looper following pheromone treatments (trap capture reductions were >90%).[89] Therefore, one must apparently expect some reproduction to be occurring in a population under pheromonal control even when trap capture data indicate highly effective reductions in mating. This again points to the need for establishing sound relationships between direct and indirect methods of assessing the impact of pheromonal control. The need for a rapid assessment following treatment is obvious. This is especially true since both the tobacco budworm and pink bollworm require only one mating to lay a full complement of fertile eggs. For example, if 10% or more of a population under pheromonal control mates each night and if virgins get the first mate during the night, then it is possible for the whole population to mate in 10 days providing they live that long. We therefore need to investigate the relationships between reductions in mating and the impact of those reductions over time on the fecundity of the population. Information of this type could help us evaluate a pheromone control program with more accuracy.

VII. CURRENT STATUS AND PROSPECTS FOR THE FUTURE

It is surely plain that, in our opinion, direct observations of behavior including mating are feasible for many if not all of the nocturnal noctuids that attack field crops. We have attempted to describe how these observations can be made and how the results can be compared with those obtained from indirect measurements of mating disruption. Since direct observations of in-field behavior are labor intensive in that they require an adequate number of observers and at least relatively high population densities to obtain adequate numbers of observations, use of direct measurements in the development of pheromones for insect control will probably be restricted to the initial phases and in obtaining behavioral information necessary for the development of simpler criteria.

Much of our presentation here is based on qualitative information. As a result, it is somewhat speculative. Nevertheless, the entire premise behind the use of pheromones to suppress insect population lies in their potential to block or alter behavioral activities of the insect. Therefore, it becomes imperative that we determine as accurately as possible how the insect is being affected. Naturally, this demands that we first understand the activities and behaviors exhibited in undisturbed populations. Their task will require the energy and ingenuity of many people and use of an array of techniques and tools including our eyes and our powers of observation. Once we can adequately qualify the complexities of insect behavior relevant to the disruption of sexual communication by pheromonal treatments, quantification then becomes the next step in the process. Hopefully, we have raised some important questions in our presentation and have provided some insight into means for obtaining answers.

REFERENCES

1. **Beroza, M.,** Insect attractants are taking hold, *Agric. Chem.*, 15, 37, 1960.
2. **Wright, R. H.,** 'Metarchon': a new term for a class of non-toxic pest control agents, *Nature (London)*, 204, 603, 1964b.
3. **Wright, R. H.,** After pesticides — what? *Nature (London)*, 204, 121, 1964a.
4. **Marshall, E.,** White House seeks to reform USDA research, *Science*, 206, 307, 1979.
5. **Phillips, J. R. and Whitcomb, W. H.,** Field behavior of adult bollworm, *Heliothis zea* (Boddie), *J. Kans. Entomol. Soc.*, 35, 242, 1962.

6. **Agee, H. R.**, Mating behavior of bollworm moths, *Ann. Entomol. Soc. Am.*, 62, 1120, 1969.
7. **Sparks, A. N., Carpenter, J. E., Klun, J. A., and Mullinix, B. G.**, Field responses of male *Heliothis zea* (Boddie) to pheromonal stimuli and trap design, *J. Ga. Entomol. Soc.*, 14, 318, 1979.
8. **Lingren, P. D., Greene, G. L., Davis, D. R., Baumhover, A. H., and Henneberry, T. J.**, Nocturnal behavior for four lepidopteran pests that attack tobacco and other crops, *Ann. Entomol. Soc. Am.*, 70, 161, 1977.
9. **Lingren, P. D., Raulston, J. R., and Sparks, A. N.**, Interception of native male tobacco budworm by barriers of released laboratory-reared sterile females, *Environ. Entomol.*, 6, 217, 1977.
10. **Lingren, P. D., Sparks, A. N., Raulston, J. R., and Wolf, W. W.**, Application for nocturnal studies of insects, *Bull. Entomol. Soc. Am.*, 24, 206, 1978.
11. **Lingren, P. D., Raulston, J. R., Sparks, A. N., and Proshold, F. I.**, Tobacco budworm: nocturnal behavior of laboratory-reared irradiated and native adults in the field, USDA/SEA, ARR-W-5, U.S. Department of Agriculture, Beltsville, Md., 1979.
12. **Raulston, J. R., Snow, J. W., Graham, H. M., and Lingren, P. D.**, Tobacco budworm: effect of prior mating and sperm content on the mating behavior of females, *Ann. Entomol. Soc. Am.*, 68, 701, 1975.
13. **Raulston, J. R., Graham, H. M., Lingren, P. D., and Snow, J. W.**, Mating interaction of native and laboratory-reared tobacco budworms released in the field, *Environ. Entomol.*, 5, 195, 1976.
14. **Raulston, J. R., Lingren, P. D., Sparks, A. N., and Martin, D. F.**, Mating interaction between native tobacco budworms and released backcross adults, *Environ. Entomol.*, 8, 349, 1979.
15. **Raulston, J. R., Sparks, A. N., and Lingren, P. D.**, Design and comparative efficiency of a wind oriented live trap for capturing *Heliothis* spp., *J. Econ. Entomol.*, 73, 586, 1980.
16. **Sparks, A. N., Raulston, J. R., Lingren, P. D., Carpenter, J. E., Klun, J. A., and Mullinix, G. B.**, Field responses of male *Heliothis virescens* to pheromonal stimuli and traps, *Bull. Entomol. Soc. Am.*, 25, 268, 1979.
17. **Goodenough, J. L. and Snow, J. W.**, Tobacco budworms: nocturnal activity as indexed by attraction to live virgin females in electric grid traps, *J. Econ. Entomol.*, 66, 543, 1973.
18. **Garcia, P. I.**, Figeros apuntes sobre la biologia del gusana rosado del algodon, *Bol. Biseccion Estudies Biol.*, Mexico, 3, 1925.
19. **Graham, H. M.**, Hourly response of laboratory-reared and native male pink bollworms to traps baited with hexalure, *Environ. Entomol.*, 2, 965, 1973.
20. **Kaae, R. S. and Shorey, H. H.**, Sex pheromones of Lepidoptera. 44. Influence of environmental conditions on the location of pheromone communication and mating in *Pectinophora gossypiella*, *Environ. Entomol.*, 2, 1081, 1973.
21. **Leppla, N. C.**, Calling behavior during pheromone release in female pink bollworm moth, *Ann. Entomol. Soc. Am.*, 65, 281, 1972.
22. **Lingren, P. D., Burton, J., Shelton, W., and Raulston, J. R.**, Night vision goggles; for design, evaluation, and comparative efficiency determination of a pheromone trap for capturing live adult male pink bollworms, *J. Econ. Entomol.*, 73, 622, 1980.
23. **Lukefahr, M. and Griffin, J.**, Mating and oviposition habits of the pink bollworm moth, *J. Econ. Entomol.*, 50, 487, 1957.
24. **McLaughlin, J. R., Shorey, H. H., Gaston, L. K., Kaae, R. S., and Stewart, F. D.**, Sex pheromones of Lepidoptera. XXXI. Disruption of sex pheromone communication in *Pectinophora gossypiella* with hexalure, *Environ. Entomol.*, 1, 645, 1972.
25. **Ouye, M. T., Graham, H. M., Richmond, C. A., and Martin, D. F.**, Mating studies of the pink bollworm, *J. Econ. Entomol.*, 57, 222, 1964.
26. **Ouye, M. T., Garcia, R. S., Graham, H. M., and Martin, D. F.**, Mating studies on the pink bollworm, *Pectinophora gossypiella* (Lepidoptera: Gelechiidae), based on presence of spermatophores, *Ann. Entomol. Soc. Am.*, 58, 880, 1965.
27. **Sharma, R. K., Rice, R. E., Reynolds, H. T., and Shorey, H. H.**, Seasonal influence and effect of trap location on catches of pink bollworm males in sticky traps baited with hexalure, *Ann. Entomol. Soc. Am.*, 64, 102, 1971.
28. **Squire, F. A.**, Nocturnal habits in *Platyedra gossypiella*, *Nature (London)*, 140, 69, 1937.
29. **Van Steenwyk, R. A., Ballmer, G. R., and Reynolds, H. T.**, Nocturnal trap catches of the pink bollworm, *Ann. Entomol. Soc. Am.*, 71, 354, 1978.
30. **Mitchell, E. R.**, Nocturnal activity of adults of three species of loopers based on collections in pheromone traps, *Environ. Entomol.*, 2, 1078, 1973.
31. **Shorey, H. H.**, Sex pheromones of noctuid moths. II. Mating behavior of *Trichoplusia ni* (Lepidoptera: Noctuidae) with special reference to the role of the sex pheromone, *Ann. Entomol. Soc. Am.*, 57, 371, 1964.
32. **Shorey, H. H.**, The biology of *Trichoplusia ni* (Lepidoptera: Noctuidae). IV. Environmental control of mating, *Ann. Entomol. Soc. Am.*, 59, 502, 1966.

33. **Shorey, H. H. and Gaston, L. K.**, Sex pheromones of noctuid moths. V. Circadian rhythm of pheromone responsiveness in males of *Autographa california, Heliothis virescens, Spodoptera exigua,* and *Trichoplusia ni*(Lepidoptera: Noctuidae), *Ann. Entomol. Soc. Am.*, 5, 597, 1965.
34. **Sower, L. L., Gaston, L. K., and Shorey, H. H.**, Sex pheromones of noctuid moths. XXVI. Female release rate, male response threshold, and communication distance for *Trichoplusia ni, Ann. Entomol. Soc. Am.*, 64, 1448, 1971.
35. **Sower, L. L., Shorey, H. H., and Gaston, L. K.**, Sex pheromones of noctuid moths. XXV. Effects of temperature and photoperiod on circadian rhythms of sex pheromone release by females of *Trichoplusia ni, Ann. Entomol. Soc. Am.*, 64, 488, 1971.
36. **Sparks, A. N., Raulston, J. R., and Lingren, P. D.**, unpublished data.
37. **Lingren, P. D., Henneberry, T. J., and Bariola, L. A.**, Nocturnal behavior of adult cotton leafperforators in cotton, *Ann. Entomol. Soc. Am.*, 73, 44, 1980.
38. **Allen, N.**, Mating habits of the tobacco hornworm, *J. Econ. Entomol.*, 48, 526, 1955.
39. **Mitchell, E. R., Webb, J. C., Baumhover, A. H., Endris, R. G., Hines, R. W., and Masuda, S.**, Activity of tobacco hornworn moths as determined by electrocutor grid traps baited with blacklight lamp and virgin females, *Environ. Entomol.*, 1, 679, 1972.
40. **Raulston, J. R.**, unpublished data.
41. **Lingren, P. D.**, Tobacco budworm nocturnal behavior — an aid in pest management, *Proc. 6th Desert Cotton Insects Symp.*, University of California Cooperative Extension Service, El Centro, Calif., 1978, Cl-3.
42. **Slusher, J. A.**, Night vision equipment developments: where we've been, where we are today, and where we're going, *Bull. Entomol. Soc. Am.*, 24, 197, 1978.
43. **Boogher, B. and Slusher, J. A.**, Successful photographic techniques through night vision devices, *Bull. Entomol. Soc. Am.*, 24, 203, 1978.
44. **Conner, W. E. and Master, W. M.**, Infrared video viewing, *Science*, 199, 1004, 1978.
45. **Hendricks, D. E.**, Tobacco budworm: disruption of courtship behavior with a component of the synthetic sex pheromone, *Environ. Entomol.*, 5, 978, 1976.
46. **Mitchell, E. R., Jacobson, M., and Baumhover, A. H.**, *Heliothis* spp: Disruption of pheromonal communication with (*Z*)-9-tetradecen-1-01 formate, *Environ. Entomol.*, 4, 577, 1975.
47. **Roelofs, W. L. and Cardé, R. T.**, Responses of Lepidoptera to synthetic sex pheromone chemicals and their analogues, *Annu. Rev. Entomol.*, 22, 377, 1977.
48. **Hartstack, A. W., Witz, J. A., and Buck, D. R.**, Moth traps for the tobacco budworm, *J. Econ. Entomol.*, 72, 519, 1979.
49. **Lingren, P. D.**, Light traps vs other physical and mechanical devices, in *Movement of Highly Mobile Insects: Concepts and Methodology in Research*, Rabb, R. L., and Kennedy, G. G., Eds., University Graphics, North Carolina State University, Raleigh, 1979, chap. 16.
50. **Snow, J. W., Raulston, J. R., and Guillot, F. S.**, Mating tables: a method of studying the mating and the competitive behavior of Lepidoptera and Diptera in the field, *Ann. Entomol. Soc. Am.*, 69, 751, 1976.
51. **Mitchell, E. R., Baumhover, A. H., and Jacobson, M.**, Reduction of mating potential of male *Heliothis* spp. and *Spodoptera frugiperda* in field plots treated with disruptants, *Environ. Entomol.*, 5, 484, 1976.
52. **Brooks, T. W., Doane, C. C., and Staten, R. T.**, Experience with the first commercial pheromone communication disruptive for suppression of an agricultural insect pest, in *Odor Communication in Animals*, Ritter, F. J., Ed., Elsevier/North Holland, New York, 1979, 375.
53. **Raulston, J. R. and Lingren, P. D.**, unpublished data.
54. **Haile, D. G., Snow, J. W., and Young, J. R.**, Movement of adult *Heliothis* released on St. Croix to other islands, *Environ. Entomol.*, 4, 225, 1975.
55. **Hendricks, D. E., Graham, H. M., and Raulston, J. R.**, Dispersal of sterile tobacco budworms from release points in northeastern Mexico and southern Texas, *Environ. Entomol.*, 2, 1085, 1973.
56. **Fenton, F. A. and Owen, W. L., Jr.**, The pink bollworm of cotton in Texas, *Texas Agric. Exp. Stn. Misc. Publ.*, No. 100, 1953, 39.
57. **Stern, V. and Sevacherian, V.**, Long-range dispersal of pink bollworn into the San Joaquin Valley, *Calif. Agric.*, 32, 4, 1978.
58. **Glick, P. A. and Noble, L. W.**, Airborne movement of the pink bollworm and other arthropods, *U.S. Dep. Agric. Tech. Bull.*, No. 1255, 20, 1961.
59. **Callahan, P. S., Sparks, A. N., Snow, J. W., and Copeland, W. W.**, Corn earworm moth: vertical distribution in flight, *Environ. Entomol.*, 1, 497, 1972.
60. **Taylor, L. R.**, Mortality and mobility of insect migrants high in the air, *Nature (London)*, 186, 410, 1960.
61. **Riley, J. R.**, Radar as an aid to the study of insect flight, in *A Handbook on Biotelemetry and Radio Tracking*, Amlaner, C. J., Jr. and Macdonald, D. W., Eds., Pergamon Press, New York, 1979, 131.
62. **Schaefer, G. W.**, Radar studies of locust, moth, and butterfly migration in the Sahara, *Proc. R. Entomol. Soc. London*, Series C, 33, 1969.

63. **Schaefer, G. W.**, Radar observations of insect flight, in *Insect Flight*, Rainey, R. C., Ed., Blackwell Scientific, Oxford, 1976, 157.
64. **Greenbank, D. O., Schaefer, G. W., and Rainey, R. C.**, Spruce budworm (Lepidoptera: Tortricidae) moth flight and dispersal: new understanding from canopy observations, radar, and aircraft, *Mem. Entomol. Soc. Can.*, No. 110, 49, 1980.
65. **Greenwalt, C. H.**, Dimensional relationships for flying animals, *Smithsonian Misc. Coll.*, 144, 49, 1962.
66. **Tumlinson, J. H., Hendricks, D. E., Mitchell, E. R., Doolittle, R. E., and Brennan, M. M.**, Isolation, identification, and synthesis of the sex pheromone of the tobacco budworm, *J. Chem. Ecol.*, 1, 203, 1975.
67. **Jones, W. A., Jacobson, M., and Martin, D. F.**, Sex attractant of the pink bollworm moth: isolation, identification, and synthesis, *Science*, 152, 1516, 1966.
68. **Sekul, A. A. and Sparks, A. N.**, Sex pheromone of the fall armyworm moth: isolation, identification and synthesis, *J. Econ. Entomol.*, 5, 1270, 1967.
69. **Sparks, A. N. and Raulston, J. R.**, unpublished data.
70. **Klun, J. A., Plimmer, J. R., Bierl-Leonhardt, B. A., Sparks, A. N., and Chapman, O. L.**, Trace chemicals: the essense of sexual communication systems in *Heliothis* species, *Science*, 204, 1328, 1979.
71. **Henneberry, T. J., Gillespie, J. M., Bariola, L. A., Flint, H. M., Lingren, P. D., and Kydonieus, A. F.**, Pink bollworm and tobacco budworm mating disruption studies on cotton, in *Management of Insect Pests with Semiochemicals-Concepts and Practices*, Knipling, E. F., Ed., Plenum Press, New York, 1981.
72. **Lingren, P. D.**, unpublished data.
73. **Bartlett, A. C. and Lingren, P. D.**, unpublished data.
74. **Lingren, P. D.**, unpublished data.
75. **Brooks, T. W. and Kitterman, R., Gossyplure H. F.** — Pink bollworm population suppression with male sex attractant pheromone released from hollow fibers — 1976 experiments, in *1977 Proc. Beltwide Cotton Prod. Mech. Conf.*, National Cotton Council of America, Memphis, Tenn., 1978, 79.
76. **Shorey, H. H., Kaae, R. S., and Gaston, L. K.**, Sex pheromones of Lepidoptera. Development of a method for pheromonal control of *Pectinophora gossypiella* in cotton, *J. Econ. Entomol.*, 67, 347, 1974.
77. **Shorey, H. H., Gaston, L. K., and Kaae, R. S.**, Air-permeation with gossyplure for control of the pink bollworm, in *Pest Management with Insect Sex Attractants*, ACS Symp. Ser. No. 23, Beroza, M. Ed., American Chemical Society, Washington, D. C., 1976, 67.
78. **Gaston, L. K., Kaae, R. S., Shorey, H. H., and Sellers, D.**, Controlling the pink bollworm by disrupting sex pheromone communication between adult moths, *Science*, 196, 904, 1977.
79. **Boness, M., Eiter, K., and Disselnkotter, H.**, Studies on sex attractants of Lepidoptera and their use in crop protection, *Pflanzenschutz-Nachrichten, Bayer*, 30, 213, 1977.
80. **Graham, H. M., Glick, P. A., and Ouye, M. T.**, Temperature effect on reproduction and longevity of laboratory-reared adult pink bollworm (Lepidoptera: Gelechiidae), *Ann. Entomol. Soc. Am.*, 60, 1211, 1967.
81. **Sevacherian, V., Toscano, N. C., Van Steenwyk, R. A., Sharma, R. K., and Sanders, R. R.**, Forecasting pink bollworm emergence by thermal summation, *Environ. Entomol.*, 6, 545, 1977.
82. **Butler, G. D., Jr. and Hamilton, A. G.**, Temperature-dependent development rates for four strains of *Pectinophora gossypiella*, *Ann. Entomol. Soc. Am.*, 69, 450, 1976.
83. **Mauney, J. R.**, Floral initiation of upland cotton *Gossypium hirsutum* L. in response to temperatures, *J. Exp. Bot.*, 17, 452, 1966.
84. **Hespeth, J. D. and Low, A.**, Effect of temperature on components of yield and fibre quality of cotton varieties of diverse origin, *Emp. Cotton Growing Rev.*, 45, 243, 1968.
85. **Mutsaers, H. J. W.**, Growth and assimilate conversion of cotton bolls (*Gossypium hirsutum* L.). 2. Influence of temperature on boll maturation period and assimilate conversion, *Ann. Bot.*, 40, 317, 1976.
86. **Constable, G. A.**, Temperature effects on the early field development of cotton, *Aus. J. Exp. Agric. Anim. Husb.*, 16, 905, 1976.
87. **Raulston, J. R.**, unpublished data.
88. **Lingren, P. D.**, unpublished data.
89. **Shorey, H. H., Kaae, R. S., Gaston, L. K., and McLaughlin, J. R.**, Sex pheromones of Lepidoptera. XXX. Disruption of sex pheromone communication in *Trichoplusia ni* as a possible means of mating control, *Environ. Entomol.*, 1, 641, 1972.

Chapter 7

MONITORING INSECTS WITH PHEROMONES: TRAPPING OBJECTIVES AND BAIT FORMULATIONS*

G. E. Daterman

TABLE OF CONTENTS

I. INTRODUCTION

Pheromone-baited traps can be extremely helpful in pest management simply by making pest populations more visible. Unless a particular pest is obviously present and causing economic injury, it may be surprisingly difficult to detect or evaluate its potential for damage. Insect populations have numerous regulating factors, and they characteristically fluctuate greatly in their densities. Numbers of moths captured by monitoring traps can reflect these changes in population densities and thereby aid in predicting outbreaks. Insects caught in traps are also good indicators of where and when adult activity is occurring, and this information can be important for timing management activities to minimize resource damage.

The objectives of monitoring may differ, and flexibility in trapping devices and strategies is required. Baited traps are used for a variety of purposes that may include both research and operational pest-management programs. Discussion here will be limited to trapping uses that are in some way directly supportive of management objectives. Available controlled release bait formulations and their applications for monitoring will also be reviewed. Trap designs have been discussed in a previous chapter.

II. MONITORING OBJECTIVES

A. Detecting Introduced Pests

Detecting an introduced pest is one of the more important uses of pheromone-baited traps. The monitoring objective is straightforward; it is simply to determine if the exotic pest has gained a foothold in a new area. When first introduced, such a pest will understandably be present in very low numbers and difficult to find. Early detection is possible with use of a pheromone bait to which the insect is highly sensitive. Obviously, the sooner an exotic pest is discovered the better the chance to eliminate the new infestation or at least minimize damage or further spread.

Effective detection surveys require reliable, long-lasting baits with maximum attractiveness, plus efficient and inexpensive traps. The baits should remain effective for at least several weeks since the pest's seasonal activity period may be unknown in the area and it would be expensive to continually recharge baits at a large number of survey locations.

Detection surveys using pheromone baits have been conducted operationally for several years for such pests as the gypsy moth, *Lymantria dispar* (L.), and the smaller European elm bark beetle, *Scolytus multistriatus* (Marsham), a principal vector of the Dutch elm disease. Numerous detections attributable to pheromone baits are listed in the Cooperative Plant Pest Report published weekly by the U.S. Department of Agriculture. An example of detection of an exotic pest by pheromone trapping was the recovery of gypsy moth males in traps in San Jose, California, in 1975 and 1976.[1]

A detection trapping system may also be used to verify that a particular location is free of a pest. The State of California,[2] for instance, will permit shipments of ornamental pines from states known to be infested with European pine shoot moth, *Rhyacionia buoliana* (Schiff.), if an official in the state of origin certifies that the source of the pines and peripheral locations were appropriately surveyed by pheromone-baited traps and no moths were captured. This reliance on the sensitivity of pheromone-baited traps has significant economic ramifications, and clearly underlines why the baits must be reliable. As in all surveys for introduced pests the capture of a single insect is significant, and maximum bait attractiveness is required.

B. Distribution Surveys

Extensive surveys with pheromone-baited traps can help define a pest's local or regional distribution. The Douglas-fir tussock moth, *Orgyia pseudotsugata* McDunnough, for example, is a well-known pest of western fir forests, but a survey with pheromone-baited traps defined the limits of its distribution more clearly. Trap captures showed the moth existed in a number of areas where it had not previously been recorded.[3] The western pine shoot borer, *Eucosma sonomana* Kearfott, is a newly recognized pest of young pines, and until recently was known from only a few locations. In one season of trapping, over 5000 males were captured, and the known range of this species was greatly expanded.[4] Probably the most dramatic increase in knowledge resulting from trapping with pheromone baits involves the Sesiidae or clear-wing moths. A majority of the 112 described species in America north of Mexico have been captured in baited traps.[5] A few years' surveys with pheromone-baited traps greatly expanded the known temporal and geographic distributions of many sesiid species,[6,7] and a further bonus was the discovery of several new species.[6]

Optimal bait formulations for distributional surveys should again be maximum strength, and have long-lasting capacity because seasonal history of a species may vary considerably from place to place.

C. Timing Seasonal Activity

Precise information on when a pest's adult activity cycle begins can be immensely important to pest management specialists. The adult activity cycle is associated with reproduction, and effective control of damage may depend on the timely application of control agents against the new generation. Baited traps can signal when adult flight begins and how long it continues. Such information greatly aids decisions that govern when and how long control agents must be applied. This approach to regulating application of control agents is inherently more efficient than a fixed spray schedule, which requires that control procedures begin on a fixed date and be repeated at fixed intervals. Several studies have shown that monitoring with pheromones permits more accurate timing of spray applications. Quantities of pesticides necessary to control the codling moth, *Laspeyresia pomonella* (L.), for example, have been reduced by this approach.[8-10] An additional advantage was the less severe buildup in mite populations with the lower dosage of pesticide.[9]

The requirements for baits used in programs to time spray applications are basically the same as for detection and distribution surveys. Baits must be long-lasting and consistent in performance. Trap design needs to be considered carefully, because the numbers of insects trapped may be a criterion for timing,[10,11] and trap capacity can depend greatly on trap design. This aspect will be elaborated in the following sections.

D. Population Assessment

Survey objectives discussed up to this point have been concerned only with the presence or absence of a pest. Monitoring with baited traps offers a further dimension, however, because the traps can also provide managers with quantitative information on the numerical trend in the population. This kind of information focuses attention on areas with impending outbreak conditions. Such predictive capability opens up opportunities for preventive management actions that can minimize losses.

Several workers have recognized the value of using sex pheromone traps for quantitative sampling. Madsen and co-workers reported that populations of certain apple pests including the codling moth, *Laspeyresia pomonella* (L.), could be estimated on the basis of numbers of captured males in pheromone-baited traps.[11,12] Riedl and Croft have related trap catches of codling moth to subsequent orchard damage,[13] and they concluded that under some population conditions monitoring with pheromone traps can provide growers with valuable predictions of damage potential later in the season. The Douglas-fir tussock moth is a defoliator pest of conifer forests in western North

America which periodically and very suddenly erupts into the outbreak condition. A monitoring system using pheromone-baited traps is being developed to provide warning against these sudden outbreaks.[14,15] The main thrust of this monitoring system is to direct attention to the locations with potential outbreak-level populations. This can be a major labor-saving device because the forest areas susceptible to tussock moth cover millions of acres and vary in accessibility. Pheromone monitoring allows managers to direct attention to the comparatively few areas with potential problems in any one year (Table 1). Average trap catches of greater than 25 males are indicative of an area potentially phasing from endemic populations into the suboutbreak stage. This suboutbreak level is still 1 year or more short of causing economic damage, so considerable time is available for developing management options.

Several variables can influence the effectiveness of monitoring with traps for predictive purposes.[13,16] The main principle emerging for predictive monitoring is that both the bait formulation and trap design must be standardized to ensure consistent performance from year to year. Only a slight change in trap design can affect the catch,[13] and it is obvious that if bait strength is substantially changed the moth catch will be directly influenced.

Problems associated with population assessment by pheromone traps vary with the pest, although a major consideration for all trapping programs is trap saturation. Most traps used with pheromone baits have a limit to the number of insects they can hold. Once this number is approached, the trap ceases to reflect differences in population densities. For codling moth, Riedl and Croft found their trapping system ceased to indicate higher infestations when the accumulative catch exceeded 100 moths per trap.[13] Thus, for the particular trap design they used, this signaled the point of saturation. Higher infestations were no longer reflected by higher capture numbers, simply because the trap holding capacity was approaching its maximum.

Several approaches can minimize or avoid the effects of trap saturation, and success of the chosen strategy can be profoundly influenced by the bait formulation.

1. Frequent Servicing of Traps

An obvious approach to minimizing trap saturation is to service traps frequently to make cumulative counts and replace the trap or remove captured specimens, thus providing expanded capacity for additional captures. Traps with high-strength baits can fill up quickly, however; sometimes within hours or days. Thus, in large or remote areas, this method of avoiding saturation would not be practical.

2. High Capacity Trap Designs

Another approach is simply to design a trap with greater holding capacity for captured insects. Historically, a drawback to this approach has been that such traps are larger, more complex, and more expensive. Also, the readily available commercial traps traditionally used with pheromones are adhesive coated and, although useful for many monitoring objectives, have obvious limits on capture capacities. Huber and Hoffman have recently demonstrated,[17] however, that a pheromone trap with high-capture capacity can be made. The Huber-Hoffman oil trap is constructed from a 240-mℓ plastic-coated paper cup with a plastic lid. The cup is partially filled with oil, which acts as a capture reservoir. When males are lured into the trap by an attractant, they become entrapped by the oil and sink to the bottom, making room for other insects. For pink bollworm, *Pectinophora gossypiella* (Saunders), the Huber-Hoffman oil trap was shown to have a higher capture capacity than traditional "sticky" traps.[17] This is clearly an example of an inexpensive, practical trapping device that could be used for large-scale monitoring. Further use and development of the Huber-Hoffman oil trap and its modifications are clearly justified for developing monitoring programs for other pest species.

Table 1
SUMMARY OF TUSSOCK MOTH POPULATION MONITORING WITH PHEROMONE-BAITED TRAPS

Season	No. plots	Plots with >20 males/trap	Plots with high larval counts following year
1977	105	7	2
1978	118	9	7
1979	246	5	0

Table 2
HYPOTHETICAL EXAMPLE OF WEIGHTING TRAP CAPTURES RELATIVE TO THE MAXIMUM POSSIBLE CATCH BY EQUATION[1]

No. insects captured	Weighting calculation	Population index
20	$\frac{(200-1)(20)}{(200-20)}$	22
170	$\frac{(200-1)(170)}{(200-170)}$	1,128
199	$\frac{(200-1)(199)}{(200-199)}$	39,601

3. Numerical Weighting of Captures

To overcome the saturation problem common to all sticky traps with a limited adhesive surface, numbers of captured moths can be weighted numerically in relation to remaining surface area available for capture. This approach is based on the assumption that each additional male captured by a trap should count more, because a lesser proportion of responding males will actually be caught as the available trapping surface decreases. Thus, subsequent captures by the same trap are arbitrarily given an exaggerated value. Presumably, the exaggerated value will provide a more meaningful index or scale that reflects the actual population density for the location. One possible equation for developing such a population index (PI) is

$$PI = \frac{(Mc-1)(Ac)}{Mc-Ac} \quad (1)$$

where Mc = maximum number of insects the trap can hold, and Ac = actual number trapped.

Table 2 shows an example of the type of population index that would result from use of Equation 1. The value for monitoring is that this approach gains 200-fold (0 to 40,000 for PI vs. 0 to 200 for actual insects trapped) in the range of trap-expressed differences for reflecting changes in population. The maximum capture capacities for particular traps and pest species would have to be carefully worked out, together with the subsequent correlations to actual population densities or damage potentials for the locality.

4. Weak Baits

Entomologists have traditionally used maximal bait strengths for pest monitoring with a philosophy of "the higher the catch the better".[17,18] For detection surveys, this is valid, but for population assessment there are clearly some advantages to using weaker baits. For one, use of weak baits reduces the possibility of capturing insects attracted over long distances, an effect that might exaggerate estimates of the local population. Secondly, weaker baits can be used to minimize the trap saturation problem. When bait strength is drastically reduced, it follows that the plume of effective attractant emanating from the trap (a sufficient concentration of pheromone to induce insect response) will also be reduced in size.[19] This in turn reduces the probability that insects flying in the vicinity will encounter the smaller plume. It follows, *a priori,* that only the higher density populations will produce a sufficiently high incidence of flying insects to result in substantial captures. This effect has been verified in tussock moth monitoring where zero captures are common in areas where populations are known to be low.[20]

Basically, this approach extends the ability of small-capacity traps to estimate higher density populations at the expense of losing sensitivity to lower densities. It is a useful strategy where low-density populations of the pest are considered harmless. Many forest insects fit into this category, and the tussock moth is a clear example; it is always present in the forest, but becomes a problem only when populations reach high levels. The high levels can be detected by the weak baits (Table 1), whereas no managerial need exists for capturing moths at lower population levels.

III. BAIT FORMULATIONS

A variety of materials has been used in attempts to control the release rate of volatile attractants. Some commercial preparations are now available, plus an array of formulations improvised by various investigators. Fortunately a variety of releaser systems is available because flexibility is clearly necessary for successful pheromone monitoring. Depending on several variables, the desired release rate can vary tremendously. For evaluation surveys of the Douglas-fir tussock moth, required release rates of pheromone are in picograms per hour,[14] but the attractants for the screwworm, *Cochliomyia hominivorax* (Coquerel),[21] and some bark beetles,[22,23] are released at a rate of milligrams per hour. Such differences of a million-fold or more could obviously strain the release capabilities of some controlled release systems. The following section reviews various types of controlled release bait systems that have been used for insect monitoring.

A. Liquid Reservoirs

Where high release rates of an attractant are required, a vial or some other vessel has generally been used to hold a reservoir of attractant. The attractant release from the reservoir is usually regulated by varying the size or number of openings through which vaporization can occur. Bark beetle traps have commonly been baited with variations of this method.[22-24] Tilden and co-workers[25] found they could regulate release rate of bark beetle pheromones by inserting a silk thread to the bottom of the reservoir and varying the extension of this wick above the vial. Burgess and Wiens[26] dispensed a flea beetle attractant by loading a glass vial with the attractant, placing a "pipe-cleaner" wick into this reservoir with the top of the pipe-cleaner fitted into the underside of a rubber stopper. The liquid was conducted upward on the wick, absorbed by the rubber stopper, and released as a vapor from the outer surface of the stopper. With the exception of the commercially available ConRel system, the release rates of liquid reservoir techniques all fall into the milligram per hour range.

Some advantages of the liquid-reservoir system are: it provides a relatively constant rate of release; reservoir-type baits are easily constructed requiring minimal laboratory facilities; and, unless large quantities are needed, it is an inexpensive system. Some limitations include: baiting stations may require frequent servicing, because at least some of the more volatile attractants will rapidly vaporize; the system is limited in versatility in that it is more suited for high-release rates and not always flexible enough to meet lower-rate applications; and, because the attractant is usually in pure form, precautions must be taken to protect it from direct sunlight, and other environmental factors which might cause degradation.

The ConRel Company (Needham Heights, Mass.) produces a line of commercial baits that are specialized versions of the liquid-reservoir system. Their approach is to use plastic capillary tubing of various lengths and diameters. The rate of release is primarily governed by the diameter of the tubing, whereas the active life of the bait also depends on the length of tubing (or fiber) filled with the attractant.

As with other open-reservoir systems, the hollow fibers can be expected to provide relatively flat release curves. The ConRel system also provides lower release rates (micrograms per hour range), which are suitable for many purposes. As with other reservoir systems, chemical stability of some attractants remains a major concern, and precautions must be taken to protect baits from exposure to sunlight and heat.

B. Fibrous Substrates

Fibrous substrates consisting of fabric, paper, and similar materials provide another commonly used releaser system. Examples of such substrates include filter paper, dental rolls, polyester fibers, and cigarette filters. A common characteristic of all such substrates is a short-lived, high rate of release. A number of tactics have been employed to slow the rate of release, including encasement of cigarette filters in impermeable containers to decrease the surface area for release.[27] Investigators have also mixed the respective attractants with various inert substances designed to inhibit vaporization of the attractant and thus prolong the period of bait effectiveness. These materials are generally referred to as extenders or "keepers", and have included such materials as cottonseed oil, wheatgerm oil, polyethylene glycol-600-distearate, and Carbowax compounds.[27-29] With few exceptions,[30] these efforts succeeded in extending attractiveness by only a matter of days. Except where high-release rates over a period of only a few days are required, better systems for release of attractants in monitoring baits are now available.

C. Polyethylene Wicks

Various types of polyethylene containers have been successfully used as dispensers for pheromones. Containers have included small polyethylene bags,[31,32] hollow caps or stoppers,[33] and tubing.[23,33] The pheromone chemicals diffuse through the polyethylene plastic and vaporize off the exterior surfaces of the container. The bait quantity may range from a few micrograms, which are absorbed directly into the polyethylene, to a liquid reservoir of material that diffuses through the container walls over a period of time.

Polyethylene stoppers, in particular, have proved to be effective controlled release trap baits. Such baits have remained attractive for as much as a year.[33] The polyethylene stopper has also shown versatility in successfully dispensing a variety of pheromone compounds, including bark beetle and moth pheromone chemicals.[23,31,33] A further advantage of the polyethylene stopper is its low cost and extreme simplicity.

Limitations of this system are relatively minor. Depending on the attractant to be dispensed, precautions must sometimes be taken with the polyethylene stoppers. Some stoppers are partially constructed of acetate plastic, which is soluble in certain pheromone chemicals, particularly some of the aromatic coleopteran compounds. Also, rate of release is partially regulated by the surface area of the polyethylene containers, and because available container sizes are limited, the range of release rates is also limited. There has been little difficulty with chemical stability in use of polyethylene wicks, although Maitlen et al.[34] reported a degradation of codling moth pheromone in a few days in polyethylene stoppers.

In summary, the polyethylene wick offers an effective bait system for many monitoring trap uses. Such baits are convenient to prepare, long lasting, and effective for numerous research purposes, as well as some operational survey applications.

D. Rubber Wicks

Numerous pheromone chemicals are absorbed into rubber substrates from which they will then slowly vaporize over time.[34,35] Rubber bands and rubber stoppers are the items most commonly used as monitoring bait substrates. Specifically, rubber septa (soft rubber sleeve stoppers) have been used most commonly and with the greatest degree of success. Flint and co-workers[36,37] have found rubber septa excellent releasers for tobacco budworm, *Heliothis virescens* (F.), and pink bollworm pheromones, and Maitlen et al.[34] used rubber septa to develop an improved survey bait for the codling moth.

Rubber septa are long lasting, reliable bait releasers which offer some flexibility in providing high and low release rates. A further advantage of rubber septa is the array of data available on their release characteristics, primarily due to the joint efforts of Butler and McDonough[38,39] who have studied the release of a series of pheromone chemicals from septa. This type of information is important for preparing lures for monitoring and research, but it may be especially important where multicomponent pheromones are involved and one or more of the chemical constituents differs in rate of release from rubber septa. Butler and McDonough's data cover the effects of chemical chain length, double bonding, and isomerism on release rate from septa.[38,39] Their findings are generally consistent with those of Kydonieus,[40] although they also illustrate some interesting effects of polymer cross-linking present in rubber. Acetate molecules with a chain length of up to 15 carbons yielded ever increasing half-lives for release. The half-life of a 16-carbon acetate, hexadecyl acetate, however, suddenly decreased to 478 days, considerably less than the half-life for the 15 carbon acetate. This apparent inconsistency (Table 3) was attributed to the effects of polymeric cross-linking in the rubber.[38] Butler and McDonough point out that cross-linking can be an advantage for prolonging release of some compounds. The reverse may be true, however, where a faster release is required from a chemical that diffuses slowly through the cross-linked matrix.

Users should be aware that not all rubber septa are the same. Some contain a chemical constituent (1,2-dianilinoethane) that can cause adverse reactions with aldehyde pheromones.[41] Perhaps a more important limitation of rubber septa as pheromone releasers is their relatively fixed shape and the related ratio of surface area to volume. Modifying the surface area for release can have major effects on release rate from a given lure volume, and rubber septa are not as readily manipulated in this way as the PVC and laminated-polymer formulations discussed below. These limitations are relatively minor, however, and rubber septa are reliable, long lasting releaser substrates suitable for many monitoring purposes.

E. Polyvinyl Chloride (PVC) Pellets

Solid plastic pellets formulated from a low-temperature-fusing polyvinyl chloride resin, a phthalate plastizer, and the attractant have proved to be effective monitoring baits.[42,43] An even mix of the attractant is achieved by thorough stirring. The plastisol

Table 3
HALF-LIVES FOR EVAPORATION OF PHEROMONES AND RELATED COMPOUNDS FROM RUBBER SEPTA

Chemical	Average half-life in days
Decyl acetate	4.97
Dodecyl acetate	37.30
Tetradecyl acetate	350.00
Pentadecyl acetate	1353.00
Hexadecyl acetate	481.00

From Butler, L. I. and McDonough, L. M., *J. Chem. Ecol.*, 5, 825, 1979. With permission.

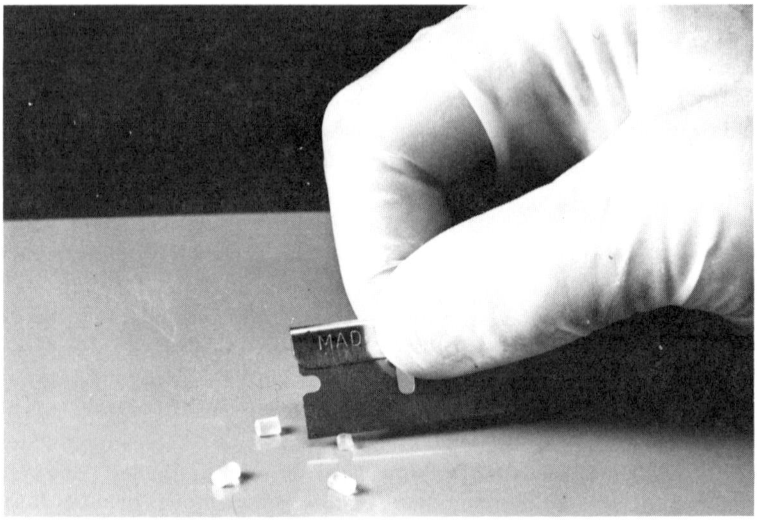

FIGURE 1. Cutting 5 mm × 3 mm polyvinyl chloride (PVC) monitoring baits to size.

mixture is then placed under a vacuum to remove air bubbles, transferred into molds, and fused into solid plastic by a few minutes exposure to the prescribed fusing temperature. Molds for PVC pellets have customarily been hollow glass tubing of different diameters. Sections of the glass tubing can be conveniently scored and broken off, and the exposed strip of PVC cut into pellet-sized baits (Figure 1).

The PVC formulation is a highly versatile controlled release system for dispensing monitoring baits. A wide range of release rates can be obtained by manipulating the concentration of formulated active ingredient, the "stiffness" of the finished polymer, or the shape (ratio of surface area to volume) of the releaser. Thus a very low (picograms per hour) release of tussock moth pheromone could be achieved for evaluation survey trapping[14] and high rates (micrograms to milligrams per hour) for point source releasers in mating disruption trials for the western pine shoot borer.[44,45]

Attractant formulations in PVC have been most widely used as trap baits for monitoring forest insects. These include the Douglas-fir tussock moth,[3,14] European pine shoot moth,[42] western pine shoot borer,[4,44] certain bark beetles,[46] and the spruce budworm,[16,47] *Choristoneura fumiferana* (Clem.). In addition to dispensing attractants,

PVC has also been successfully used for the controlled release of pesticides.[48] Excellent controlled release baits can be formulated from PVC. They are reliable, long lasting, and most important, highly flexible for meeting release prescriptions for different applications of various chemicals. This system would probably see greater use if the simplicity and ease of PVC formulation were more widely known.

F. Laminated Plastic Wicks

The Hercon Products Group of the Herculite Protective Fabrics Corp. in New York produces laminated plastic devices for controlling the release of both pesticides and behavioral chemicals.[40] The formulations for the behavioral compounds can be adapted for either control or monitoring. The release mechanism for the Hercon system has been described in detail by Kydonieus.[40]

The Hercon releasers consist of a three-layered membrane system with the active ingredient contained in the inner layer, which acts as a reservoir. The outer layers are initially inert, but become permeated by the active ingredient as it diffuses to the outer surfaces where it is released. The concentration of active ingredient in the reservoir layer, the membrane thickness, type of polymer, polymer stiffness, and configuration of baits (ratio of surfacearea to volume) are the variables that can be readily modified for regulating release rates.

The outstanding feature of the Hercon system is its versatility in meeting release rate requirements with different types of compounds. By varying one or more of the factors that affect release through polymer membranes, the desired rate can usually be obtained. A wide range of reservoir capacities are possible with the Hercon system; Kydonieus reported effective formulations that contained from 0.5 to 40% active ingredient by weight.[40] Although 40% may be near a maximum formulation capacity, the lower end of the scale undoubtedly could be reduced to a much lower concentration as might be necessary to achieve the lower release rates needed for some monitoring purposes.[14]

Because of the variables available for modification, the laminated plastic wick system is probably the most versatile of the controlled release systems discussed in this chapter. It has been successfully used with a wide variety of active ingredients including pheromones of high and low volatility, plus conventional pesticides. This system could likely be adapted for controlling the release of most pheromones or other insect behavior-modifying chemicals that may be used in monitoring applications.

IV. FACTORS INFLUENCING RELEASE RATE

The different factors governing release rate generally fall under one of two distinct categories: the structural and chemical characteristics of the release substrate and active ingredient; and the environmental conditions associated with field application. The first set of factors can be manipulated somewhat, thus modifying the rate of release. The environmental aspects usually cannot be controlled, but considerations for local environmental conditions at the time of formulation may help optimize effectiveness in the field.

A. Structural and Chemical Characteristics

The effects of polymer formulations on release rate have been described in considerable detail.[38-40] Table 4 summarizes formulation characteristics most readily changed for modifying release rate of a given chemical. All these factors can be changed somewhat for customizing a particular monitoring bait system.

Changing release rates by changing the concentration of pheromone in the formulation is an obvious approach that has been demonstrated for most systems.[40] Figure

Table 4
FORMULATION CHARACTERISTICS OF POLYMER SYSTEMS ASSOCIATED WITH *INCREASING* RELEASE RATES OF VOLATILES

Increasing concentration of attractant
Increasing polymer flexibility (decreasing "stiffness")
Increasing ratio of surface area to volume
Decreasing membrane thickness

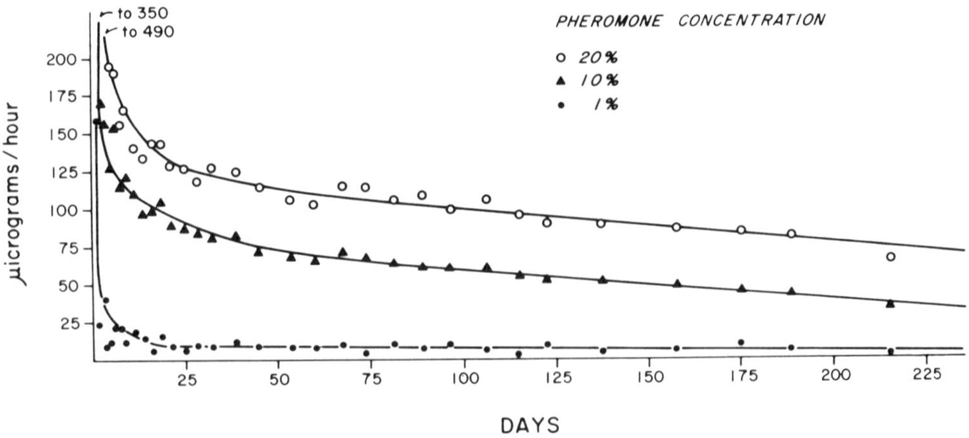

FIGURE 2. Release rates of *Z*-7-dodecenyl acetate formulated in polyvinyl chloride (PVC) pellets at different concentrations of active agent and held at 26°C. (From Fitzgerald, T. D., St. Clair, A. D., Daterman, G. E., and Smith, R. G., *Environ. Entomol.*, 2, 607, 1973. With permission.)

2 illustrates the effect of this approach for modifying release rate for a PVC system. Changing the thickness of membranes through which the active component must diffuse is a means of modification most appropriate for the Hercon laminated plastic system. Figure 3 illustrates this effect for two Hercon preparations.[40] Polymer "stiffness" can be changed in some releaser substrates by the use of co-diffusants[40] or greater quantities of plasticizers to make formulations more flexible which also causes them to be more permeable and release faster.[40,48] Changing the ratio of surface area to volume modifies the proportion of surface area of volatilization. Obviously a given volume of formulation will have less surface area available for release if shaped into a compact cube or ball than if it were flattened into sheets or strung out into thin strands. Lloyd and Matthyse clearly demonstrated this effect when they were able to increase the release rate of a PVC-pesticide formulation 4 to 13 times by changing the pellet diameter.[48]

Cross-linking of polymers, as discussed for rubber septa,[38,39] has a decided effect on release rate, which may be a desirable feature for some materials.

Two major aspects of chemical structure of the active ingredients can be expected to influence release rate. The first item is molecular weight or size. Increased molecular weight of the diffusant generally results in decreasing rates of release. This effect has been appropriately demonstrated for the Hercon system by Kydonieus for pheromone molecules ranging in size from 12 to 20 carbons (Figure 4).[40] This general effect probably holds true for most controlled release systems, although exceptions may exist as evidenced with the rubber septa wicks (Table 3). The second characteristic of the dif-

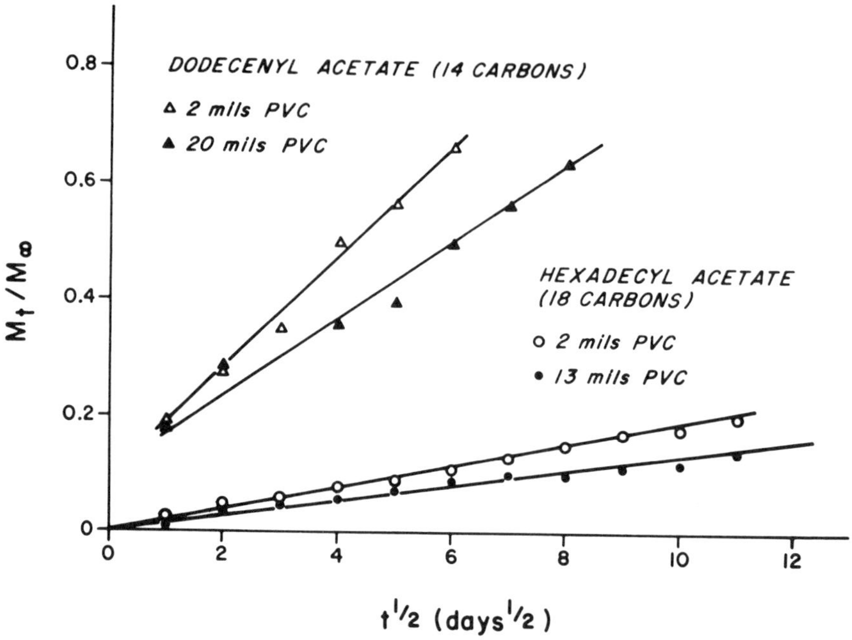

FIGURE 3. Effect of polymer film thickness on release rate. (From Kydonieus, A. F., *Controlled Release Pesticides*, Scher, H. B., Ed., American Chemical Society, Washington, D.C., 1977. With permission.)

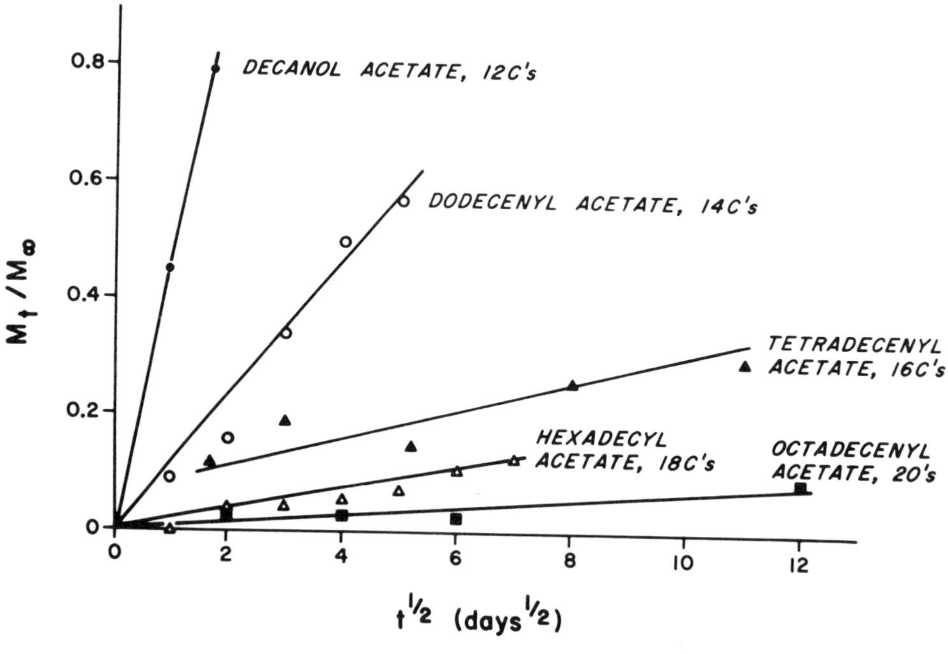

FIGURE 4. Effect of molecular weight of diffusant on release rate. (From Kydonieus, A. F., *Controlled Release Pesticides*, Scher, H. B., Ed., American Chemical Society, Washington, D.C., 1977. With permission.)

fusant molecule that affects release is chemical functionality. Compounds with the same length carbon chain but with different chemical functionalities may differ substantially in rate or release from the same polymer matrix.[39,40] For example, the release rates of 14-carbon acetates, alcohols, aldehydes, and epoxides are all likely to differ from one another.[40]

For particular release systems and applications, other chemical characteristics of the diffusant molecule may also be important. Examples might be the effects of isomerism or the degree of unsaturation in the attractant molecules. These considerations amount to subtleties which must be worked out for individual release systems and field applications.

B. Environmental Effects

The major environmental factors governing release are temperature and wind, with temperature being most important. Other factors may also influence release under certain circumstances. Very high humidities or rain could slow release, for example, and direct sunlight could cause chemical degradation of the active chemical agent — leading to a major reduction in effectiveness.

Temperature is the major environmental factor regulating release rate. Higher temperatures result in higher rates, and the performance of a formulation can definitely be influenced by local conditions. Flint et al.,[37] for instance, found the half-life of a 16-carbon acetate in rubber septa was 159 days at a constant 23°C laboratory temperature, whereas it averaged 120 days in the high temperatures of Arizona cotton fields. Figure 5 illustrates the effects of a range of constant temperatures on an array of pheromone chemicals with varying molecular weights.[49] Temperature obviously has profound effects, with the magnitude of effect particularly important for materials of the lower molecular weights.[49] Such curves have considerable value for designing bait releasers to be used in monitoring programs.

C. The Question of "Constant" Release Rates

Many researchers and management practitioners are concerned with the shape of the release curve over time. Theoretically this is a valid concern since most if not all monitoring lures are first order delivery systems,[40] which result in ever-decreasing increments of release material per unit time. For further elaboration of first order release and description of equations to calculate release at different time intervals, readers should consult the literature.[38,40,50-52] From a practical standpoint, for most monitoring objectives, there is either no cause for concern or lure modifications can be made so that release rate will be nearly constant over the time of application.

As one example, Figure 6 shows the release of European pine shoot moth pheromone from PVC. The curve is steep, but if baits were first "cured" at room temperature for 7 to 10 days before field use, they would emit pheromone at a rate of about 5 to 35 ng/min for at least an additional 40 days. Actually, because of the cool night temperatures associated with this insect's habitat and seasonal flight occurrence, the release curve would likely be extended from that shown in Figure 6 so that release rate per minute would drop only to 10 or 15 ng by day 50. In any event, this range of release is well within the optimal dosage to attract the insect,[42] and because the monitoring objective concerns basic detection, the drop in rate over time would have no adverse effect.

If a decreasing release rate over time continues to be a concern for a particular monitoring system, a strategy that may resolve the problem is simply to alter the ratio of surface area to volume of the lure. With some controlled release systems this may not be possible, but for both the Hercon and PVC systems it is very simple to do. Using a PVC formulation, Lloyd and Matthyse showed an increasing rate of release

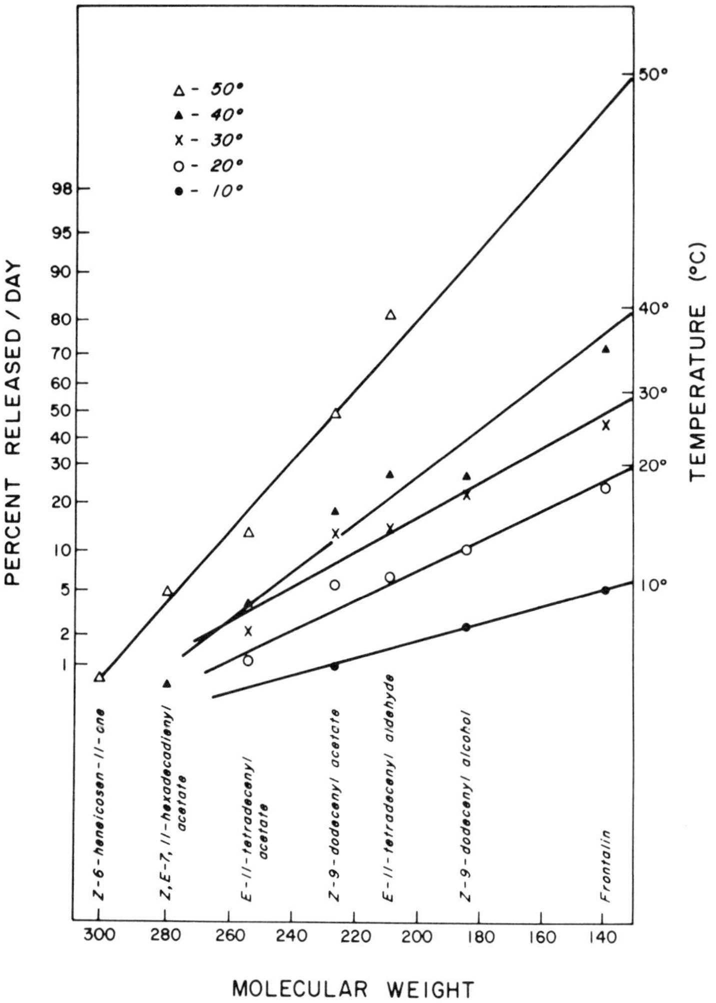

FIGURE 5. Temperature effects on release rates of pheromones with varying molecular weights of active agent. (From Sower, L. and Daterman, G.[51]).

with increased surface area. Thus, to prolong release and flatten the release curve over time, the approach would be to decrease the surface area, relative to the volume, for a given lure. Figure 2 demonstrates this effect for a PVC formulation of a 12-carbon acetate with a surface area:volume ratio of 1:2.4). For comparison, consider that the ratio of the sample illustrated in Figure 6 was nearly reversed (1.7:1). After 270 days the pheromone losses of the bulkier samples (Figure 2) averaged 63%.[43] This contrasts markedly with the 15-day half-life of the bait formulation with higher surface area illustrated in Figures 1 and 6. Most significant for monitoring purposes are the flatter release curves that were obtained when the relative surface area was reduced. For practical purposes, after the initial 20 to 25 days of release, the release rates remained nearly constant for the next 200 days or more (Figure 2). For the comparatively short period of an insect's seasonal flight, the effect would amount to constant rate of pheromone release.

V. SUMMARY AND RECOMMENDATIONS

Two major points are emphasized in this review: first, that variations exist in pest-

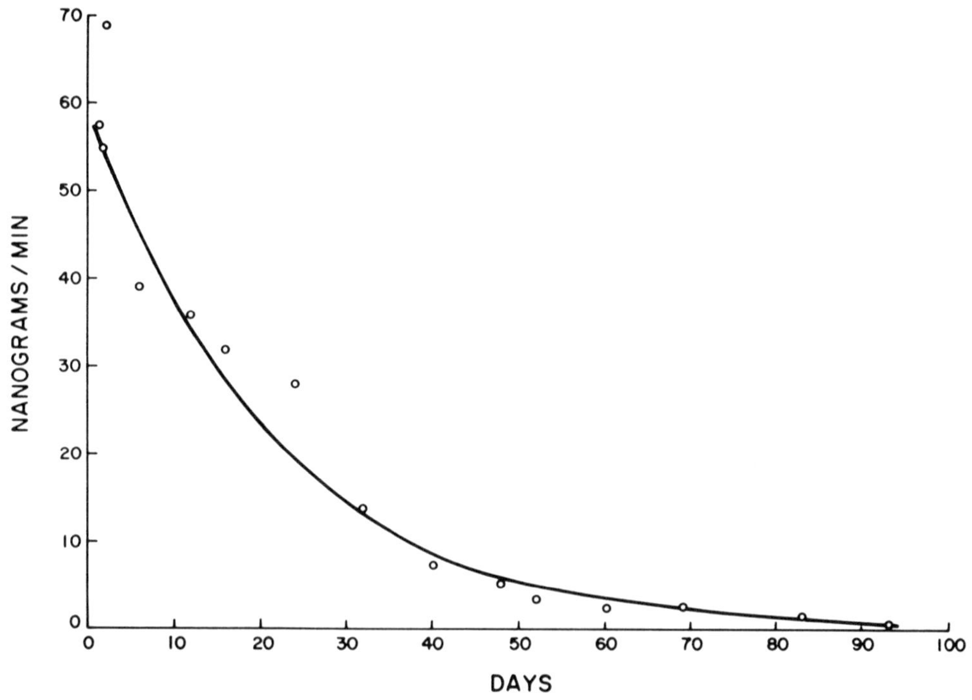

FIGURE 6. Emission of *E*-9-dodecenyl acetate from 5 mm × 3 mm polyvinyl chloride (PVC) pellet containing 5% pheromone at 25.5°C. (From Daterman, G., *U.S.D.A. For. Serv. Res. Pap. PNW*, 180, 1974.

monitoring objectives and associated release requirements of the attractants; and, second, that a variety of controlled release bait systems are now available, several of which can be tailored to a particular release requirement for a given attractant and monitoring objective.

Monitoring programs are concerned with either simple detection of a pest or population assessment. Detection monitoring usually seeks to determine where a particular pest occurs or when it is active. Population assessment is concerned with measuring populations to predict outbreaks or potential damage. For population assessment, trap captures must reflect proportionate population levels. For this purpose, bait formulations must be designed for capturing insects at population levels significant to the monitoring objective. Bait strength can be adjusted, for example, so that only higher level populations, capable of causing damage, will be reflected in trap catches.

Many pest-management specialists have experience with detection trapping. Few are acquainted with monitoring for population assessment. A growing number of studies have shown, however, that the predictive information provided by population assessment monitoring can be very useful to pest management programs. Versatile controlled release baits are available that can be adapted to the specific needs of a monitoring program. Pest managers should carefully review their monitoring needs and be flexible in considering the options in monitoring approaches.

The various controlled release bait systems fall into two categories: those improvised by various investigators and commercial products. The more versatile improvised systems include polyethylene wicks (mainly polyethylene caps), rubber septa, and polyvinyl chloride (PVC) pellets. The commercial systems are mainly the ConRel hollow fiber and the Hercon laminated polymer process. In addition to the ConRel and Hercon systems, the Zoecon Company of Palo Alto, California, also produces monitoring baits commercially, and Bend Research Inc., Bend, Oregon, has custom-designed insect lures, including a system for the Douglas-fir tussock moth.[53] Pheromone researchers and practitioners should encourage commercial production and application of monitoring systems for the general purpose of promoting the use of pheromones on an operational scale. It is true that some improvised bait systems are useful for operational applications, but it is necessary that monitoring materials be made readily available in large quantity and at reasonable costs for managers to conveniently incorporate them into large-scale management programs. Production by commercial sources can probably best meet such needs.

All the dispensing systems discussed in this chapter might work to a degree for a given application. Some dispensing systems, however, are more adaptable than others and where release requirements are highly specific, use of one of the more versatile methods is recommended. The two most versatile dispensing systems, from the standpoint of regulating rate of release of a given chemical, are the Hercon laminated polymer and PVC pellet systems. This is mainly because all three of the major substrate factors affecting release can be readily modified for both systems. The three major factors for substrate modification are: concentration of the active ingredient, stiffness of the polymer, and ratio of surface area to volume of the bait.

A final recommendation for investigators and practitioners is to be flexible in considering monitoring strategies and choice of controlled release systems for lures. With calibrated traps and bait strength, different options in monitoring objectives are possible. The appropriate bait formulation may depend on many variables including nature of the attractant chemical, required rate of release, local environmental conditions, and the monitoring objective. Users must be flexible enough to consider available options in choosing the most appropriate formulation system for meeting their monitoring objective.

REFERENCES

1. U.S. Department of Agriculture, Gypsy moth—California, Animal and Plant Health Inspection Service, *Coop. Plant Pest Rep.*, 1, 491, 1976.
2. **Dilley, D.**, California Dept. of Food and Agriculture, Sacramento, Calif., personal communication, 1978.
3. **Livingston, R. L. and Daterman, G. E.**, Surveying for Douglas-fir tussock moth with pheromone, *Bull. Entomol. Soc. Am.*, 23, 172, 1977.
4. **Sartwell, C., Daterman, G. E., Koerber, T. W., Stevens, R. E., Sower, L. L., and Medley, R. D.**, Distribution and hosts of *Eucosma sonomana* in the Western United States as determined with pheromone-baited traps, *Ann. Entomol. Soc. Am.*, 73, 1980.
5. **Nielsen, D. G.**, Clearwing moth pheromone research: a perspective, in Proc. Symp. Pheromones of the Sesiidae (formerly Aegeriidae), Washington, D.C., ARR-NE-6, U.S. Department of Agriculture, Science and Education Administration, 1979, 75.
6. **Sharp, J. L. and Eichlin, T. D.**, Distribution and seasonal occurrence of Sesiidae (Lepidoptera) attracted to E,Z and Z,Z acetate and alcohol, in Proc. Symp. Pheromones of the Sesiidae (formerly Aegeriidae), U.S. Department of Agriculture, Science and Education Administration, ARR-NE-6, Washington, D.C., 1979, 75.
7. **Duckworth, W. D. and Eichlin, T. D.**, A classification of the Sesiidae of America north of Mexico (Lepidoptera: Sesiidae), *Calif. Dep. Agric. Bur. Entomol. Occas. Pap.*, No. 27, 1977.
8. **Batiste, W. C.**, Integrated control of codling moth on pears in California — effects of varying dosage of Azinphosmethyl and oil with individual cover sprays, *Environ. Entomol.*, 1, 503, 1972.

9. **Batiste, W. C.**, Integrated control of codling moth on pears in California: a practical consideration where moth activity is under surveillance, *Environ. Entomol.*, 1, 213, 1972.
10. **Madsen, H. F. and Vakenti, J. M.**, Codling moth: use of Codlemone®-baited traps and visual detection of entries to determine need of sprays, *Environ. Entomol.*, 2, 677, 1973.
11. **Madsen, H. F. and Vakenti, J. M.**, Codling moths: female-baited and synthetic pheromone traps as population indicators, *Environ. Entomol.*, 1, 554, 1972.
12. **Madsen, H. F. and Peters, F. E.**, Pest management: monitoring populations of *Archips argyrospilus* and *Archips rosanus* (Lepidoptera: Tortricidae) with sex pheromone traps, *Can. Entomol.*, 108, 1281, 1976.
13. **Riedl, H. and Croft, B. A.**, A study of pheromone trap catches in relation to codling moth (Lepidoptera: Olethreutidae) damage, *Can. Entomol.*, 106, 525, 1974.
14. **Daterman, G. E.**, Monitoring and early detection, in *The Douglas-Fir Tussock Moth: A Synthesis*, Brookes, M. H., Stark, R. W., and Campbell, R. W., Eds., *U.S. For. Serv. Tech. Bull.*, 1585, 99, 1978.
15. **Daterman, G. E., Livingston, R. L., Wenz, J. M., and Sower, L. L.**, How to use pheromone traps to determine outbreak potential, *Douglas-fir Tussock Moth Ag. Handb.*, No. 546, U.S. Department of Agriculture, Washington, D.C., 1979.
16. **Sanders, C. J.**, Evaluation of sex attractant traps for monitoring spruce budworm populations (Lepidoptera: Tortricidae), *Can. Entomol.*, 110, 43, 1978.
17. **Huber, R. T. and Hoffman, M. P.**, Development and evaluation of an oil trap for use in pink bollworm pheromone mass trapping and monitoring programs, *J. Econ. Entomol.*, 72, 695, 1979.
18. **Foster, R. N., Staten, R. T., and Miller, E.**, Evaluation of traps for pink bollworm, *J. Econ. Entomol.*, 70, 289, 1977.
19. **Bossert, W. H. and Wilson, E. O.**, The analysis of olfactory communication among animals, *J. Theor. Biol.*, 5, 443, 1963.
20. **Daterman, G. E.**, unpublished data.
21. **Broce, A. G., Davey, R. B., and Snow, J. W.**, Plastic wicks as dispensers of the screwworm attractant, Swormlure-2, *J. Econ. Entomol.*, 72, 115, 1979.
22. **Payne, T. L., Coster, J. E., Richerson, J. V., Edson, L. J., and Hart, E. R.**, Field response of the southern pine beetle to behavioral chemicals, *Environ. Entomol.*, 7, 578, 1978.
23. **Hedden, R. L. and Pitman, G. B.**, Attack density regulation: a new approach to the use of pheromones in Douglas-fir beetle population management, *J. Econ. Entomol.*, 71, 633, 1978.
24. **Rudinsky, J. A., Furniss, M. M., Kline, L. N., and Schmitz, R. F.**, Attraction and repression of *Dendroctonus pseudotsugae* (Coleoptera: Scolytidae) by three synthetic pheromones in traps in Oregon and Idaho, *Can. Entomol.*, 104, 815, 1972.
25. **Tilden, P. E., Bedard, W. D., Wood, D. L., Lindahl, K. Q., and Rauch, P. A.**, Trapping the western pine beetle at and near a source of synthetic attractive pheromone. Effects of trap size and position, *J. Chem. Ecol.*, 5, 519, 1979.
26. **Burgess, L. and Wiens, J. E.**, Dispensing allyl isothiocyanate as an attractant for trapping crucifer-feeding flea beetle, *Can. Entomol.*, 112, 93, 1980.
27. **Hendricks, D. E., Hartstack, A. W., and Shaver, T. N.**, Effect of formulations and dispensers on attractiveness of virelure to the tobacco budworm, *J. Chem. Ecol.*, 3, 497, 1977.
28. **Daterman, G. E. and McComb, D.**, Female sex attractant for survey trapping European pine shoot moth, *J. Econ. Entomol.*, 63, 1406, 1970.
29. **Hedlin, P. A., McKibben, G. H., Mitchell, E. B., and Johnson, W. L.**, Identification and field evaluation of the compounds comprising the sex pheromone of the female boll weevil, *J. Chem. Ecol.*, 5, 617, 1979.
30. **Beroza, M., Bierl, B. A., Tardif, J. G. R., Cook, D. A., and Paszek, E. C.**, Activity and persistence of synthetic and natural sex attractants of the gypsy moth in laboratory and field trials, *J. Econ. Entomol.*, 64, 1499, 1971.
31. **St. Clair, A., Goulding, R. L., and Rudinsky, J. A.**, Controlled-release dispensing system for 3,2-MCH, Antiaggregative pheromone of the Douglas-fir beetle (Col.: Scolyt.), *Z. Ang. Ent.*, 83, 297, 1977.
32. **Toba, H. H., Kishaba, A. N., and Wolf, W. W.**, A polyethylene bag for dispensing the synthetic sex pheromone of the cabbage looper, *J. Econ. Entomol.*, 62, 517, 1969.
33. **Glass, E. H., Roelofs, W. L., Arn, H., and Comeau, A.**, Sex pheromone trapping red-banded leaf roller moths and development of a long-lasting polyethylene wick, *J. Econ. Entomol.*, 63, 370, 1970.
34. **Maitlen, J. C., McDonough, L. M., Moffitt, H. R., and George, D. A.**, Codling moth sex pheromone: baits for mass trapping and population survey, *Environ. Entomol.*, 5, 199, 1976.
35. **Rothchild, G. H. L.**, A comparison of methods of dispensing synthetic sex pheromone for the control of oriental fruit moth, *Cydia molesta* (Busck) (Lepidoptera: Tortricidae), in Australia, *Bull. Entomol. Res.*, 69, 115, 1979.

36. **Flint, H. M., McDonough, L. M., Salter, S. S., and Walters, S.**, Rubber septa: a long lasting substrate for (Z)-11-hexadecenal and (Z)-9-tetradecenal, the primary components of the sex pheromone of the tobacco budworm, *J. Econ. Entomol.*, 72, 798, 1979.
37. **Flint, H. M., Butler, L., McDonough, L. M., Smith, R. L., and Forey, D. E.**, Pink bollworm: response to various emission rates of gossyplure in the field, *Environ. Entomol.*, 7, 57, 1978.
38. **Butler, L. I. and McDonough, L. M.**, Insect sex pheromones, evaporation rates of acetates from natural rubber septa, *J. Chem. Ecol.*, 5, 825, 1979.
39. **Butler, L. I. and McDonough, L. M.**, personal communication.
40. **Kydonieus, A. F.**, The effect of some variables on the controlled release of chemicals from polymeric membranes, in *Controlled Release Pesticides*, ACS Symp. Ser. 23, Scher, H. B., Ed., American Chemical Society, Washington, D.C., 1977, chap. 14.
41. **Steck, W. F., Bailey, B. K., Chisholm, M. D., and Underhill, E. W.**, 1,2-Dianilinoethane, a constituent of some red rubber septa which reacts with aldehyde components of insect sex attractants and pheromones, *Environ. Entomol.*, 8, 732, 1979.
42. **Daterman, G. E.**, Synthetic sex pheromone for detection survey of European pine shoot moth, *U.S. For. Serv. Res. Pap. PNW*, 180, 1974.
43. **Fitzgerald, T. D., St. Clair, A. D., Daterman, G. E., and Smith, R. G.**, Slow release plastic formulation of the cabbage looper pheromone *cis*-7-dodecenyl acetate: release rate and biological activity, *Environ. Entomol.*, 2, 607, 1973.
44. **Daterman, G. E., Sartwell, C., and Sower, L. L.**, Prospects for controlling forest Lepidoptera with controlled-release pheromone formulations, in *Proc. 6th Int. Symp. on Controlled Release of Bioactive Materials*, Baker, R. W., Ed., Academic Press, New York, 1980.
45. **Sartwell, C., Daterman, G. E., Sower, L. L., Overhulser, D. L., and Koerber, T. W.**, Mating disruption with synthetic sex attractants controls damage by *Eucosma sonomana* (Lepidoptera: Tortricidae, Olethreutinae) in *Pinus ponderosa* plantations. I. Manually applied polyvinyl chloride formulation, *Can. Entomol.*, 112, 159, 1980.
46. **Pitman, G. B.**, personal communication, 1978.
47. **Schmidt, J. D. and Seabrook, W. D.**, Mating of caged spruce budworm moths in pheromone environments, *J. Econ. Entomol.*, 72, 509, 1979.
48. **Lloyd, J. E. and Matthysse, J. G.**, Polymer-insecticide systems for use as livestock feed additives, *J. Econ. Entomol.*, 59, 363, 1966.
49. **Sower, L. L. and Daterman, G. E.**, unpublished data, 1980.
50. **Baker, R. W. and Lonsdale, H. K.**, *Controlled Release of Biologically Active Agents*, Tanquary, A. C. and Lacey, R. E., Eds., Plenum Press, New York, 1974.
51. **Crank, J.**, *The Mathematics of Diffusion*, Oxford University Press, London, 1956.
52. **Barrer, R. M.**, *Diffusion in Polymers*, Academic Press, London, 1968.
53. **Daterman, G. E., Peterson, L. J., Robbins, R. G., Sower, L. L., Daves, G. D., Jr., and Smith, R. G.**, Laboratory and field bioassay of the Douglas-fir tussock moth pheromone, (Z)-6-heneicosen-11-one, *Environ. Entomol.*, 5, 1187, 1976.

Chapter 8

LAMINATED STRUCTURE DISPENSERS

Alberto R. Quisumbing and Agis F. Kydonieus

TABLE OF CONTENTS

I. INTRODUCTION

Excessive use of insecticides and sole reliance on these toxicants in pest control have contributed to serious environmental problems. These include resistance to insecticides in an increasing number of insect pest species; harmful effects on nontarget species such as fish and wildlife as well as beneficial insects (e.g., honey bees and other pollinators); increased outbreaks of secondary pests caused by the destruction of insect parasites and predators; and the contamination of air, water, and soil with the persistent residues of insecticides.

These environmental problems and the accompanying potential hazards to human health have spurred scientists to search for and evaluate alternatives to the use of insecticides for insect pest control. In the U.S., the Environmental Protection Agency (EPA), the federal organization that regulates pest control products, is endorsing the use of such alternatives and facilitating their introduction as major or supplemental components of integrated pest management (IPM) programs. "Biorational pesticides", such as insect pheromones, are currently considered IPM tools. Thus, pheromones and other insect behavior-modifying chemicals represent the farmer's newest weapon in his war against insect pests.

The Hercon multilayered, laminated controlled release system has been used to dispense these revolutionary behavior-modifying chemicals. Such use signals a marked departure from sole reliance on conventional insecticides for agricultural pest control.

This chapter will discuss the Hercon patented controlled release system and describe the various pheromone formulations that are now used or are being evaluated for specific insect pest management programs.

A. Hercon — The Organization

The Hercon Division of Health-Chem Corporation has been involved in pheromone research since 1973. Working with U.S. Department of Agriculture and state department of agriculture researchers, university and agricultural chemical industry scientists, the Hercon Division has formulated and evaluated pheromone dispensers targeted for use against over 30 different insect species. A partial list of the insects is shown in Table 1. Some of the pheromone research has been discussed in previous review articles.[1-6]

Formerly a research group within Herculite Products, Inc. (previously Herculite Protective Fabrics Corporation), a subsidiary of the Health-Chem Corporation, the Hercon Division now functions as part of the parent corporation. The Hercon executive and research offices are at corporate headquarters in New York City. Manufacturing, production, and quality control facilities are in York, Pennsylvania.

B. The Hercon System: A General Description

The mathematics affecting transport, and factors affecting the release rates of materials dispensed through Hercon multilayered laminates were discussed in two prior publications.[6,7] Basically, the Hercon dispenser consists of several layers of laminated polymeric material with the active ingredient implanted and protectively sealed in a layer between outer plastic layers. Figure 1 is a schematic illustration of a typical Hercon multilayered system.

The specially formulated inner layer serves as a reservoir for the active ingredient. In addition to behavior-modifying chemicals, other active agents have been implanted in this sealed reservoir, such as insecticides,[6,8-12] antibacterial agents,[13,14] and fragrances and perfumes for air fresheners.[6,9,15] Some of these multilayered dispensers have now been marketed for several years, e.g., Insectape® insecticidal strips, Staph-Chek® antibacterial fabrics, and Scentstrip® air fresheners.

Table 1
LIST OF INSECTS WHOSE BEHAVIOR-MODIFYING CHEMICALS HAVE BEEN INCORPORATED INTO HERCON MULTILAYERED, CONTROLLED RELEASE DISPENSERS (AS OF AUGUST 1980)

ORDER:	Family	Common name	Scientific name[a]	Application	Ref.
COLEOPTERA:	Curculionidae	Boll weevil	*Anthonomus grandis* Boheman	M—MT	30, 31, 32
	Dermestidae	Khapra beetle	*Trogoderma granarium* Everts		
	Scarabaeidae	Japanese beetle	*Popillia japonica* Newman	M—MT	33
	Scolytidae	Ambrosia beetles	*Gnathotrichus sulcatus* (Le Conte) and *G. retusus*		
	Scolytidae	Smaller European elm bark beetle	*Scolytus multistriatus* Marsham	M—MT-TT	34, 35, 36
	Scolytidae	Southern pine beetle	*Dendroctonus frontalis* Zimmermann		
	Scolytidae	Spruce bark beetle	*Ips typographus* (L.)	MT	37, 38
DIPTERA:	Muscidae	House fly	*Musca domestica* L.		10, 39
	Tephritidae	Mediterranean fruit fly	*Ceratitis capitata* (Wiedemann)	M	40
HOMOPTERA:	Pseudococcidae	Comstock mealybug	*Pseudococcus comstocki* (Kuwana)		
HYMENOPTERA:	Vespidae	Western yellowjacket	*Vespula pensylvanica* (Saussure)		
LEPIDOPTERA:	Gelechiidae	Angoumois grain moth	*Sitotraga cerealella* (Walker)	DM	41
	Gelechiidae	Peach twig borer	*Anarsia lineatella* Zeller		
	Gelechiidae	Pink bollworm	*Pectinophora gossypiella* (Saunders)	M—DM	42, 43, 44
	Lymantriidae	Douglas-fir tussock moth	*Orgyia pseudotsugata* (McDunnough)	M	45
	Lymantriidae	Gypsy moth	*Lymantria dispar* (L.)	M—MT—DM	46, 47, 48, 49, 50, 51, 52, 53
	Noctuidae	Black cutworm	*Agrotis ipsilon* (Hufnagel)	M	54
	Noctuidae	Fall armyworm	*Spodoptera frugiperda* (J. E. Smith)	DM	55
	Noctuidae	Egyptian cotton leafworm	*Spodoptera littoralis* (Boisduval)	DM	
	Noctuidae	Cabbage looper	*Trichoplusia ni* (Hübner)	M	56
	Noctuidae	Soybean looper	*Pseudoplusia includens* (Walker)	M	
	Noctuidae	Bollworm/corn earworm/tomato fruitworm	*Heliothis zea* (Boddie)	M	57, 58
	Noctuidae	Tobacco budworm	*Heliothis virescens* (Fabricius)	M	4, 21, 40, 54, 58, 59, 60

Table 1 (continued)
LIST OF INSECTS WHOSE BEHAVIOR-MODIFYING CHEMICALS HAVE BEEN INCORPORATED INTO HERCON MULTILAYERED, CONTROLLED RELEASE DISPENSERS (AS OF AUGUST 1980)

Noctuidae	American (= old world) bollworm	*Heliothis armigera* (Hübner)	M	
Olethreutidae	Codling moth	*Laspereysia pomonella* (L.)	M—DM	
Olethreutidae	Oriental fruit moth	*Grapholitha molesta* (Busck)	DM	61, 62
Olethreutidae	Western pineshoot borer	*Eucosma sonomana* Kearfott	DM	63
Pterophoridae	Artichoke plume moth	*Platypilia carduidactyla* (Riley)	M	
Pyralidae	Asiatic rice borer	*Chilo suppressalis* (Walker)	DM	
Pyralidae	Indian meal moth	*Plodia interpunctella* (Hübner)	DM	
Pyralidae	Navel orangeworm	*Amyelois transitella* (Walker)	DM	64
Sesiidae	Grape root borer	*Vitacea polistiformis* (Harris)	DM	65
Sesiidae	Lesser peachtree borer	*Synanthedon pictipes* (Grote & Robinson)	DM	66, 67, 91
Sesiidae	Peachtree borer	*Synanthedon exitiosa* (Say)	DM	66, 67, 91
Tortricidae	European grape(vine) moth	*Eupoecilia ambiguella* Hübner	DM	68
Tortricidae	Spruce budworm	*Choristoneura fumiferana* (Clemens)	DM	
Tortricidae	Western spruce budworm	*Choristoneura occidentalis* Freeman	DM	

Note: Unless specified, Hercon dispensers of specific pheromones or attractants were formulated for preliminary evaluations. Legend: M represents Monitoring or Survey, MT Mass Trapping, DM Disruption of Mating or confusion technique, and TT Trap Tree system.

[a] Scientific names used are from "Common Names of Insects and Related Organisms" (1978 Revision).[29]

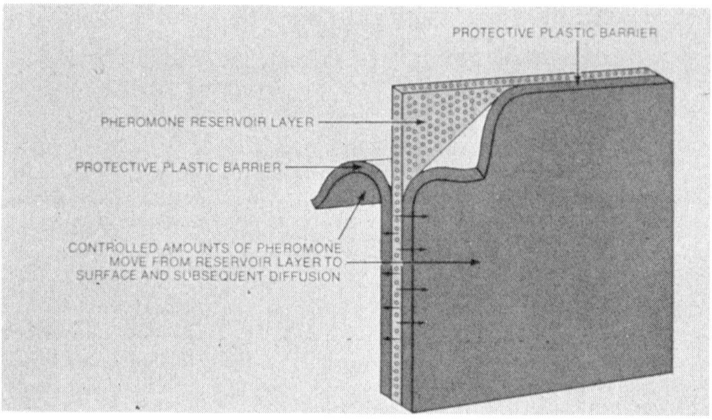

FIGURE 1. The Hercon pheromone dispenser.

The Hercon commercial insect pheromone dispensers have been used since the early 1970s in various IPM programs. Table 2 lists Hercon formulations which have received EPA registrations. Hercon pheromone products are used in monitoring or survey (to determine presence or level of infestation of an insect species), mass trapping (the dispensers serve as baits in traps set out for pest population suppression), and disruption of mating (as additional point sources of pheromone, the dispensers confuse mating communication between insects, thereby reducing the number of successful matings). Pheromone products for monitoring and survey do not require EPA registrations.

The multilayered dispensers come in various forms (Figure 2). The hand-applied formulations are called Luretape® (e.g., strips and "squares") while the equipment-applied form (e.g., "flakes" or confetti) is called Disrupt^{TM}. Disrupt is applied with specially built aerial or ground equipment. The flakes are mixed with an adhesive so they adhere to plant surfaces on contact.

Essentially, the laminated active agent in the dispenser migrates continually, due to an imbalance of chemical potential, from the reservoir layer through one or more outer layers to the exposed surface and it thereby becomes available for biological action. At the surface, the active chemical is removed by volatilization, ultraviolet light degradation, alkaline or acid hydrolysis, or mechanical contact by humans, insects, rainfall, wind, or other agents.

The construction of the dispenser and its composition may be varied depending on the properties of the active ingredient and the desired release rate and effective life span. One or both outer surfaces may be made active by using one or two permeable outer layers.

1. Factors Affecting Release Rates

The permeation of the behavior-modifying chemical through the barrier membranes is governed primarily by Henry's Law and Fick's First Law.[16,17] The concentration and composition of the stored chemical (diffusant) and/or construction of the barrier membranes control the release rate of the dispensed chemical. Equations for diffusion of chemicals through laminated structures under both "steady" and "unsteady state" conditions have been described by Kydonieus and Quisumbing,[6] and Kydonieus.[7] Reservoir concentration, membrane thickness, polymer stiffness, codiffusants, molecular weight of diffusant, and chemical functionality were factors affecting the transport of pheromones and other active ingredients through Hercon membranes.

Table 2
LIST OF HERCON PHEROMONE FORMULATIONS/PRODUCTS AND STATUS OF EPA REGISTRATION (DECEMBER 1980)

Product name	EPA Reg. No. or file symbol	Product use
Luretape with Grandlure	8730-15, on August 20, 1979	Mass trapping
Luretape with Disparlure	8730-17, on August 28, 1979	Mating disruption, mass trapping
Luretape with Japonilure	8730-18, on April 8, 1980	Mass trapping
Disrupt/Gypsy Moth	8730-19, on May 2, 1980	Mating disruption
Disrupt/Western Pineshoot Borer	8730-20, on November 14, 1980	Mating disruption
Disrupt/Pink Bollworm	8730-ER (registration pending)	Mating disruption
Luretape with Multilure	8730-EG (registration pending)	Mass trapping
Disrupt/H.V. flakes[a]	8730-EUP-5	Mating disruption of *H. virescens*
Disrupt/Virelure flakes[b]	8730-EUP-8	Mating disruption of *H. virescens*
Disrupt/Peachtree Borer	8730-EUP-10	Mating disruption

[a] Active ingredient of Disrupt/H.V. flakes is *n*-tetradecyl formate.
[b] Active ingredient of Disrupt/Virelure flakes is two-component virelure.

FIGURE 2. Types of Hercon dispensers.

a. Reservoir Concentration

The release rate of a pheromone from the Hercon dispenser depends on its concentration in the reservoir layer. Within a given time, the mass of pheromone released from a dispenser with a higher concentration of the chemical will be greater than a similarly constructed dispenser with a lower pheromone concentration. The effect of reservoir concentration on the release rates of the insecticide chlordane, and the insect repellent deet, were illustrated in previous publications.[6,7]

In Chapter 10 of this publication, Bierl-Leonhardt[18] further illustrates the influence of pheromone concentration on the emission rate. She reported that a Hercon dispenser with a virelure concentration of 17 $\mu g/mm^2$, had a half-life, i.e., time to halve the pheromone content, of over 60 days. (Virelure is a two-component mixture of (*Z*)-11-hexadecenal and (*Z*)-9-tetradecenal,[19-21] and is present in the seven-component *Heliothis virescens*, tobacco budworm, sex pheromone.[22,23] A similarly constructed dispenser with 25 $\mu g/mm^2$ virelure concentration, on the other hand, had a half-life of 25 days.)

b. Membrane Thickness

The mass of the pheromone released is inversely proportional to the thickness of the outer membranes.[6,7] Within a given time, relatively faster release rates of dodecenyl acetate and hexadecyl acetate were observed in the thinner dispensers. Bierl-Leonhardt[18] also showed that dispensers with thinner outer films have shorter half-lives than those with thicker ones: a 9-mil thick dispenser of gossyplure had a longer half-life (45 days) than another with the same pheromone concentration but with a 4-mil polyvinyl chloride (PVC) outer film. The latter had a 22-day half-life. (Gossyplure is a 1:1 mixture of (*Z,Z*)-7,11-hexadecadienyl acetate and (*Z,E*)-7,11-hexadecadienyl acetate, and is the sex pheromone of the pink bollworm, *Pectinophora gossypiella*.[24,25])

c. Polymer Stiffness

The amount of active ingredient transported into and through the membrane becomes progressively smaller as the polymer stiffness increases. In previous studies[6,7] with an antibacterial agent, an insecticide, and a fungicide, transport or release was fastest through the flexible PVC barrier film, followed by rigid PVC, acrylic, polypropylene, nylon, and polyester. Polymer backbone stiffness is an important consideration when controlling the release rate of certain pheromone components, e.g., aldehydes, which are normally volatile and unstable.

Table 3 lists polymers of varying thickness and stiffness which may be used as barrier films in the Hercon multilayered, laminated dispensing system.

d. Codiffusants

These are chemical agents capable of altering the structure of barrier films by swelling, softening and/or dissolving the polymer matrix. Since some codiffusants, e.g., phenol and ethylene glycol phenyl ether, can soften polymer stiffness, their presence can increase emission rates, thereby shortening the half-life of a dispenser.

e. Molecular Weight

This factor is also directly related to the diffusivity. Relative release rates of five pheromone acetates ranging in chain length from 12- to 20-carbon atoms have been discussed previously.[6,7] In this study, the 12-carbon decanol acetate diffused through barrier membranes the fastest, followed by dodecenyl acetate (14-C), tetradecenyl acetate (16-C), hexadecyl acetate (18-C), and octadecenyl acetate (20-C) in this order. Thus, release rates decreased as molecular weight increased.

f. Chemical Functionality

The rule that "like dissolves like" also holds true with polymers. Ability of the diffusing molecules to dissolve in the polymer matrix is important in the transport process because it increases the magnitude of the distribution coefficient. In the earlier articles on multilayered, laminated controlled release dispensers,[6,7] a study comparing the emission rates of two pheromones with 18 carbon atoms (octadecane and hexadecyl acetate), and three 16-carbon pheromones (epoxyhexadecane, hexadecanone, and hexadecenal) showed that functional groups substantially affect the release rate.

Table 3
LIST OF PLASTIC FILMS WHICH MAY BE USED AS BARRIER MEMBRANES IN HERCON MULTILAYERED LAMINATES

Acrylic	Polyethylene
Cellophane	Polypropylene
Mylar	Rigid polyvinyl chloride
Nylon	Flexible polyvinyl chloride
Polycarbonate	Polymeric polyvinyl chloride
Polyester	Urethane

II. LAMINATED PHEROMONE DISPENSERS

In addition to insecticides and perfumes or fragrances, behavior-modifying chemicals such as aggregation and sex pheromones, food attractants, and pheromone analogs ("mimics") have been incorporated and released by the Hercon laminated structures (Table 1).

In the remaining part of this chapter, we will review various pheromone studies in which Hercon dispensers were used against important insect pests. In several instances, a more detailed discussion of these research efforts is given elsewhere in this volume; in these cases, the particular chapter will be identified.

In 1975, Inscoe and Beroza[26] and Mayer and McLaughlin,[27] having searched the literature, listed the behavior-modifying chemicals according to their general structure. An updated list is presented by Inscoe[28] in Volume II, Chapter 15 of this book. Table 4 classifies the target species of a sampling of Hercon formulations, according to chemical type (i.e., aldehyde, alcohol, ester, or ketone) of the insect's behavior-modifying chemical.

A. Aldehydes

The pheromones or pheromonal components of some of the most important insect pests in the world, e.g., tobacco budworm, corn earworm, old world bollworm, boll weevil, Asiatic rice borer, and spruce budworm, have been reported to be aldehydes. And while conventional laboratory-made dispensers have been used in research trials, the Hercon multilayered laminate is the only commercially available dispenser of aldehydes that has been used effectively, to date, in large-scale programs. The construction of the Hercon controlled release dispenser is unique in its ability to protect aldehydes effectively. Field trials with the boll weevil, spruce budworm, and *Heliothis* pheromones have shown that the polymeric layers of the dispenser effectively protect the unstable aldehyde active ingredients from oxidation, ultraviolet light degradation, and weather.

1. Grandlure, Boll Weevil Pheromone

Grandlure is the sex and aggregation pheromone[69] of the boll weevil, the insect considered by many to be the worst pest of cotton. Produced by the male weevil, the pheromone attracts both male and female weevils. Grandlure is a four-component mixture of two aldehydes, (*Z*)-3,3-dimethyl-Δ^1,α-cyclohexaneacetaldehyde (Compound III) and (*E*)-3,3-dimethyl-Δ^1,α-cyclohexaneacetaldehyde (Compound IV); and two alcohols, (R,*Z*)-2-isopropenyl-1-methylcyclobutaneethanol (Compound I) and (*Z*)-3,3-dimethyl-Δ^1,β-cyclohexaneethanol (Compound II). The most attractive combination of the four components is a 40:30:15:15 ratio of Compounds I:II:III:IV.[70]

The Hercon formulation of grandlure is used in mass trapping, monitoring, and

Table 4
LISTING OF INSECT SPECIES CLASSIFIED ACCORDING TO THE CHEMICAL TYPE OF COMPOUNDS EVALUATED AS BEHAVIOR-MODIFYING CHEMICALS AND FORMULATED IN THE HERCON CONTROLLED RELEASE SYSTEM

Chemical type	Insect species
Aldehydes	*Amyelois transitella*
	Anthonomus grandis (2 of 4 components)
	Chilo suppressalis
	Choristoneura fumiferana
	Choristoneura occidentalis
	Heliothis armigera
	Heliothis virescens (6 of 7)
	Heliothis zea
	Platypilia carduidactyla
Alcohols	*Anarsia lineatella*
	Anthonomus grandis (2 of 4)
	Gnathotrichus sulcatus
	Heliothis virescens (1 of 7)
	Laspereysia pomonella
Esters	
Acetates	*Agrotis ipsilon*
	Anarsia lineatella
	Eucosma sonomana
	Eupoecilia ambiguella
	Grapholitha molesta
	Pectinophora gossypiella
	Plodia interpunctella
	Pseudococcus comstocki
	Spodoptera frugiperda
	Spodoptera littoralis
	Synanthedon exitiosa
	Synanthedon pictipes
	Vitacea polistiformis
Formates	*Heliothis virescens*
Carboxylate	*Ceratitis capitata*
Ketones	*Orgyia pseudotsugata*
	Popillia japonica
Miscellaneous	*Dendroctonus frontalis*
	Gnathotrichus sulcatus
	Ips typographus
	Lymantria dispar
	Scolytus multistriatus

survey programs. In 1978, the Hercon Division received EPA registration for its Hercon® Luretape with Grandlure; this formulation contained approximately 25 mg grandlure active ingredient per square inch, and was designed to last 4 weeks in the field. Both top and bottom barrier films utilize a 16-mil polymeric PVC material.

This construction was chosen after a series of field tests by U.S. Department of Agriculture researchers in Mississippi.[30-32] In 1975, Hardee, McKibben, and Huddleston[30] found the multilayered laminates equal in effectiveness to the standard cigarette filter-in-vial formulations but longer lasting than the standard's 14 days. In another series of tests conducted during the winter of 1974-1975 throughout the cotton belt in the United States and Mexico, Johnson et al.[31] reported that Luretape, applied once, was as effective or more effective for eight weeks than the standard filter-in-vial which was changed weekly. Other formulations, such as impregnated cigarette filters in plastic tubes or gel inside a plastic straw, hollow fiber dispensers, and grandlure-containing vials with a flexible foam barrier, were not as effective as the standard or the Hercon product.

The commercial Luretape formulation is recommended for use in traps set out from time of germination of cotton through harvest. Up to 10 Luretape-baited traps per acre is suggested, with a minimum of two traps in low infestations. Baited boll weevil traps are placed along the perimeter of cotton fields, in sites of previous severe boll weevil damage, and in overwintering areas favored by the boll weevil.

In Volume II, Chapter 5 of this book, Hardee discusses in detail a boll weevil mass trapping program, and the potential of trap cropping using grandlure dispensers to lure the weevils into a specific area which will be treated with conventional insecticides.

2. Tobacco Budworm Sex Pheromone

Hercon formulations of the two-component virelure, previously reported to be the sex pheromone of *Heliothis virescens*,[19,20] and the seven-component lure, later reported to be the complete pheromone,[22,23] were bioassayed in field tests and evaluated in mating disruption studies. A four-component mixture consisting of the major ingredients of the pheromone was also evaluated in separate studies.[93] In all cases, plastic films less permeable than PVC were chosen as protective barriers because of the high volatility and unstable nature of the aldehyde components.

Virelure consists of (*Z*)-11-hexadecenal and (*Z*)-9-tetradecenal, as reported in 1974[19] and 1975.[20] Five years later, Klun et al.[22,23] announced that the *H. virescens* pheromone consisted of the following: (*Z*)-7-hexadecenal (2%), (*Z*)-9-hexadecenal (3%), (*Z*)-11-hexadecenal (75%), hexadecanal (7%), (*Z*)-9-tetradecenal (5%), tetradecanal (2%), and (*Z*)-11-hexadecen-1-ol (6%).

a. Virelure Studies

Hendricks et al.[56] reported that a 16:1 ratio of (*Z*)-11-hexadecenal and (*Z*)-9-tetradecenal caught significantly more male moths than ratios of 24:1, 32:1, or 40:1. The Hercon multilayered laminate was tested against a filter-in-glass vial formulation and it proved to be superior by being attractive for 21 days, with attractiveness not regressing significantly for the first 7 days. The cigarette filter-in-vial, on the other hand, readily released most of its pheromone content during the first two nights, and fewer male moths were trapped after the third night. The Hercon formulation contained approximately 10 mg virelure per square inch, with the 6-mil polymeric outer layers helping prevent the rapid oxidation of the virelure aldehydes.

The virelure formulation bioassayed as bait in tobacco budworm cone traps was also used in a preliminary 9.3-acre mating disruption study conducted in 1979 in Mississippi.[94] The 10 mg/in.^{2} Luretape units were manually applied 10 ft apart on cotton plants every 4th row across a cotton field. The number of tobacco budworm eggs, larvae, and percent damage were determined for the virelure-treated field as well as for 6 10-acre fields within a 3-mile radius that were conventionally treated, i.e., sprayed with insecticides when needed. The encouraging results obtained suggested that disruption of mating resulted from the Luretape treatment. While pretreatment observations showed no significant infestation or damage differences among the fields, 2 months later the fields that did not receive any virelure treatment had to be sprayed at least 5 times for *Heliothis* suppression. The virelure-treated field had 1.6% damage; the six conventionally treated fields had an average of 5.5% damage.

These promising results stimulated further testing with both manually and aerially applied virelure formulations in Mississippi and Arizona in 1980. The preliminary research performed in Arizona is discussed by Henneberry et al.[42] in Volume II, Chapter 6 of this book.

b. Seven-Component Pheromone Studies

Hartstack and Lopez in College Station, Texas, and their colleagues conducted most of the field evaluations of Hercon dispensers containing the seven components of the *H. virescens* pheromone.[57,58] The purpose of the tests was to develop a long lasting *H. virescens* bait for use in conjunction with a specially designed cone trap, the Texas pheromone (TP) trap.[71]

Hartstack et al.[58] reported that there was no significant difference in catch of tobacco budworm males in TP traps baited with five virgin females, Luretape with virelure, Luretape with the seven-component mixture, and virelure in rubber septa. Hartstack et al.[58] also stated that there was no significant difference in trap catches of male *H. virescens* moths when dosages of the seven-component mixture were reduced from 20 mg to 5 mg per Luretape unit bait. When dosage was reduced to 2.5 mg, however, significantly fewer males were caught than at a dosage of 10 mg or greater.

3. Corn Earworm Sex Pheromone

The sex pheromone of the corn earworm, *Heliothis zea,* is composed of four aldehydes, all of which are components of the *H. virescens* attractant. Klun and his colleagues[22,23] reported that the four components are (*Z*)-7-hexadecenal (2%), (*Z*)-9-hexadecenal (3%), (*Z*)-11-hexadecenal (87%), and hexadecanal (8%).

In 1980, Hartstack et al.[58] and Lopez et al.[57] evaluated several *H. zea* pheromone dispensers, including Hercon Luretape formulations, for use in monitoring and research activities. Trapping studies were conducted in corn fields near Tampico, Mexico with the TP cone trap at temperatures from 18° to 27°C.

The Lopez study[57] indicated that a Hercon 34-mil-thick dispenser containing 2.5 mg/in.^{2} was most attractive; aging of this product for 11 days in the greenhouse did not significantly alter its efficacy based on trap catch. This dispenser was fabricated with 16-mil polymeric films as outer layers.

Rubber septa containing 1.25, 2.5, and 5 mg pheromone per unit bait were also effective both as unaged and aged dispensers, but their efficacy, in terms of trap catches, was lower than that of the Hercon formulations. Traps baited with a cotton dental roll each containing 133.6 μg of the four-component *H. zea* pheromone (in 10 $\mu\ell$ heptane) caught the most moths but had to be replaced nightly to be effective.

4. Spruce Budworm Pheromone

The pheromone for both the spruce budworm and western spruce budworm, *Choristoneura fumiferana* and *C. occidentalis,* have been reported to be (*E*)-11-tetradecenal.[72] A discussion of spruce budworm mating disruption studies is given in Volume II, Chapter 13 by Sanders and Seabrook.[87]

In 1979, Daterman[95] aerially applied a Hercon formulation of the western spruce budworm pheromone and succeeded in disrupting mating communication based on a reduction of spruce budworm moth trap catches in the pheromone-treated areas. However, damage reduction was not obtained due to excessively heavy adult infestations in the test sites; in other words, the successful disruption of mating communication was not reflected in the incidence of actual mating. Additional studies are planned in areas with isolated and moderately low infestations of spruce budworm.

The Hercon formulation containing spruce budworm pheromone was designed for long-lasting effectiveness, i.e., for 6 to 8 weeks. Analyses of dispensers by Hercon researchers showed that the product emitted the pheromone throughout the entire spruce budworm moth flight. This was made possible by allowing the active ingredient to be emitted only through one outer film layer.

5. Navel Orangeworm Pheromone

In 1979, disruption of mating studies were conducted in an attempt to minimize navel orangeworm damage to almonds.[64] The active ingredient used was (*Z,Z*)-11,13-hexadecadienal, an aldehyde reported to be a component of the sex pheromone of adult *Amyelois transitella* moths.[73] Hercon dispensers of this behavior-modifying chemical were manually applied to trees in a preliminary study reported by Landolt et al.[64] They reported a 99% reduction of trap catch for 39 days. These results indicated that disruption of mating communication by permeating the orchard atmosphere with the pheromone component had occurred. The traps were baited with live female navel orangeworm moths as the complete sex pheromone has not yet been identified.

The navel orangeworm aldehyde component is protected in the Hercon laminate by 8-mil thick plastic outer layers. In 1980, the Hercon formulation was tested both as a flake and as hand-applied dispensers, to determine their potential in reducing adult moth mating and egg laying in almond orchards. Early reports from this study suggest damage reduction and increased crop yields in areas treated with the mating disruptant (personal communication).

B. Alcohols

The peach twig borer, ambrosia beetles, codling moth, boll weevil, and tobacco budworm all have pheromones or pheromonal components which are alcohols. Hercon pheromone dispensers have been prepared for these insects in the past for use in monitoring and research programs.

The dispensers of the boll weevil and tobacco budworm pheromones have already been described under Section A. Aldehydes; the alcohols are only part of the multicomponent pheromone systems.

The alcohol component of the peach twig borer, *Anarsia lineatella,* is (*E*)-5-decenol, and it is found together with (*E*)-5-decenol acetate in female moths.[74] The Hercon dispenser used in research studies and preliminary disruption of mating work was fabricated with PVC films as outer layers.

Ambrosia beetles, *Gnathotrichus sulcatus* and *G. retusus,* are important pests in the western U.S. and Canada where they bore into the sapwood of many coniferous tree species, particularly western hemlock and Douglas-fir. The tunnels or "pinholes" result in lower lumber and plywood quality.

Hercon laminates of sulcatol and lineatin, behavior-modifying chemicals of the ambrosia beetles, were prepared in 1979 for evaluations in Canada.[96] Sulcatol is the common name of the attractant, 6-methyl-5-hepten-2-ol[88] while lineatin is 3,3,7-trimethyl-2,9-dioxotricyclo [3.3.$10^{4.7}$] nonane. Sample dispensers contained 180 mg sulcatol and 5 mg lineatin per unit.

The pheromone of *Laspeyresia pomonella* is (*E,E*)-8,10-dodecadien-1-ol.[75] The multilayered laminates containing this alcohol have been used in preliminary disruption of mating studies in California and Washington in the U.S., and in Australia. Two 8-mil plastic films are used to sandwich the codling moth attractant.

Rothschild[89] describes the codling moth disruption of mating studies in Volume II, Chapter 9 of this book. He conducted his work in Australia.

C. Esters

Several ester-type pheromones have been formulated in Hercon multilayered systems and tested as baits in survey programs or as point pheromone sources in disruption of mating studies. Most of the chemicals have been acetates, e.g., those that attract the pink bollworm, western pineshoot borer, European grape moth, black cutworm, peachtree borer, and Angoumois grain moth.

1. Gossyplure, Pink Bollworm Pheromone

The sex pheromone of the pink bollworm, a 1:1 mixture of (*Z,Z*)- and (*Z,E*)-7,11-hexadecadienyl acetate, is called gossyplure. The insect is a major pest of cotton in the southwestern U.S. The larvae attack the squares (buds), flowers, or bolls of the cotton plants. There are several generations in a growing season.

In addition to Luretape with gossyplure, which is used as bait in delta-type traps set out in a field to determine the level of infestation and/or degree of effectiveness of a mating disruption program, Hercon multilayered flakes are available for aerial application over large areas. The latter, commercially named Disrupt/Pink Bollworm, consists of 1/8- by 1/8-in. units of 8-mil thick polymeric films on both sides of a reservoir containing 2.8% active ingredient by weight.

The label for this product calls for the application of 53.6 to 71.4 g of flakes per acre which is equivalent to 1.5 to 2.0 g Gossyplure per acre. An application is effective for 21 days, depending on local conditions.

The use of gossyplure-containing flakes in large scale mating disruption programs is discussed in detail by Henneberry et al.[42] in Volume II, Chapter 6 of this book. The authors report on the successful reduction of boll damage by pink bollworm larvae through the disruption of mating of the adult moths. They found that damage in pheromone-treated fields was less than the damage in fields treated with conventional insecticides.

2. Western Pineshoot Borer Pheromone

In 1980, the Hercon Division obtained from the Environmental Protection Agency product registration for its controlled release formulation of a 4:1 *Z:E* mixture of 9-dodecenyl acetate, the pheromone of the western pineshoot borer. The product, Disrupt/Western Pineshoot Borer, is designed to lower the incidence of the adult *Eucosma sonomana* mating by disrupting the normal male orientation to females. This treatment reduces damage caused by the larvae to ponderosa, lodgepole, and Jeffrey pine plantations in the western U.S. Daterman[63] discusses the use of this behavior-modifying chemical in Volume II, Chapter 11.

Like the previously discussed Disrupt/Pink Bollworm product, the western pineshoot borer disruptant is aerially applied before emergence of the adult moths, i.e., early in the spring or immediately after spring thaw or snow melt. The EPA-accepted label for this product recommends the application of 4 to 8 g active ingredient per acre, depending on the level of infestation. This application rate is equivalent to 47.6 to 95.2 g of flakes per acre. The Hercon formulation provides optimum release of the pheromone throughout the moth flight.

3. European Grape Moth Pheromone

Eupoecilia (Clysia) ambiguella is an important tortricid moth that attacks grapes in Europe. Field experiments to control this insect by disrupting their mating communication have been discussed by Arn et al.[76] Since the European grape moth and *Lobesia botrana* Schiff are the only insect pests of European vineyards, pheromones are desired to eliminate the use of insecticides to control these pests. An effective mating-suppression program can minimize, if not eliminate, the problem of insecticide residues in grapes and its by-products, as well as prevent the emergence of secondary pests by allowing beneficial insects and predators to survive.[76]

The *E. ambiguella* sex pheromone, (*Z*)-9-dodecenyl acetate,[77] has been used to disrupt mating. A Hercon formulation was prepared using a 1:5 mixture of the sex pheromone and dodecyl acetate, the latter being a synergist for both trapping[78] and disruption of mating.[79] The dispensers, each 2- × 5-cm in size, were prepared by laminating the reservoir layer containing approximately 13.2 mg active ingredient per unit between two permeable barriers. The dispensers were placed alternately at 1- and 1.5-m heights on fences surrounding the test vineyards which were located in a wine-growing area near Lake Geneva, Switzerland.

Disruption of mating was measured by trap catch suppression, oviposition damage to grapes, and by examining the mating status of moths, both male and female, lured to traps baited with a food attractant. Arn et al.[68,76] reported that most of the trapped moths were found in the untreated (no pheromone treatment) area; percent disruptions for the 3-month long season were 99.2% for the first flight and 98.2% for the second. Damage due to oviposition was also generally greater in grapes from the untreated area. Another indication of successful mating suppression was the observation that, based on spermatophore counts on female moths caught in food traps, only 33% of the females in the treated area were mated; counts were 51% in the untreated area. Also, most of the female moths with multiple matings were collected from the untreated area.

4. Black Cutworm Pheromone

The larvae of *Agrotis ipsilon* is a serious pest of a wide variety of crops, e.g., corn, beans, tomatoes, cabbage, and other plants that do not have hard, woody stems. The cutworms cause damage by feeding and cutting off many plants just above, at, or a short distance below, the soil surface. The degree of infestation varies from year to year; frequently, the cutworm can cause so much damage to corn that replanting becomes necessary. Bait stations, burlap traps, and bait traps are currently used to detect larval populations before they reach economic injury levels. Following the identification of the black cutworm sex attractant pheromone components, Hercon prepared dispensers containing the pheromone for testing as baits in adult moth traps. The development of an effective bait and a trap for adult black cutworm moths can significantly improve monitoring programs for this pest.

The two components, (*Z*)-7-dodecenyl acetate and (*Z*)-9-tetradecenyl acetate (3:1 ratio),[80] were evaluated in the field in Wooster, Ohio.[54] The Hercon controlled release system was tried after the researchers failed to obtain positive results with a commercial hollow fiber (an Albany International Conrel formulation) product, even though the two components were diluted with 2-methyloctadecane to slow the pheromone release rate.

In a series of tests,[54] the mean number of male moths caught in traps baited with the Hercon dispensers did not differ significantly from catches of traps baited with the hand-prepared rubber septa. The hollow fibers, on the other hand, did not appear to be as attractive as the standard rubber septa. The availability of the commercially produced laminated dispensers makes possible the establishment of large-scale monitoring programs for predicting black cutworm infestations.

The best Luretape bait, prepared with PVC/polyester films as outer layers, contained 0.39 mg pheromone per square inch. No volatility retardant was required.

5. Peachtree Borer Pheromone

Both manually applied Luretape and aerially distributed Disrupt dispensers of the peachtree borer pheromone, (*Z,Z*)-3,13-octadecadien-1-ol acetate, were evaluated as a mating disruptant for both peachtree borer, *Synanthedon exitiosa,* and lesser peachtree borer, *S. pictipes,* moths. The peachtree borer pheromone also disrupts the mating communication of the lesser peachtree borer which has the (*E,Z*) isomer of 3,13-octadecadien-1-ol acetate as the female sex pheromone.[90]

Yonce and Gentry[66] discuss the disruption of mating studies in Volume II, Chapter 7.

In a trapping study, Gentry et al.[91] reported that male peachtree borer catches in traps baited with either Hercon laminated dispensers or hollow fibers were the same, but catches of male lesser peachtree borer moths were greatest when Luretape^{TM} baits were used to dispense the pheromone.

6. Angoumois Grain Moth Pheromone

A Hercon multilayered dispenser of (*Z,E*)-7,11-hexadecadien-1-ol acetate, was found to effectively disrupt the mating of *Sitotroga cerealella* in preliminary studies conducted by Vick et al.[41] The tests were conducted in small enclosures simulating grain warehouses where the Angoumois grain moth is often an important pest.

The Hercon dispenser had a release rate at 26°C of approximately 0.31 $\mu g/cm^2/hr$. The researchers also stated that Angoumois grain moth mating frequency is dependent on population density, that the presence of the behavior-modifying chemical in the atmosphere of the enclosure can reduce the incidence of mating, with significant mating inhibition failing at high population levels.

7. Oriental Fruit Moth Pheromone

Successful disruption of mating and reduction of damage by *Grapholitha molesta* in peaches had been reported previously by Gentry, Bierl, and Blythe.[61] In tests conducted from 1973 to 1975 at the Southeastern Fruit and Tree Nut Research Station in Byron, Georgia, Hercon strips containing (*Z*)-8-dodecenyl acetate, the oriental fruit moth sex pheromone, were placed on every other tree in a 2-acre peach orchard. This resulted in disruption of mating communication, in terms of trap catches of male moths, for 12 weeks. Damage to twigs was also reduced from 78 to 100% (compared to an untreated orchard) during the same 12-week period.

A formulation containing the pheromone with a dodecyl alcohol synergist was also evaluated; performance of this modified dispenser was not as good as with pheromone alone.

Additional information on the disruption of mating of the oriental fruit moth is given by Gentry, Yonce, and Bierl-Leonhardt[62] in Volume II, Chapter 8.

D. Ketones

The pheromone of the Japanese beetle, *Popillia japonica,* is an ideal example of a ketone-type, behavior-modifying chemical that is effectively dispensed by the Hercon dispenser.

The insect was introduced into North America about 1916 and is now well-established in the eastern part of the U.S. The beetles are serious pests. The adults defoliate trees, ornamentals and vegetables, and the larvae (grubs) feed on roots of turf and grasses.

1. Japonilure, Japanese Beetle Pheromone

The sex pheromone of *Popillia japonica,* commonly called japonilure, was isolated from virgin females and identified as (*Z*)-5-(1-decenyl)dihydro-2(3H)-furanone by Tumlinson et al.[69] in 1977. The isolated chemical attracted male beetles in field bioassays. These workers also reported that the pure synthetic (R,*Z*)-5-(1-decenyl)dihydro-2(3H)-furanone was as attractive as live females, and the chemical isolated from female beetles. The synthesized racemic mixture failed to attract the beetles, apparently because the (S,*Z*)-isomer in the chemical inhibited male response.

A controlled release dispensing system is recommended for japonilure because the stereospecific isomer is difficult and expensive to synthesize. As a result, the amount of pheromone used in pest management programs — from monitoring and survey to mass trapping — is very limited.

Tests were conducted at the Japanese Beetle Research Laboratory in Wooster, Ohio,

Table 5
SYNERGISTIC EFFECT OF JAPONILURE SEX ATTRACTANT AND PHENETHYL PROPIONATE + EUGENOL FOOD ATTRACTANT WHEN USED AS BAITS IN JAPANESE BEETLE TRAPS[33]

	Mean weight of beetles (grams)			
Bait used	Day 1	Day 15	Day 22	Total
Japonilure	35.1	15.7	5.2	56.0
Japonilure & PEP + eugenol	47.3	81.0	19.6	147.9
PEP + eugenol	10.3	30.1	10.0	50.4

in 1978 and 1979 to find an effective dispensing system, one that required the least amount of the active ingredient. These tests showed that the Hercon dispenser containing 1 mg japonilure was effective for the season-long, 3-month period of beetle flight. This Luretape product has been registered by the EPA and is commercially available as the additional bait in the Japanese beetle traps for homeowners. The Luretape with Japonilure releases the sex pheromone on the exposed (top) side; the bottom side is impermeable and has a pressure-sensitive adhesive backing for easy attachment to the traps.

The addition of the japonilure bait makes the trap at least three times more attractive than the trap containing only the phenethyl propionate (PEP) + eugenol (3:7) mixture, previously the standard bait in traps used by the U.S. Department of Agriculture Animal and Plant Health Inspection Service for Japanese beetle survey. Table 5 illustrates the synergism resulting from the addition of the japonilure bait to the standard apple-smelling PEP-eugenol mixture.[33] Traps with japonilure and PEP + eugenol caught 147.9 g of beetles vs. 56.0 and 50.4 g for traps baited only with either japonilure or PEP + eugenol. Table 5 also shows that the sex attractant was more effective than the food lure early in the season, and that this attractancy reversed in mid-season.[33]

Ladd and Klein,[92] in Volume II, Chapter 4, discuss the trapping of Japanese beetles with synthetic female sex pheromone and various food-type lures.

E. Miscellaneous

1. Smaller European Elm Bark Beetle Attractant

Hercon dispensers of the elm bark beetle, gypsy moth, and spruce bark beetle attractants are examples of products containing important compounds other than aldehydes, alcohols, esters, or ketones. The attractant of the elm bark beetle, called multilure, consists of three compounds, namely: (S)-α-5-ethyl-2,3-dimethyl-6,8-dioxabicyclo(3.2.1) octane (called multistriatin); (S)-4-methyl-3-heptanol, an alcohol; and (3aα, 3bβ, 4α, 7β, 7aR*)-(S)-3a, 3b, 4,5,6,7-hexahydro-3,7-dimethyl(4-(1-methylethyl)-cyclopenta(1,3)cyclopropa(1,2)benzene (or α-cubebene).[82] The first two compounds are beetle-produced components, while α-cubebene is host-produced.[83]

The Hercon Luretape dispensers of multilure are being used in monitoring-survey traps, to determine the level of infestation of the smaller European elm bark beetle, *Scolytus multistriatus*, and in mass trapping, to reduce the incidence of Dutch elm disease, of which the insect is a major vector.

In 1980, the Hercon Division submitted for EPA registration a Hercon Luretape product consisting of a two-bait system for the elm bark beetle sticky traps. One bait contained 6.5% 4-methyl-3-heptanol, 4.2% multistriatin and 3.6% α-cubebene. The other bait contained 10.1% by weight of the host-produced attractant, α-cubebene. The pair of baits is effective for one trapping season.

Table 6
ELM BARK BEETLE TRAP DENSITY TABLE SHOWING RECOMMENDED NUMBER OF TRAPS TO PLACE IN AREAS OF MASS TRAPPING. THE NUMBER OF TRAPS PER GIVEN AREA SHOULD NOT EXCEED THE NUMBER SHOWN BELOW, BECAUSE AN EXCESSIVE NUMBER OF TRAPS MAY DISORIENT THE BEETLES AND THUS REDUCE TRAP CATCHES

Number of elm trees in area	Area to be trapped (acres)								
	5	12	24	60	120	140	600	1200	2400
1-5	2	3							
10	2	4	6	7				Decrease Trapping	
25	3	5	7	9	10	14		Area	
50		6	8	11	14	18	30		
100			9	13	18	22	35	70	
250		Increase		15	22	26	40	80	160
500		Trapping Area			26	30	45	90	180
1,000						34	50	100	200

In one study by Peacock and Cuthbert,[84] the number of male and female beetles lured by the aggregating multilure released from aged and fresh baits were similar over an 11-week period.

The label of the commercial Hercon multilure dispenser recommends that baited traps be deployed immediately after elm tree bud break. The traps should be at least 50 m from the nearest elm tree, and at least 50 to 200 m apart. Recommended trap height is 8 to 10 ft above the ground, with placement on utility poles, buildings, fences, and trees other than elm.

To prevent the possibility of trap catch reduction resulting from beetle disorientation, Lanier[97] has recommended that the number of traps in a given area not exceed a certain number. Table 6, which appears on the product label, shows the number of traps to be used in areas with different numbers of elm trees. Lanier indicates that trap efficiency increases as the incidence of Dutch elm disease and competing sources of attractant (e.g., diseased trees and other multilure-baited traps) decrease.

2. *Spruce Bark Beetle Attractant*

Like multilure, the aggregating attractant of the spruce bark beetle, *Ips typographus*, consists of a combination of chemicals produced by the insect and the host tree. The use of the insect's behavior-modifying chemicals in an integrated control program, through mass trapping and trap-tree practices, is discussed in detail later in Volume II, Chapters 1 and 2.[37,38]

The Hercon Division formulates the multicomponent attractant in one dispenser 25 cm by 100 cm (1 m) long. Each unit contains 1,500 mg methyl butenol, 70 mg *z*-verbenol, and 10 mg ipsdienol.

In 1979, over 600,000 *Ips typographus* baits, manufactured by the Hercon Division, were used in mass trapping programs in Norway and Sweden. A budget of $20 million was set for this massive project which involved 1% of the total population of Norway in the placement of traps in the forests during winter and early spring. It is believed that the mass trapping program saved more than three million trees from destruction by the spruce bark beetle.

3. Disparlure, Gypsy Moth Pheromone

The Hercon Division has three products containing disparlure which are commercially available for use in integrated pest management programs against the gypsy moth, *Lymantria dispar*. The Hercon disparlure dispenser, initially made and tested in 1973, was the first pheromone dispenser ever produced by Hercon. This product was used by the U.S. Department of Agriculture in its gypsy moth monitoring and survey programs.[46,48]

Today, Hercon disparlure formulations are used as baits for traps, as hand-applied point sources for disruption of mating, and as an aerially applied flake dispenser, also for disruption of mating.

a. Luretape with Disparlure for Traps

An EPA-registered Hercon formulation of the racemic disparlure is used as bait to lure gypsy moth males into traps originally designed by the U.S.D.A. Animal and Plant Health Inspection Service for gypsy moth monitoring programs. The bait, a 1-in.^{2} plastic strip, contains 22.4 mg active ingredient (13.0% by weight), and is made of 2-mil thick plastic films.

To prevent trapped moths from escaping, another Hercon formulation — the Insectape® insecticidal strips containing propoxur (Baygon®) — is used inside the trap to provide the toxicant that kills the moths. This insecticidal product is also registered for use in residences and commercial establishments as a cockroach control agent.[39]

The success of a trapping program depends on many factors, including the level of infestation, attractiveness of the lure, effectiveness of the trap design, number of traps per given area, and the behavior of the target insect. Webb[85] discusses these and other factors in Volume II, Chapter 3 of this book.

Both racemic and the optically active disparlure have been used as baits in the gypsy moth traps. The Hercon formulation of the optically active ("plus") (*Z*)-7,8-epoxy-2-methyloctadecane, contains 0.5 mg pheromone per bait; the latter is used in U.S.D.A. monitoring programs. Unpublished observations have indicated that the plus-disparlure is seven to ten times more attractive than the racemic bait, i.e., 22.4 mg per bait, but because of the very limited supply of the optically active enantiomer, racemic disparlure is still commonly used in gypsy moth trap baits.

b. Luretape with Disparlure for Mating Disruption

In addition to its use in traps, the EPA-registered Luretape with Disparlure may be used for disruption of mating of the gypsy moth. The Luretape label suggests the application of 40 2-in.^{2} dispensers per acre, which is equivalent to 100 Luretape strips per hectare, or about 2 g disparlure per acre.

The Hercon dispenser used in mating disruption has the same construction as that of the bait. The Luretape strips are attached to trees in a grid pattern; the dispensers continue to release the disparlure for at least two seasons. Volume I, Chapter 10 and Volume II, Chapter 10 by Bierl-Leonhardt[18] and Plimmer,[86] respectively, discuss the release rates of disparlure from the Hercon formulations, and disruption of mating programs with disparlure.

c. Disrupt/Gypsy Moth for Mating Disruption

In 1979, the Hercon Division received an EPA registration for another disparlure formulation, one that is applicable by aircraft or ground equipment. The tradename for this product is Disrupt/Gypsy Moth. As with the Luretape, Disrupt is also designed to lower incidence of gypsy moth mating by disrupting normal male orientation to the nonflying females, thus suppressing the larval population that causes defoliation of trees the following spring.

FIGURE 3. Hercon application equipment.

Disrupt contains 9.4% disparlure by weight or 19.1 mg active ingredient per square inch of the 3/32- by 3/32-in. flakes.

The EPA-accepted label allows the application of 4 to 8 g disparlure per acre before emergence of adult gypsy moths. Disrupt may be used in forested areas and wooded residential neighborhoods. Only one application is needed per season.

Disparlure is another tool for use in integrated pest management programs against the gypsy moth. The Hercon Division recommends the use of this behavior-modifying chemical to lower the incidence of mating in low-level populations. Since the effectiveness of disparlure is dependent on the density of the pest population, in areas of heavy infestations, disparlure is recommended as a follow-up to larvicidal treatments to help prevent the surviving moths from mating, a practice that will delay the build up of future infestations, or maintain the population below the need-to-spray (with conventional insecticides) levels.

Disrupt is applied with Phero-Tac^{TM}, an inert sticker that causes the flakes to adhere to foliage or other plant parts. Equipment specially built to apply Disrupt is now available from Hercon, and is pictured in Figure 3.

REFERENCES

1. **Kydonieus, A. F., Smith, I. K., and Beroza, M.,** Controlled release of pheromones through multi-layered polymeric dispensers, in *Controlled Release Polymeric Formulations,* ACS Symp. Ser. 33, Paul, D. R. and Harris, F. W., Eds., American Chemical Society, Washington, D.C., 1976, 283.
2. **Kydonieus, A. F. and Beroza, M.,** Insect control with the multi-layered Luretape dispenser, in *Proc. 1977 Controlled Release Pesticides Symp.,* Goulding, R., Ed., Oregon State University, Corvallis, 1977, 78.
3. **Kydonieus, A. F. and Beroza, M.,** The Hercon dispenser: formulation and recent test results, in *Management of Insect Pests with Semiochemicals-Concepts and Practices,* Mitchell, E.R. Ed., Plenum Press, New York, 1981.
4. **Kydonieus, A. F., Bierl-Leonhardt, B. A., Plimmer, J. R., Barry, M. W., and Quisumbing, A. R.,** Cotton Insects: dispenser development and disruption of mating trials, in *Proc. 6th Int. Symp. Controlled Release Bioactive Materials,* Baker, R., Ed., Academic Press, New York, 1980, IV-13.

5. **Kydonieus, A. F. and Quisumbing, A. R.**, Hercon commercial pheromone products, in *Abstr. 7th Int. Symp. Controlled Release Bioactive Materials,* Lewis, D. H., Ed., Fort Lauderdale, Florida, 1980, 27.
6. **Kydonieus, A. F. and Quisumbing, A. R.**, Multilayered laminated structures, in *Controlled Release Technologies: Methods, Theory and Applications,* Vol. 1, Kydonieus, A. F., Ed., CRC Press, Boca Raton, Fla., 1980, 183.
7. **Kydonieus, A. F.**, The effect of some variables on the controlled release of chemicals from polymeric membranes, in *Controlled Release Pesticides,* ACS Symp. Ser. 53, Scher, H. B., Ed., American Chemical Society, Washington, D.C., 1977, 152.
8. **Kydonieus, A. F., Smith, I. K., and Hyman, S.**, A polymeric delivery system for the controlled release of pesticides: Hercon Roach-Tape, in *Proc. 1976 Int. Controlled Release Pesticides Symp.,* Cardarelli, N. F., Ed., University of Akron, Ohio, 1976, 3.40.
9. **Kydonieus, A. F., Quisumbing, A. R., and Hyman, S.**, Application of a new controlled release concept in household products, in *Controlled Release Polymeric Formulations,* ACS Symp. Ser. 33, Paul, D. R. and Harris, F. W., Eds., American Chemical Society, Washington, D.C., 1976, 295.
10. **Quisumbing, A. R., Kydonieus, A. F., Calsetta, D. R., and Haus, J. B.**, Hercon "Lure 'N Kill" Flytape: a non-fumigant insecticidal strip containing attractants, in *Proc. 1976 Int. Controlled Release Pesticides Symp.,* Cardarelli, N. F., Ed., University of Akron, Ohio, 1976, 3.40.
11. **Quisumbing, A. R., Kydonieus, A. F., and Gauthier, N. L.**, Use of laminated controlled release contact-action insecticidal strips for cluster fly control, in *Proc. 5th Int. Symp. on Controlled Release of Bioactive Materials,* Brinckman, F. E. and Montemarano, J. A., Eds., Gaithersburg, Maryland, 1978, 5.62.
12. **Kydonieus, A. F., Baldwin, S., and Hyman, S.**, Hercon granules and powders for agricultural applications, in *Proc. 1976 Int. Controlled Release Pesticides Symp.,* Cardarelli, N. F., Ed., University of Akron, Ohio, 1976, 4.23.
13. **Hyman, S., Bernstein, B. S., and Kapoor, R.**, U.S. Patent 3,705,938, 1972.
14. **Kydonieus, A. F., Rofheart, A., and Hyman, S.**, Marketing and economic considerations for Hercon consumer and industrial controlled release products, in *Chemical Marketing and Economic Preprints of Symposium on Economics and Market Opportunities for Controlled Release Products,* ACS Chemical Marketing and Economics Division, American Chemical Society, Washington, D.C., 1976, 140.
15. **Kydonieus, A. F.**, Hercon controlled release system — a novel way to dispense fragrances and room deodorizers, *Soap Cosmet. Chem. Spec.,* May 1978.
16. **Crank, J. and Park, G. S.**, Methods of measurement, in *Diffusion of Polymers,* Crank, J. and Park, G. S., Eds., Academic Press, New York, 1968, 1.
17. **Richards, R. W.**, The Permeability of Polymers to Gases, Vapours and Liquids, NTIS AD-767 627, U.S. Department of Commerce, Springfield, Va., March 1973.
18. **Bierl-Leonhardt, B. A.**, Release rates from formulations and quality control methods, in *Insect Suppression with Controlled Release Pheromone Systems,* Vol. 1, Kydonieus, A. F. and Beroza, M., Eds., CRC Press, Boca Raton, Fla., 1982, chap. 10.
19. **Roelofs, W. L., Hill, A. S., Cardé, R. T., and Baker, T. C.**, Two sex pheromone components of the tobacco budworm moth, *Heliothis virescens, Life Sci.,* 14, 1555, 1974.
20. **Tumlinson, J. H., Hendricks, D. E., Mitchell, E. R., Doolittle, R. E., and Brennan, M. M.**, Isolation, identification and synthesis of the sex pheromone of the tobacco budworm, *J. Chem. Ecol.,* 1, 203, 1975.
21. **Hendricks, D. E., Hartstack, A. W., and Shaver, T. N.**, Effect of formulations and dispensers on attractiveness of virelure to the tobacco budworm, *J. Chem. Ecol.,* 3, 497, 1977.
22. **Klun, J. A., Plimmer, J. R., Bierl-Leonhardt, B. A., Sparks, A. N., and Chapman, O. L.**, Trace chemicals: the essence of sexual communication systems in *Heliothis* spp., *Science,* 204, 1328, 1979.
23. **Klun, J. A., Bierl-Leonhardt, B. A., Plimmer, J. R., Sparks, A. N., Primiani, M., Chapman, O. L., Lepone, G., and Lee, G. H.**, Sex pheromone chemistry of the female tobacco budworm moth, *Heliothis virescens, J. Chem. Ecol.,* 6, 177, 1980.
24. **Hummel, H. E., Gaston, L. K., Shorey, H. H., Kaae, R. S., Byrne, K. J., and Silverstein, R. M.**, A clarification of the chemical status of the pink bollworm sex pheromone, *Science,* 181, 873, 1973.
25. **Bierl, B. A., Beroza, M., Staten, R. T., Sonnet, P. E., and Adler, V. E.**, The pink bollworm sex attractant, *J. Econ. Entomol.,* 67, 211, 1974.
26. **Inscoe, M. N. and Beroza, M.**, Insect-behavior chemicals active in field trials, in *Pest Management with Insect Sex Attractants and Other Behavior-Controlling Chemicals,* ACS Symp. Ser. 23, Beroza, M., Ed., American Chemical Society, Washington, D.C., 1976, 145.
27. **Mayer, M. S. and McLaughlin, J. R.**, An annotated compendium of insect sex pheromones, *Univ. Fla. Agric. Exp. Stations Monogr. Ser.,* 6, 1975.

28. **Inscoe, M. N.,** Insect attractants, attractant pheromones, and related compounds, in *Insect Suppression with Controlled Release Pheromone Systems,* Vol. 2, Kydonieus, A. F. and Beroza, M., Eds., CRC Press, Boca Raton, Fla., 1982, chap. 15.
29. **Sutherland, D. W. S., Ed.,** *Common Names of Insects and Related Organisms,* (1978 Revision), Entomological Society of America Spec. Publ. 78-1, College Park, Md., 1978.
30. **Hardee, D. D., McKibben, G. H., and Huddleston, P. M.,** Grandlure for boll weevils: controlled release with a laminated plastic dispenser, *J. Econ. Entomol.,* 68, 477, 1975.
31. **Johnson, W. L., McKibben, G. H., Rodriguez, J., and Davich, T. B.,** Boll weevil: increased longevity of grandlure using different formulations and dispensers, *J. Econ. Entomol.,* 69, 263, 1976.
32. **Davich, T. B. and McKibben, G. H.,** Belt-wide Grandlure formulation test, presented at the Entomological Society of America annual meeting, New Orleans, November 30 to December 1, 1975.
33. **Michelotti, F. W.,** A new Japanese beetle trap containing pheromone and floral lure as synergistic attractants, presented at the Entomological Society of America annual meeting, Denver, November 25 to 29, 1979.
34. **Lanier, G. N.,** Protection of elm groves by surrounding them with multilure-baited sticky traps, *Bull. Entomol. Soc. Am.,* 25, 109, 1979.
35. **Wollerman, E. H.,** Attraction of European elm bark beetles, *Scolytus multistriatus,* to pheromone-baited traps, *J. Chem. Ecol.,* 5, 781, 1979.
36. **Cuthbert, R. A. and Peacock, J. W.,** The Forest Service program for mass-trapping *Scolytus multistriatus, Bull. Entomol. Soc. Am.,* 25, 105, 1979.
37. **Bakke, A. and Riege, L.,** Mass trapping of the spruce bark beetle, *Ips typographus,* in Norway as part of an integrated control program, in *Insect Suppression with Controlled Release Pheromone Systems,* Vol. 2, Kydonieus, A. F. and Beroza, M., Eds., CRC Press, Boca Raton, Fla., 1982, chap. 1.
38. **Bakke, A.,** The pheromone of the spruce bark beetle, *Ips typographus,* and its potential use in the suppression of beetle populations, in *Insect Suppression with Controlled Release Pheromone Systems,* Vol. 2, Kydonieus, A. F. and Beroza, M., Eds., CRC Press, Boca Raton, Fla., 1982, chap. 2.
39. **Kydonieus, A. F. and Quisumbing, A. R.,** Multilayered laminated structures, in *Controlled Release Technologies: Methods, Theory, and Applications,* Vol. 1, Kydonieus, A. F., Ed., CRC Press, Boca Raton, Fla., 1980, 183.
40. **Nakagawa, S., Harris, E. J., and Urago, T.,** Controlled release of trimedlure from a three-layer laminated plastic dispenser, *J. Econ. Entomol.,* 72, 625, 1979.
41. **Vick, K. W., Coffelt, J. A., and Sullivan, M. A.,** Disruption of pheromone communication in the Angoumois grain moth with synthetic female sex pheromone, *Environ. Entomol.,* 7, 528, 1978.
42. **Henneberry, T. J., Gillespie, J. M., Bariola, L. A., Flint, H. M., Butler, G. D., Jr., Lingren, P. D., and Kydonieus, A. F.,** Mating disruption as a method of controlling pink bollworm and tobacco budworm on cotton, in *Insect Suppression with Controlled Release Pheromone Systems,* Vol. 2, Kydonieus, A. F. and Beroza, M., Eds., CRC Press, Boca Raton, Fla., 1982, chap. 6.
43. **Henneberry, T. J., Gillespie, J. M., Bariola, L. A., Flint, H. M., Butler, G. D., Jr., Lingren, P. D., and Kydonieus, A. F.,** Gossyplure in laminated plastic formulations for mating disruption and pink bollworm control, *J. Econ. Entomol.,* in press.
44. **Henneberry, T. J., Gillespie, J. M., Bariola, L. A., Flint, H. M., Butler, G. D., Jr., Lingren, P. D., and Kydonieus, A. F.,** Recent Progress with Gossyplure for Pink Bollworm Mating Disruption, presented at 8th Desert Cotton Insect Symp., El Centro, California, March 12, 1980.
45. **Daterman, G. E. and Sower, L. L.,** Douglas-fir tussock moth pheromone research using controlled release system, in *Proc. 1977 Int. Controlled Release Pesticides Symp.,* Goulding, R., Ed., Oregon State University, Corvallis, 1977, 68.
46. **Beroza, M., Paszek, E. C., Mitchell, E. R., Bierl, B. A., McLaughlin, J. R., and Chambers, D. L.,** Tests of a 3-layer laminated plastic bait dispenser for controlled emission of attractants from insect traps, *Environ. Entomol.,* 3, 926, 1974.
47. **Beroza, M., Hood, C. S., Trefrey, D., Leonard, D. E., Knipling, E. F., and Klassen, W.,** Field trials with disparlure in Massachusetts to suppress mating of the gypsy moth, *Environ. Entomol.,* 4, 705, 1975.
48. **Beroza, M., Paszek, E. C., DeVilbiss, D., Bierl, B. A., and Tardif, J. G. R.,** A 3-layer laminated plastic dispenser of disparlure for use in traps for gypsy moths, *Environ. Entomol.,* 4, 712, 1975.
49. **Plimmer, J. R., Schwalbe, C. P., Paszek, E. C., Bierl, B. A., Webb, R. E., Marumo, S., and Iwaki, S.,** Contrasting effectiveness of (+) and (±) enantiomers of disparlure for trapping native populations of gypsy moth in Massachusetts, *Environ. Entomol.,* 7, 815, 1978.
50. **Bierl, B. A., DeVilbiss, E. D., and Plimmer, J. R.,** Use of pheromones in insect control programs: slow release formulations, in *Controlled Release Polymeric Formulations,* ACS Symp. Ser. 33, Paul, D. R. and Harris, F. W., Eds., American Chemical Society, Washington, D.C., 1976, 265.

51. **Plimmer, J. R., Bierl, B. A., and Schwalbe, C. P.,** Controlled release of pheromone in the gypsy moth program, in *Controlled Release of Pesticides,* ACS Symp. Ser. 53, Scher, H. B., Ed., American Chemical Society, Washington, D.C., 1977, 168.
52. **Webb, R. E., Dull, C. W., McComb, C. W., Bierl, B. A., and Plimmer, J. R.,** Gypsy moth mating suppressed by disparlure emitted from laminated plastic dispenser, unpublished.
53. **Webb, R. E., Schwalbe, C. P., Leonhardt, B. B., and McLane, W.,** Aerial application of disparlure dispensers, unpublished.
54. **Clement, S. L., Schmidt, B. S., and Levine, E.,** Monitoring the flight activities of the black cutworm with blacklight and pheromone traps, presented at the Entomological Society of America annual meeting, Denver, November 25 to 29, 1979.
55. **Tingle, F. C. and Mitchell, E. R.,** Controlled release plastic strips containing (*Z*)-9-dodecen-1-ol acetate for attracting *Spodoptera frugiperda, J. Chem. Ecol.,* 4, 41, 1978.
56. **Hendricks, D. E., Hartstack, A. W., and Raulston, J. R.,** Compatibility of virelure and looplure dispensed from traps for cabbage looper and tobacco budworm survey, *Environ. Entomol.,* 6, 566, 1977.
57. **Lopez, J. D., Jr., Shaver, T. N., and Hartstack, A. W., Jr.,** Evaluation of dispensers for the pheromone of *Heliothis zea,* in press.
58. **Hartstack, A. W., Jr., Lopez, J. D., Klun, J. A., Witz, J. A., Shaver, T. N., and Plimmer, J. R.,** New trap designs and pheromone bait formulations for *Heliothis, Proc. Beltwide Cotton Prod. Res. Conf.,* National Cotton Council, Memphis, 1980, 132.
59. **Hollingsworth, J. P., Hartstack, A. W., Buck, D. R., and Hendricks, D. E.,** Electric and Non-Electric Moth Traps Baited with the Synthetic Sex Pheromone of the Tobacco Budworm, U.S. Department of Agriculture Publ. ARS-S-173, Feb. 1978.
60. **Hendricks, D. E., Hartstack, A. W., and Shaver, T. N.,** Effect of formulations and dispensers on attractiveness of virelure to the tobacco budworm, *J. Chem. Ecol.,* 3, 497, 1977.
61. **Gentry, C. R., Bierl, B. A., and Blythe, J. L.,** Air permeation field trials with the oriental fruit moth pheromone, in *Proc. 1976 Int. Controlled Release Pesticides Symp.,* Cardarelli, N. F., Ed., University of Akron, Ohio, 1976, 3.22.
62. **Gentry, C. R., Yonce, C. E., and Bierl-Leonhardt, B. A.,** Oriental fruit moth: mating disruption trials with pheromone, in *Insect Suppression with Controlled Release Pheromone Systems,* Vol. 2, Kydonieus, A. F. and Beroza, M., Eds., CRC Press, Boca Raton, Fla., 1982, chap. 8.
63. **Daterman, G. E.,** Control of western pineshoot borer damage by mating disruption — a reality, in *Insect Suppression with Controlled Release Pheromone Systems,* Vol. 2, Kydonieus, A. F. and Beroza, M., Eds., CRC Press, Boca Raton, Fla., 1982, chap. 11.
64. **Landolt, P. J., Curtis, C. E., Coffelt, J. A., Vick, K. W., and Doolittle, R.,** Disruption of sexual communication in the navel orangeworm, *Amyelois transitella,* in almond orchards, presented at the Entomological Society of America annual meeting, Denver, November 25 to 29, 1979.
65. **Johnson, D. T.,** Development of control strategies for the grape root borer, presented at the Entomological Society of America, Southeastern Branch meeting, Biloxi, Miss., January 28-31, 1980.
66. **Yonce, C. E. and Gentry, C. R.,** Disruption of mating of peachtree borer, in *Insect Suppression with Controlled Release Pheromone Systems,* Vol. 2, Kydonieus, A. F. and Beroza, M., Eds., CRC Press, Boca Raton, Fla., 1982, chap. 7.
67. **Gentry, C. R., Bierl-Leonhardt, B. A., and McLaughlin, J. R.,** Trapping and air permeation with pheromone to combat peach orchard pests, presented at the Entomological Society of America, Southeastern Branch meeting, Biloxi, Miss., January 28 to 31, 1980.
68. **Arn, H., Schmid, A., Jaccard, C., Bierl-Leonhardt, B. A., and Rauscher, S.,** Mating reduction in free-living grape moths (*Eupoecilia ambiguella* Hb.) by disruption of pheromone communication with (Z)-9-dodecenyl acetate, Compte-rendu de la reunion sur los pheromones sexuelles des insectes et mediateurs chimiques, Institut Nacional de Recherche Agronomique, Antibes, 1978.
69. **Tumlinson, J. H., Hardee, D. D., Gueldner, R. C., Thompson, A. C., Hedin, P. A., and Minyard, J. P.,** Sex pheromone produced by male boll weevil: isolation, identification and synthesis, *Science,* 166, 1010, 1969.
70. **Hardee, D. D., McKibben, G. H., Rummel, D. R., Huddleston, P. M., and Coppedge, J. R.,** Response of boll weevils to component ratios and doses of the pheromone, grandlure, *Environ. Entomol.,* 3, 135, 1974.
71. **Hartstack, A. W., Witz, J. A., and Buck, D. R.,** Moth traps for the tobacco budworm, *J. Econ. Entomol.,* 72, 519, 1979.
72. **Weatherston, J., Roelofs, W. L., Comeau, A., and Sanders, C. J.,** Studies of physiologically active arthropod secretions: X. Sex pheromone of the eastern spruce budworm, *Choristoneura fumiferana* (Lepidoptera: Tortricidae), *Can. Entomol.,* 103, 1741, 1971.
73. **Coffelt, J. A., Vick, K. W., Sonnet, P. E., and Doolittle, R. E.,** Isolation, identification, and synthesis of a female sex pheromone of the navel orangeworm, *Amyelois transitella* (Lepidoptera: Pyralidae), *J. Chem. Ecol.,* 5, 971, 1979.

74. **Roelofs, W. L., Kochansky, J., Anthon, E., Rice, R., and Cardé, R.**, Sex pheromone of the peach twig borer moth (*Anarsia lineatella), Environ. Entomol.*, 4, 580, 1975.
75. **Roelofs, W. L., Comeau, A., Hill, A., and Milicevic, G.**, Sex attractant of the codling moth: characterization with electroantennogram technique, *Science*, 174, 297, 1971.
76. **Arn, H., Schmid, A., Bierl-Leonhardt, B. A., Rauscher, S., and Jaccard, C.**, Field experiments to develop control of the grape moth, *Eupoecilia ambiguella*, by communication disruption, in *Management of Insect Pests with Semiochemicals-Concepts and Practices*, Mitchell, E.R., Ed., Plenum Press, New York, 1981.
77. **Arn, H., Rauscher, S., Buser, H. R., Roelofs, W. L.**, Sex pheromone of *Eupoecilia ambiguella: cis*-9-dodecenyl acetate as a major component, *Z. Naturforsch.*, 31, 499, 1976.
78. **Arn, H.**, Developing insect control by disruption of sex pheromone communication: conclusions from programs on lepidopterous insects in Switzerland, in *Chemical Ecology: Odour Communication in Animals*, Ritter, F. J., Ed., Elsevier, Amsterdam, 1979.
79. **Arn, H., Rauscher, S., and Schmid, A.**, Sex attractant formulations and traps for the grape moth, *Eupoecilia ambiguella* Hb., *Mitt. Schweiz. Ent. Ges.*, 52, 49, 1979.
80. **Hill, A. S., Rings, R. W., Swier, S. R., and Roelofs, W. L.**, Sex pheromone of the black cutworm moth, *Agrotis ipsilon, J. Chem. Ecol.*, 5, 439, 1979.
81. **Hardee, D. D.**, Mass trapping and the trap cropping of the boll weevil, *Anthonomus grandis* Boheman, in *Insect Suppression with Controlled Release Pheromone Systems*, Vol. 2, Kydonieus, A. F. and Beroza, M., Eds., CRC Press, Boca Raton, Fla., 1982, chap. 5.
82. **Pearce, G. T., Gore, W. E., Silverstein, R. M., Peacock, J. W., Cuthbert, R. A., Lanier, G. N., and Simeone, J. B.**, Chemical attractants for the smaller European elm bark beetle, *Scolytus multistriatus* (Coleoptera: Scolytidae), *J. Chem. Ecol.*, 1, 115, 1975.
83. **Lanier, G. N., Silverstein, R. M., and Peacock, J. W.**, Attractant pheromone of the European elm bark beetle (*Scolytus multistriatus*): isolation, identification, synthesis and utilization studies, in *Perspectives in Forest Entomology*, Academic Press, New York, 1976, 149.
84. **Peacock, J. W. and Cuthbert, R. A.**, Field and laboratory evaluations of controlled release pheromone dispensers for *Scolytus multistriatus* pheromone, in *Proc. 1975 Int. Controlled Release Pesticides Symp.*, Harris, F. W., Ed., Wright State University, Dayton, 1975, 216.
85. **Webb, R. E.**, Mass trapping the gypsy moth, in *Insect Suppression with Controlled Release Pheromone Systems*, Vol. 2, Kydonieus, A. F. and Beroza, M., Eds., CRC Press, Boca Raton, Fla., 1982, chap. 3.
86. **Plimmer, J. R.**, Disruption of mating in the gypsy moth, in *Insect Suppression with Controlled Release Pheromone Systems*, Vol. 2, Kydonieus, A. F. and Beroza, M., Eds., CRC Press, Boca Raton, Fla., 1982, chap. 10.
87. **Sanders, C. J. and Seabrook W. D.**, Disruption of mating in the spruce budworm, in *Insect Suppression with Controlled Release Pheromone Systems*, Vol. 2, Kydonieus, A. F. and Beroza, M., Eds., CRC Press, Boca Raton, Fla., 1982, chap. 13.
88. **Byrne, K. J., Swigar, A. A., Silverstein, R. M., Borden, J. H., and Stokkink, E.**, Sulcatol: population aggregation pheromone in the scolytid beetle, *Gnathotrichus sulcatus, J. Insect Physiol.*, 20, 1895, 1974.
89. **Rothschild, G. L.**, Suppression of mating in codling moths with synthetic sex pheromone and other compounds, in *Insect Suppression with Controlled Release Pheromone Systems*, Vol. 2, Kydonieus, A. F. and Beroza, M., Eds., CRC Press, Boca Raton, Fla., 1982, chap. 9.
90. **Tumlinson, J. H., Yonce, C. E., Doolittle, R. E., Heath, R. R., Gentry, C. R., and Mitchell, E. R.**, Sex pheromones and reproductive isolation of the lesser peachtree and the peachtree borer, *Science*, 185, 614, 1974.
91. **Gentry, C. R., Bierl-Leonhardt, B. A., Blythe, J. L., McLaughlin, J. R., and Plimmer, J. R.**, Air permeation tests with (Z,Z)-3,13-octadecadien-1-ol acetate for reduction in trap catch of peachtree and lesser peachtree borer moths, in preparation.
92. **Ladd, T. L. and Klein, M. G.**, Trapping Japanese beetles with synthetic female sex pheromone and food-type lures, in *Insect Suppression with Controlled Release Systems*, Vol. 2, Kydonieus, A. F. and Beroza, M., Eds., CRC Press, Boca Raton, Fla., 1982, chap. 4.
93. **Mitchell, E. R.**, unpublished data.
94. **Hardee, D. D.**, private communication.
95. **Daterman, G.**, personal communication.
96. **Oehlschlager, A. C.**, private communication.
97. **Klein, M. G.**, personal communication.
98. **Lanier, G.**, personal communication.

Chapter 9

CONTROLLED RELEASE FROM HOLLOW FIBERS*

E. Ashare, T. W. Brooks, and D. W. Swenson

TABLE OF CONTENTS

* Reprinted with permission from *Proc. 1975 Int. Controlled Release Pesticide Symp.*, Harris, F.W., Ed., Wright State University, Dayton, 1975, 42.

I. INTRODUCTION

A new controlled release dispenser has been developed, called CONREL (for "controlled release") denoting a concept and range of versatile precision vapor dispenser systems made from hollow fibers. The concept features precise control and design flexibility resulting in high efficiency and economical utilization of active materials, and convenient physical form.

In one convenient-to-use form now in pilot production, the CONREL vapor dispenser comprises a parallel array of hollow polymeric fibers fixed to an adhesive-surfaced tape. Filled with the liquid material to be dispensed, the fibers are sealed at regular intervals along the length of the tape. Vapor release is activated by cutting the tape, thereby opening the fiber ends. The vaporizable liquid is kept from running out of the hollow fibers by capillary forces. Only the vapor is released (see Figure 1).

Hollow fibers function both as a reservoir and as a means of control over the dissemination of a vaporizable substance into the atmosphere. The controlled vapor release flexibility of CONREL is achieved by judicious selection of the number, length, and diameter of the fibers used. Design elements of CONREL dispensers are variable and are determined by the physical properties of the material to be dispensed, the desired rate of release, and the optimum time span of release. The instantaneous delivery rate is dependent upon the number and inside diameter of each hollow fiber, while the effective life is a function of the length.

This system, capable of dispensing any vaporizable material, is extremely versatile as to possible applications. Areas presently being investigated include controlled release of insect pheromones, insecticides, animal repellents, and fragrances.

Preliminary results for some of these applications will be discussed after considering the mass transport theory for vapor dispensing from hollow fibers.

II. THEORETICAL ANALYSIS

The mechanism for dispensing of volatile materials consists of three steps:

1. Evaporation at the liquid-vapor interface
2. Diffusion from the liquid-vapor interface to the open end of the hollow fiber
3. Convection away from the open end

Generally, the diffusion step will be the rate-controlling factor. The rate of diffusion can be predicted mathematically (Reference 1, p. 522).

The molar flux, N_a (Figure 2) is given by:

$$N_a = -\frac{cD}{1-x_a}\frac{dx_a}{dz} \qquad (1)$$

where x_a is the mole fraction of the volatile material, c is its molar density, and D is the diffusion coefficient.

A mass balance, $dN_a/dz = 0$, combined with Equation 1 and assuming ideal gas behavior gives:

$$-\ln(1-x_a) = C_1 z + C_2 \qquad (2)$$

where C_1 and C_2 are constants which must be determined by the boundary conditions.

$$\text{B.C. 1} \quad \text{at } z = 0, x_a = 0 \qquad (3)$$

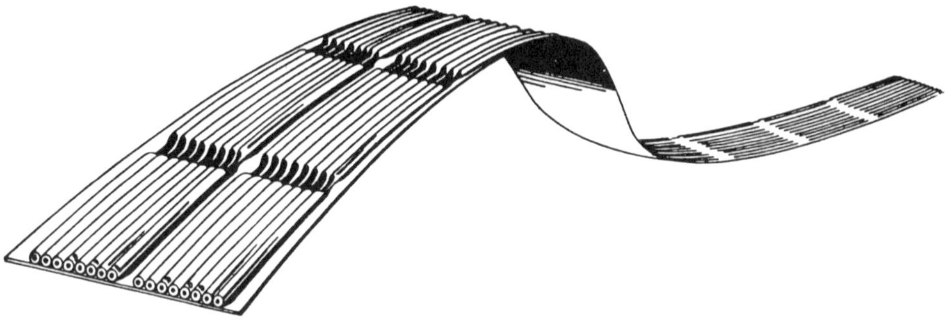

FIGURE 1. Schematic diagram of hollow fiber dispenser.

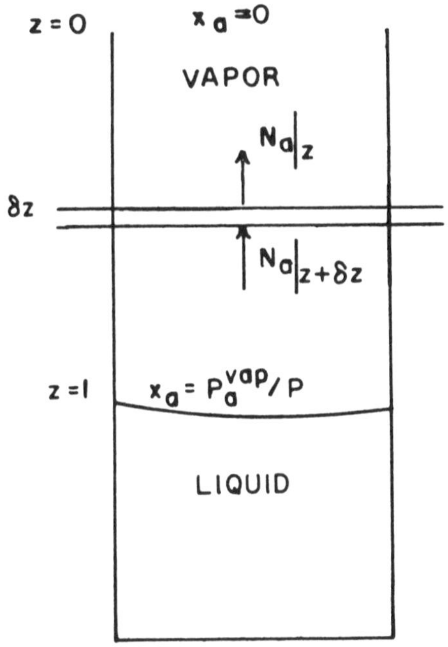

FIGURE 2. Diffusion through a stagnant gas column.

$$\text{B.C. 2} \quad \text{at } z = l,\ x_a = P_a^{vap}/P \tag{4}$$

The first boundary condition indicates that there is zero concentration at the open end of the fiber, i.e., convection is sufficient to remove the volatile material away from the fiber. The second boundary condition implies vapor-liquid equilibrium at the liquid-vapor interface, and P_a^{vap} is the vapor pressure of the volatile material.

Applying these boundary conditions gives:

$$(1 - x_a) = (1 - P_a^{vap}/P)^{z/\ell} \tag{5}$$

The liquid-vapor interface is continuously moving away from the fiber open end and the rate of this movement can be related to the molar flux, N_a

$$-N_a = \frac{\rho}{M}\frac{d\ell}{dt} \tag{6}$$

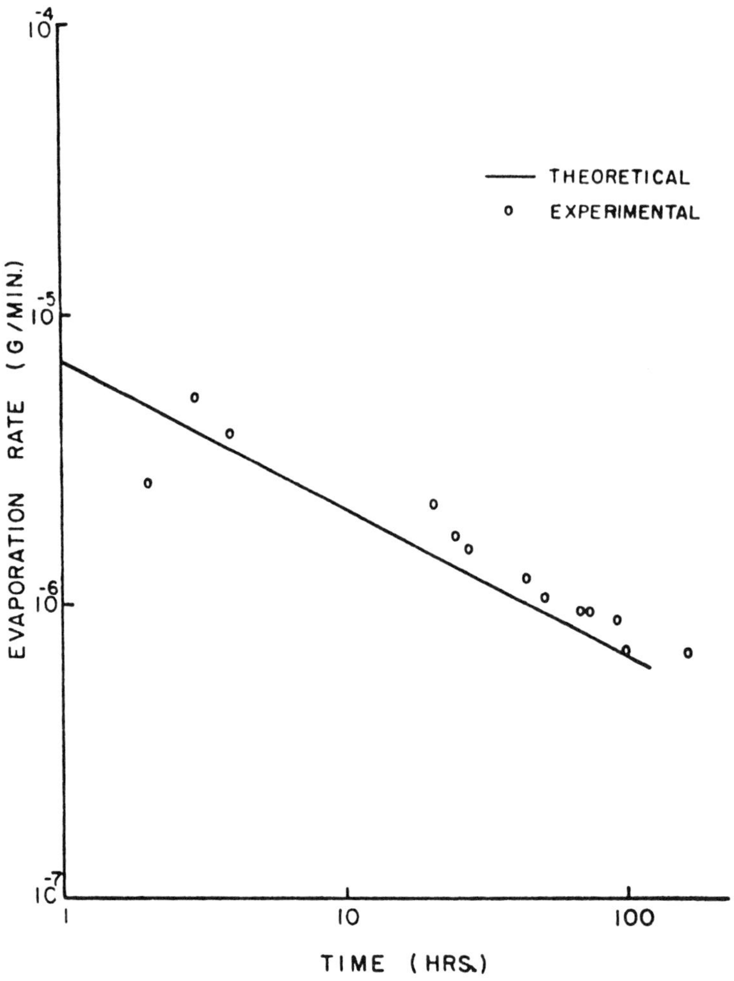

FIGURE 3. Rate of release of *o*-dichlorobenzene.

where ϱ is the liquid density, M the molecular weight, and ℓ is the distance between the fiber open end and the liquid-vapor interface.

Combining Equations 1, 5, and 6 gives:

$$\frac{d\ell}{dt} = -\frac{cMD}{\rho\ell} \ln\left(1 - \frac{P_a^{vap}}{P}\right) \qquad (7)$$

This then leads to

$$\frac{d\ell}{dt} = \left[\frac{-McD}{2\rho} \ln\left(1 - \frac{P_a^{vap}}{P}\right)\right]^{1/2} t^{-1/2} \qquad (8)$$

For most materials and temperatures considered $P_a^{vap} \ll P$. A Maclaurin expansion of the ln term gives:

$$\ln\left(1 - \frac{P_a^{vap}}{P}\right) \approx -P_a^{vap}/P$$

so that

$$\frac{d\ell}{dt} = \left[\frac{McD}{2\rho}\ \frac{P_a^{vap}}{P}\right]^{1/2} t^{1/2} \tag{9}$$

Thus, the change in meniscus level with time, $d\ell/dt$, which is related to the dispensing rate, is predicted to be inversely proportional to the half power of time. The validity of this relationship was checked by measuring the movement of the meniscus for *o*-dichlorobenzene. The results are shown in Figure 3 where the solid line represents Equation 9 and the points are measured values for two different fiber diameters. The agreement is quite good considering that the diffusion coefficient, D, was calculated by using an empirical equation (see Reference 1, Equation 16.3-1).

The rate of release is dependent on temperature since the material properties, P_a^{vap}, D, and c are temperature dependent. The major contribution to this dependence is the vapor pressure and is given by the Clausius-Clapeyron equation,

$$\frac{d\ \ln P^{vap}}{dT} = \frac{\Delta H^{vap}}{RT^2} \tag{10}$$

A 30°F temperature change (e.g., from 70°F to 100°F) will result in a vapor pressure increase for pheromone-type materials of about 200%, while the diffusivity, D, increases by only about 10% and the molar gas density, c, decreases by about 5%. Thus, from Equation 9 it is noted that the dispensing rate will about double for a 30°F temperature rise.

III. EVAPORATION STUDIES

The rate of release of volatile materials can be determined in the laboratory by following the meniscus movement in the hollow fiber via a cathetometer. These menisci readings can be made to ±0.002 cm with the equipment used, and when the hollow fiber area is considered, this is equivalent to about ±1 μg (for a fiber with 8 mil inside diameter). This type of measurement has been shown to give approximately the same result as by cold trapping the effluent vapors and analyzing by gas chromatography.[2]

Figure 4 is a typical plot of total material dispensed as a function of time for several different pheromones at 70°F. The slope of each curve is the dispensing rate. Note that there is an initial high slope (or dispensing rate) which corresponds to the time in which the meniscus is near the open end. As the meniscus recedes down the fiber with time, the dispensing rate becomes more steady as is noted by the straight line through the data points. This relatively steady dispensing rate occurs after only a very short time and is usually the value to which our system is designed.

Typical evaporation curves for other materials such as insecticide (Naled), insect attractant (trimedlure), and fragrance (cedar oil) are shown in Figure 5. These curves have the same general shape as for the pheromones. These curves are normalized by dividing by the fiber area. (For an 8 mil fiber the area is about 3×10^{-4} cm^2.)

Once the release curve for a material is determined, a CONREL dispenser can be designed for a specified application. The desired dose rate is established by selecting the proper number and diameter of fiber, and the length of the fiber is then selected to meet the lifetime specification, subject to chemical stability limitations of the active material.

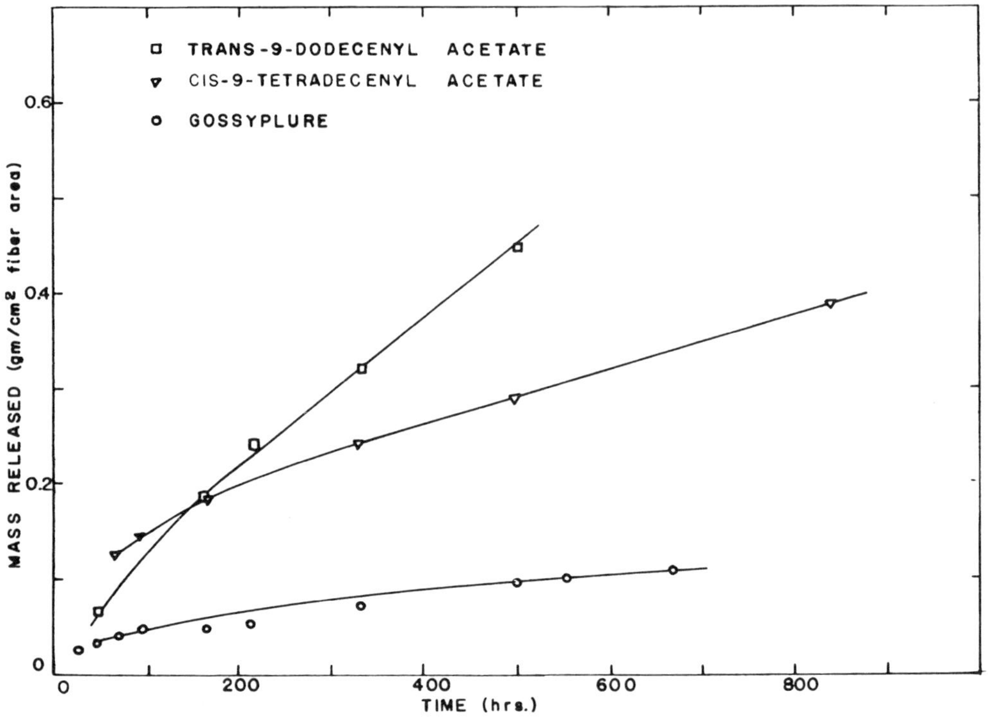

FIGURE 4. Release of insect pheromones.

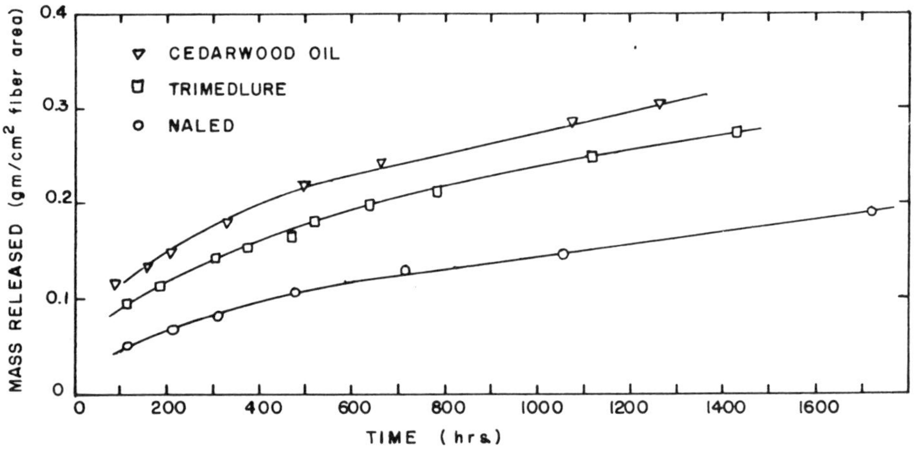

FIGURE 5. Release of vaporizable materials.

IV. FIELD TESTS

CONREL dispensers have been designed for insect pheromones for field evaluation. The evaluation includes survey, mass trapping, and disruption approaches for pheromone application. The pheromones being investigated include those for elm bark beetle, pink bollworm, cabbage looper, peachtree borer, lesser peachtree borer, gypsy moth, tobacco budworm, redbanded leaf roller, Southwestern pine tip moth, oriental fruit moth, boll weevil and corn earworm. In addition, several different insect attractants and insecticides are under investigation.

Table 1
ELM BARK BEETLE MASS TRAPPING DATA USING CONREL DISPENSERS

(Showing the date of examination and the number of beetles [unsexed] caught around the multilure baits)

Trap No.	Trap installed on tree species	6.2.75	12.2.75	26.2.75	17.3.75	Total
[a]3	*Archontophoenix cunninghamiana*	—	—	—	—	—
4	*Archontophoenix cunninghamiana*	15	37	15	143	210
7	*Ulmus procera*	3	35	53	112	203
[a]9	*Prunus* spp.	—	—	—	—	—
11	*Ulmus procera*	18	144	8	220	390
[a]12	*Quercus* spp.	—	—	—	—	—
14	*Ulmus procera*	16	302	130	200	648
17	*Ulmus procera*	210	786	209	990	2195
[a]21	*Pinus ponderosa*	—	—	—	—	—
22	*Ulmus procera*	54	235	34	160	483
[a]23	*Ulmus procera*	—	—	—	—	—

[a] Control trees, traps installed but no baits.

To date, the largest evaluation of the CONREL dispenser is for the elm bark beetle pheromone. This work is being done by the U.S. Forest Service, Delaware, Ohio in their mass trapping field experiments in Ft. Collins, Colo. and Detroit, Mich. In addition, the CONREL system is being used with the elm bark beetle pheromone in a study to determine the optimum dose rate and ratio of pheromone components.

The versatility of the CONREL dispenser is illustrated by this study. The elm bark beetle pheromone is a mixture of three components — multistriatin, 4-methyl 3-heptanol, and α-cubebene. The desired dose rate of this system is obtained by supplying each component in separate fibers, the number of fibers of each being dependent on the relative delivery rate of each component. The length of fibers necessary for each component is made equal by using an inert, volatile diluent such as hexane. This approach of using separate fibers eliminates any problem of a distillation effect, that is, the changing concentration of a mixture due to the different evaporation rates of each component of the mixture.

Table 1 presents trapping results for the CONREL dispenser for elm bark beetle pheromone.[3] These data were obtained in Australia in January and February of this year. It should be noted that these results were obtained on a small number of samples as a preliminary evaluation of the system for tests this summer in the U.S.

Early results from the Detroit trapping sites indicate that the CONREL system is performing to design specifications.[4]

The CONREL dispenser is also being evaluated in the field for the pheromones of cabbage looper (looplure) and lesser peachtree borer by investigators at the U.S.D.A. Insect Attractants and Basic Biology Laboratory in Gainesville, Fla. Preliminary results indicate the system shows promise for both survey traps and disruption.[5,6]

The generally favorable response to preliminary evaluations has stimulated us to bring the CONREL concept to commercial availability. In addition, we are also involved in developing a system for aerial application. This will consist of individual fibers sealed in the middle and with two open ends. This system would be suitable for the disruption approach. We feel that the advantages of this system are

1. The potential to provide a controlled release of vapor which could last an entire season, thus eliminating the need for multiple applications
2. The ability to provide a sufficient number of fibers per acre to give the desired daily pheromone dose rate
3. The size of the fibers (i.e., the length) would make it easier for them to adhere to the foliage, especially if the fibers are coated with an adhesive

V. CONCLUSIONS

1. Both experiment and theory indicate that the CONREL approach is a viable method of controlled release of vaporizable materials.
2. Where the concept has been tested under field conditions with insect pheromones, early results indicate dispensers perform to the design specifications.
3. Independent determinations of release rate indicate agreement with the specified release rate.

ACKNOWLEDGMENTS

We wish to acknowledge the assistance of the following with the initial field evaluation of the CONREL system: Drs. John Peacock and Roy Cuthbert, U.S. Forest Service, Northeastern Forest Experiment Station, Delaware, Ohio; Drs. J. H. Tumlinson and J. R. McLaughlin, U.S.D.A. Insect Attractants and Basic Biology Laboratory, Gainesville, Fla.; Roger Kitterman, FRL, and The Wilbur Ellis Company, Phoenix, Arizona.

REFERENCES

1. **Bird, R. B., Stewart, W. E., and Lightfoot, E. N.,** *Transport Phenomena,* John Wiley & Sons, New York, 1960.
2. **Gore, W. E.,** private communication, February 17, 1975.
3. **Peacock, J. W.,** private communication, March 31, 1975.
4. **Cuthbert, R.,** private communication, June 3, 1975.
5. **McLaughlin, J. R.,** private communication, May 19, 1975.
6. **Tumlinson, J. H.,** private communication, June 24, 1975.

Chapter 10

RELEASE RATES FROM FORMULATIONS AND QUALITY CONTROL METHODS

B. A. Bierl-Leonhardt

TABLE OF CONTENTS

I. INTRODUCTION

Pheromones have become a very popular as well as potentially vital tool in insect management programs, as evidenced by the discussion in many chapters of this book. Much effort has been directed at using these biologically important chemicals not only in survey programs to detect and monitor insect populations but also in attempts to suppress mating of a wide variety of insects. In almost all cases, these field uses of the pheromones require chemicals that are formulated in such a way that they remain biologically active over a prolonged period. Without this formulation, activity generally drops rapidly because the materials volatilize or degrade. For this reason, controlled or slow release formulations that extend the activity of the pheromone over weeks or months are highly important.

II. TYPES OF FORMULATION

Types of formulation have varied widely over the years as specific needs dictated by insect behavior are met and as new substrates are developed. Marks[1] tested polyethylene vials, cigarette filters, and rubber septa for dispensing C_{12} acetates and found that a combination of corn oil, an extender, and BHT (2,6-di-*tert*-butyl-*p*-cresol), an antioxidant, extended the period of attractiveness beyond 25 to 30 days. Bull et al.[2] used cigarette filters impregnated with grandlure and polyol extenders in experiments with the boll weevil, *Anthonomus grandis* Boheman, and Hardee et al.[3] reported development of a gel formulation that further extended the effectiveness of the synthesized pheromone. Polyethylene vials have been used to dispense attractants for the "Egyptian cotton leafworm", *Spodoptera littoralis* (Boisduval),[4] and the "red bollworm", *Diparopsis castanea* (Hampson),[5] and polyethylene caps were used[6] to dispense pheromone-type compounds in studies with *Heliothis* spp. A solid polyvinyl chloride dispenser prolonged the release of *(Z)*-7-dodecen-1-ol acetate in experiments with the cabbage looper, *Trichoplusia ni* (Hübner),[7] and of the *(E)*-9-isomer in trapping tests with the European pine shoot moth, *Rhyacionia buoliana* (Schiffermüller).[8] Other types of controlled release dispensers include: impregnated resin for control of the Douglas-fir tussock moth, *Orgyia pseudotsugata* (McDunnough),[9] plastic tubes filled with *(Z)*-8-dodecen-1-ol acetate for control of the "plum fruit moth", *Grapholitha funebrana* Treitschke,[10] or with the *(Z)*-9 isomer for suppression of the European grape berry moth, *Eupoecilia ambiguella* (Hübner);[11] and cork,[12] molecular sieve,[12] wax,[13] and paper chips[14] coated with disparlure for mating disruption experiments with the gypsy moth, *Lymantria dispar* (L.). Other substrates have included sand,[15] cotton dental rolls,[16] and nylon mesh.[17] A somewhat different approach was the inclusion complexation of urea with tetradecadienyl acetates, which yielded tablet-type dispensers that were used in tests with *Spodoptera litura* (F.);[18] these tablets remained attractive longer than impregnated natural rubber caps.[19] Others[20,21] have reported that rubber septa are useful in dispensing pheromones.

Microcapsules of various wall types, including gelatin,[12,13,22-25] polyurea,[13] and nylon,[5,12] have been used to encase solutions of pheromones and these have been used extensively in mating disruption programs involving aerial applications. Plastic laminate films containing an inner reservoir layer of pheromone have been applied in a wide variety of insect programs to control the release of pheromones[26-30] in both trapping and air permeation experiments. Hollow fibers filled with the pheromone, alone or in a solvent, constitute another innovative commercial development that has been used in a variety of applications.[13,25,31-35] Thus the variety of controlled release formulations is limited only by the ingenuity of scientists; the needs of a given program will determine the choice among the many controlled release formulations.

III. RELEASE RATES

Emission of pheromone from a controlled release formulation is generally limited by a diffusion process, that is, the rate of emission is dependent on the parameters of the barrier to free emission and the concentration gradient across that barrier.[36] If the concentration gradient remains constant, the rate follows zero order kinetics since it does not depend on how much material has been released or how much remains. Some pheromone dispensers that contain a large reservoir of phermone do follow zero order kinetics; an example of one such dispenser is a tube containing neat attractant and covered with a porous membrane.[36]

Most formulations, however, tend to follow first order kinetics, in which the rate of release is proportional to the amount of pheromone remaining; this amount declines significantly as lure is released. An expression for a first order rate is given as $\ln C_o/C = kt$ where C_o is the initial pheromone concentration, C is the residual lure content, k is the release rate constant, and t is the time period. The half-life, $t_{1/2}$, of the treatment, i.e., when $C = \frac{1}{2}C_o$, is expressed as 0.693/k. The theoretical basis for treatment of release rates has been discussed in detail elsewhere[37-40] and will not be reviewed in this chapter.

Many formulations depend upon passage of the pheromone through a membrane or wall by a process that includes absorption, solution, and diffusion.[41] With other formulations, the lure diffuses through pores or holes in the matrix and escapes to the air. In all these cases, the rate at which lure leaves the formulation can be varied by changes in the manufacturing process of the controlled release matrix, for example, by varying parameters such as film thickness, polymer stiffness, particle size, pore dimensions, solvent, etc. The effect of each of these variables on the release rate of pheromone must be determined in order to obtain the optimum formulation. Thus, as a new formulation is designed, there is a concurrent need to develop a procedure for determining the release characteristics. Field experiments are very costly and normally cannot be used to screen a wide variety of formulations. This must be done in a laboratory. The following sections contain a description of the way release rates are measured and the manner in which they are used in efforts to design formulations that will deliver effective concentrations of pheromone to the target insect over the desired period.

A. Methods of Measurement of Release Rates

Perhaps the two most critical items in evaluating the performance of controlled release formulations are the method of aging the formulations to simulate approximate field conditions and the analytical procedures used to determine the nature and amount of lure released as a function of time under a reproducible set of conditions.

The optimum method of determining release rates is, of course, monitoring the amount of lure being emitted as the formulation ages in the field. Caro et al.[42,43] have measured pheromone contents in air over areas treated with formulations; this work is fully discussed in another chapter of this book. However, this procedure is costly, difficult and time consuming and cannot be applied to routine screening of formulations. A compromise can be achieved by aging the formulations in a chamber maintained at conditions that roughly approximate those in the field. The pheromone is swept through the chamber by a measured flow of air and is trapped in the exit port on an adsorbant such as Porapak. The lure is then extracted from the adsorbant and measured chromatographically. With such a device, it is impossible to simulate the temperature fluctuation and air movement that is experienced in a field situation, but the procedure is useful for a comparison of formulations under the same conditions. An example[44] of a small-scale chamber is a 2-ℓ flask fitted with a port near the top where N_2 can be added at 175 mℓ/min and a port at the bottom where a Porapak trap can be attached; the formulation is deposited on the bottom of the flask. The amount of emitted lure is determined by high pressure liquid chromatographic analyses of the extract of the Porapak trap. The disadvantage of such a controlled-environment chamber for aging formulations is the potential need to dedicate a chamber to the aging of a single formulation for a period of weeks or months; if this has to be done, one could not screen a variety of formulations at the same time.

In view of the limitations of small-scale chambers, modified procedures have been developed to aid in determining the release rates of formulations used for survey or mating disruption in field programs. One of the easiest methods is to measure weight loss of a formulation aged in the laboratory or in the field and ascribe the difference in weight to the amount of pheromone lost in a given aging period. However, this procedure gives accurate release rates only if no components of the matrix are simultaneously volatilized, if no water is absorbed or released during the aging process, if no decomposition of the pheromone occurs, and if the weight loss is large enough to be determined accurately. Also, with formulations aged outdoors in changing humidities, the absorption and desorption of water can cause sizable inaccuracies in weight loss measurements. Nevertheless, Rothschild[45] successfully used the weight-loss method to compare controlled release dispensers containing *(Z)*-8-dodecen-1-ol acetate and showed that hollow-fiber and polymeric laminate dispensers were superior to polyethylene and silicone rubber designs in providing a relatively constant release of pheromone. The weight-loss method was also used[7] to measure the loss of *(Z)*-7-dodecen-1-ol acetate from polyvinyl chloride dispensers though the weight losses had to be corrected for the concurrent loss of plasticizers, etc. Similarly, release rates from polyethylene caps baited with 25-mg quantities of candidate compounds were obtained[6] by the weight-loss method.

Another method that has similar limitations in regard to accuracy is the measurement of the meniscus of the liquid pheromone in hollow fibers; by using a cathetometer, the meniscus can be read to ±0.002 cm, which is equivalent to ±1 μg for a fiber with an 8-mil I.D.[31] This procedure, of course, is uniquely applicable to the hollow-fiber formulations. A comparison of this meniscus-measurement technique with air-flow and static-air measurement techniques was reported by Weatherston et al.[46] for hollow fibers containing gossyplure: the release rates ranged from 0.04 to 2.86 μg/fiber a day. This 70-fold variation in measured rate illustrates the need to simulate field conditions for a realistic determination of lure loss.

Maitlen et al.[47] and Butler and McDonough[48] used an extraction procedure followed by chromatographic analysis to determine the release characteristics of rubber stoppers or septa impregnated with pheromone components. As indicated elsewhere in this chapter, this method gives reliable data only when it has been established that the amount of pheromone left in the dispenser is a true indication of the amount that was released and does not include decomposed pheromone. Radiochemical techniques have been used to measure the release of pheromone-type chemicals from plastic stoppers under field conditions[49] and from hollow fibers containing gossyplure,[46] the synthesized pheromone of the pink bollworm, *Pectinophora gossypiella* (Saunders), in a laboratory air-flow system in which the lure is trapped on ground glass.

Another procedure that has been used[30] involves measuring the release rate from pheromone dispensers in still air at a specific temperature. The dispenser is suspended in a flask for 2 hr and removed; the flask is then rinsed with ether to dissolve the condensed pheromone, and the lure content to the concentrated extract is determined by gas chromatography. This procedure provides a convenient means of comparing formulations in a static-air situation, but it may not give a realistic release rate for field applications. Cross[50] determined the release rate of dispensers suspended in a glass chamber; 1000 mℓ/min of air were pulled through the chamber, and the pheromone was retained by a Tenax GC® trap. The volatiles were then thermally desorbed from the trap for analysis. Other adsorbants such as charcoal[51] and Porapak Q®[44,52,53] have been used to trap volatile pheromones from air streams.

Bioassay in field cages has been used[26] to evaluate dispensers containing virelure, an attractant for tobacco budworm, *Heliothis virescens* (F.). The controlled release dispensers were aged in the field and then bioassayed in field cages; the numbers of insects trapped gave a comparison of the various dispenser types. McKibben and Johnson[40] described a mathematical model based on regression analysis of insect capture data for evaluation of formulations; they found that the decay model gave the effective life of the formulation in a more exact manner than a simple comparison of the number captured vs. time. Schwalbe et al.[13] and Plimmer et al.[24] evaluated nine formulations of disparlure in extensive field tests in Massachusetts and Maryland found that gelatin microcapsules gave better overall reduction of mating performance than other formulations including hollow fibers, a wax matrix material, and other capsule types. This type of test gives an indication of effectiveness in the field, but it does not indicate the release rate of the lure nor does it provide data on stability.

At USDA, we evaluate controlled release formulations by aging the materials outdoors during the period when the insect is active or in a greenhouse with summer-like temperature. Then at preselected intervals, duplicate samples of the aged formulations are analyzed; the rate of release of the lure is measured by a procedure that will be described later; the residual lure is extracted and measured by gas chromatography (GC). By determining both the amount of lure released and the residual content, one can determine whether the diffusion process that carries the lure to the evaporating surface is still operating and also whether decomposition products are being formed or emitted. Since some labile chemicals such as aldehydes can decompose in the formulation during the aging process, analysis of the residual pheromone and accompanying reaction products should reveal this problem. Also, some microcapsule formulations contain a solvent that is released more quickly than the pheromone; as a result, when solvent is no longer available to carry the lure through the capsule wall, the emission rate of pheromone drops to a low level. Much of the lure is thus trapped within the capsule walls. Figure 1 (microcapsules) illustrates such a case. The emission rate (μg/hr) dropped very quickly over a 42-day period; yet the residual lure content (dotted line) of the capsules remained near the initial level. In fact, a formulation of this type generates a burst of pheromone soon after field application, but the subsequent release rate is very low relative to the amount applied, and much of the lure is not emitted during the target period. Figure 2 (plastic laminate) reflects the performance of a controlled release formulation which shows a more consistent pattern; the emission rate dropped gradually over a 120-day aging period, and the residual lure showed a corresponding decline. With a formulation of this type, most, if not all, of the lure is eventually released. Thus, if the release rate data and the residual content of lure are consistent, decomposition of the pheromone is not a problem, and only residual lure content needs to be measured; this expedites the evaluation process in subsequent testing.

An apparatus designed[54,55] to measure the rate at which lure is emitted from samples of aged formulations traps the volatiles emitted from the formulations held at a constant temperature and airflow. A laboratory oven houses six assemblies as shown in Figure 3. Copper lines conduct air through holes in the oven wall and into glass tubes containing the samples. The air stream passes over the sample, then through a delivery tube, and into hexane to trap the emitted pheromones. After passage of air for a fixed time (usually 2 hr), the hexane solution of the emitted material and the washing from the delivery tube were concentrated, as described by Beroza et al.[56] Then the formula-

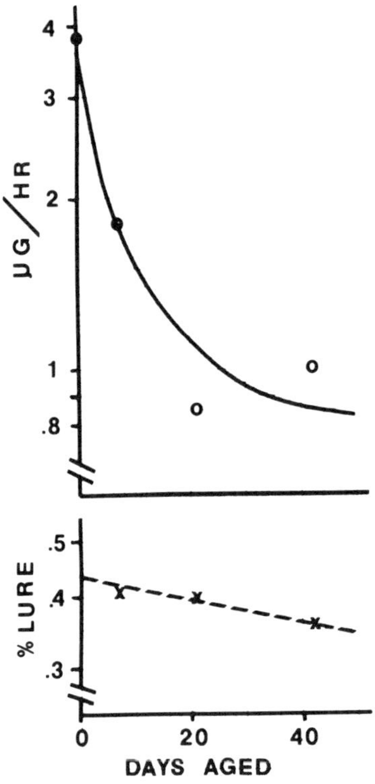

FIGURE 1. Emission rate (top) and residual lure (bottom) 50-250 μm gelatin microcapsules (25% plastic coated) initially containing 2% disparlure in 1:3 pentyl acetate:xylene.

tions were removed for further aging or for extraction and analysis of residual pheromone. The emitted pheromone was then determined on a measured aliquot of the hexane by GC with a flame ionization instrument and calculated as μg/hr. Replicate tubes were used for each formulation. Similar GC analysis of pheromone extracted from the tested formulations gave the content of residual lure. These values for different aging periods should show the same pattern as the emission rates if the formulation is behaving ideally. The major disadvantage of the procedure is the fact that the oven assembly is not at the same temperature and air-flow conditions as outdoors. Nevertheless, the procedure offers a way to expose formulations to field conditions and to determine how this exposure affects the rate at which they emit lure under fixed and reproducible conditions.

B. Temperature Effect on Release Rates

The effect of temperature on the release rate is a major consideration in the design of formulations. For example, a 30°F change in temperature (e.g., from 70°F to 100°F) causes an increase in vapor pressure of about 200% for pheromone-type materials, but diffusivity and molar gas density are affected only slightly.[31] As a result, a 30°F temperature rise will cause the release rate to double. In addition, the porosity and other physical parameters of controlled release matrices such as polymer films, rubber dispensers, microcapsule walls, etc., will vary with changing temperature. Therefore the release rate of a formulation destined for field use must be evaluated at temperatures approximating those experienced in the field situation. For example, at temperatures ranging from 82° to 100°F (28 to 38°C) in the oven assembly, the emis-

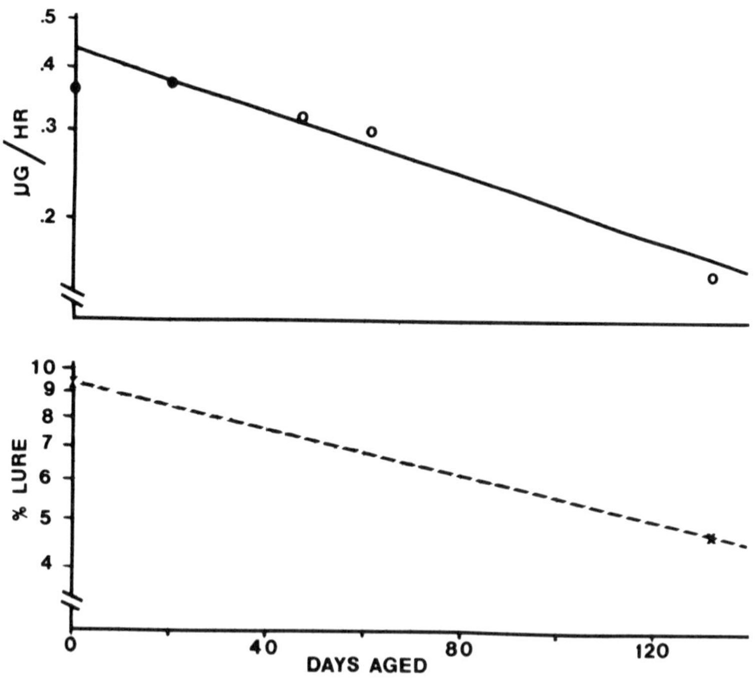

FIGURE 2. Emission rate (top) and residual lure (bottom) of 2-mil plastic laminate dispensers initially containing disparlure, aged outdoors.

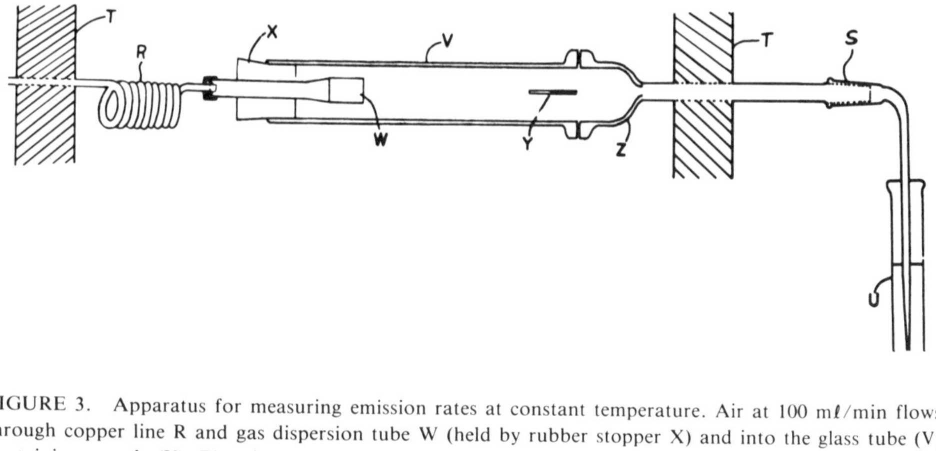

FIGURE 3. Apparatus for measuring emission rates at constant temperature. Air at 100 mℓ/min flows through copper line R and gas dispersion tube W (held by rubber stopper X) and into the glass tube (V) containing sample (Y). The air containing the volatiles then flows through adapter Z and delivery tube S into collection tube U containing 10 mℓ of hexane.

sion rate of disparlure from a Hercon® laminated dispenser ranged from 0.28 to 0.84 μg/hr.[57] Also, the release from gelatin microcapsules containing disparlure ranged[57] from 0.59 to 4.2 μg/hr for temperatures of 80 to 100°F (27 to 38°C). Rothschild,[45] too, measured the loss rate of *(Z)*-8-dodecen-1-ol acetate from five types of controlled release dispensers exposed to temperatures of 15, 20, and 29°C; in every case, the lure loss at 29°C was significantly greater than at the other two temperatures. Over 3 times as much lure was lost from polyethylene microcentrifuge tubes (closed lid) and polyethylene tube[10] dispensers in 40 days of aging at 29° as was lost at 15°C; Hercon® laminate, Conrel hollow-fiber and silicone rubber dispensers all showed less than twice as much lure lost for the same two temperatures. Likewise, Sower and Daterman[9] studied the release of *(Z)*-6-heneicosen-11-one, the synthesized sex pheromone of the Douglas-fir tussock moth, from dispensers made from films of impregnated resin backed with Mylar® film. They found that the release rate increased about twofold for each 5°C increase in temperature.

All research cited shows clearly that release rates measured at temperatures other than those experienced in the field will not clearly define the performance of a given formulation when used in insect control programs. However, if all other factors are equivalent, a formulation showing a high temperature coefficient would be best for day-flying insects: the greater amount of lure emitted at the higher daytime temperatures would capture the insects or disrupt their mating; the reduced rate of loss during the cooler evenings would conserve lure. With night-flying insects, an opposite pattern would be preferred so a formulation with a more constant release pattern at various temperatures would be desired.

C. Other Factors Affecting Release Rates

Temperature is not the only factor that influences the rate at which lure is released. Wind or air movement increases loss from the surfaces of many types of controlled release formulations; as an example, the emission of disparlure from plastic laminates was found to increase in proportion to airflow.[54] It therefore follows that such a dispenser should release lure more slowly when it is protected inside a trap than when it is stapled on a tree or exposed to wind in some other way. In contrast, the release rate from hollow fibers is virtually unaffected by wind[58] since diffusion from the liquid-vapor interface to the open end of the fiber is the rate-determining step. Thus once again, the test formulations must be aged under field conditions to obtain a true evaluation. Campion et al.[52] reported a faster release rate for polyurea-polyamide capsules in the field than in the laboratory and suggested that a combination of temperature, wind velocity, and sunlight may have produced unpredictable changes in the lifetime of these formulations.

IV. EFFECT OF VARIABLES IN FORMULATION DESIGN

As has been shown, environmental conditions affect the release rate of lure from controlled release formulations. However, the construction parameters of almost all such formulations can be varied so as to significantly alter the release pattern. These would include the permeability of capsule walls, the porosity, thickness, and cross-linking of polymers, the concentration of lure within the formulation, the nature of solvents or volatility suppressants, and many other factors.

One of the first used and simplest formulations of pheromone is a cotton wick on which pheromone is deposited. The addition of volatility suppressants (or "keepers") to the pheromone reduced the release rate.[59] Thus the half-life of 100 μg of disparlure deposited on a 6-mm wick of dental roll was increased from 2 to 3 days to 6 days by adding 15 mg of methyl dodecanoate; a similar amount of paraffin oil or trioctanoin extended the half-life to 24 and 46 days, respectively.

Also the release rates of disparlure from gelatin microcapsules depend on a number of parameters including concentration of lure, conditions for forming the capsules, and extent of plastic coating[60] as well as changes in the capsule size and type of solvent within the capsule. Data obtained at 30°C and 100 mℓ/min airflow in the laboratory apparatus previously described are summarized in Table 1. The capsules were aged outdoors where they were shielded from rain but exposed to sun, wind, and temperature variations.

Campion et al.[52] reported that the release rate from polyurea-polyamide capsules

Table 1
EMISSION RATE OF DISPARLURE FROM GELATIN MICROCAPSULES[a]

Solvent	Size of capsule (μm)	Emission (in μg/hr)[b] for indicated weeks aging: 0	1	3	6
1:3 Pentyl acetate:xylene	10—40	2.0	2.2	2.1	2.1
1:3 Pentyl acetate:xylene	50—250	3.7	1.9	0.9	1.0
1:3 Me dodecanoate:xylene	50—250	0.7	0.9	0.6	0.5
1:3 1-Hexanol:xylene	50—250	2.5	1.9	1.2	0.7
1:3 Nonane:xylene	50—250	2.0	1.5	1.1	0.5
Nonane	50—250	2.2	1.7	1.5	1.6
1:3 Me hexanoate:xylene	50—250	3.0	2.1	0.5	0.2

[a] All capsules 25% plastic coated.
[b] Emission rate at 30°C and air flow of 100 mℓ/min.

containing *(Z)*-9-tetradecen-1-ol acetate could be reduced by increasing the wall thickness of the capsules or by achieving higher level of cross-linking in the polymer.

Release rates of other formulations can similarly be affected by changes in variables of construction. The Hercon® plastic laminate formulation can be conveniently modified to yield effective dispensers for a wide variety of pheromones ranging from the more volatile C_{10}-C_{12} compounds to less volatile materials such as disparlure (a C_{19} epoxide). For example, changes in reservoir concentration, membrane thickness, polymer stiffness, and co-diffusants will alter the release rate from the laminate.[41] Results have shown that the release rate from the laminate dispenser can generally be determined by measuring the amount of lure remaining since decomposition of the lure in these formulations has not been a factor.

The emission rate from a laminate is dependent upon the concentration of lure in the inner layer when the inner and outer layers have approximately the same glass transition temperature (polymer stiffness); in this case, the release is a first order process. Figure 4 shows the lure content of three laminate disparlure dispensers as they were aged outdoors. The top curve illustrates the performance of a 5 × 25 mm dispenser made of 2-mil vinyl film that contained an initial loading of 4.3 mg; the half-life ($t_{1/2}$) of this dispenser was estimated as 12 weeks. The middle and bottom curves show that similar dispensers had half-lives of 8 and 6.5 weeks, respectively, for loadings of 1.2 and 0.22 mg of lure.

The thickness of the polymer film also affects the release rate. Some of the results produced by changes in dispenser thickness are shown in Table 2. For example, when 3 × 3 mm squares of laminate containing disparlure were aged in a greenhouse, the $t_{1/2}$ for those made of 2-mil vinyl was 29 days vs. 54 days for those made of 5-mil vinyl. Likewise, the type of plastic used has an effect. This is illustrated by the data of Table 2. For a given thickness of polymer film, lure is released fastest from vinyl and slowest from acrylic plastic. Also, a 0.5-mil Mylar® coating on vinyl greatly reduces the release rate; this coating on 2-mil vinyl gave approximately the same release rates as were given by uncoated 9-mil vinyl in 13 × 13 mm dispensers. Since the Mylar® is essentially nonpermeable to pheromone chemicals, the lure released from such a formulation escapes primarily from the edges. In fact, the edge effect is particularly important when one is comparing release rates from various sizes of dispensers that have different ratios of top plus bottom surface to amount of edge.[54] Table 2 gives the comparison of $t_{1/2}$ values for 3 × 3 and 13 × 13 mm squares of 2-mil vinyl containing

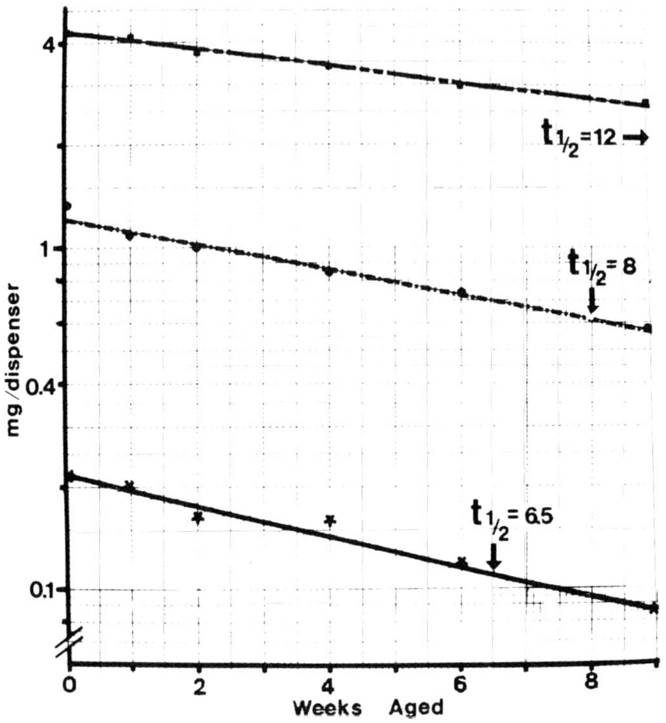

FIGURE 4. Residual lure of 5 × 25 mm dispensers made of 2-mil vinyl film and containing initial loadings of 4.3 mg (top), 1.2 mg (middle), and 0.22 mg (bottom) of disparlure.

disparlure, 4-mil vinyl containing virelure, and 6-mil acrylic also containing virelure. The lure was released roughly twice as fast from the smaller dispensers, which had the higher ratio of perimeter to top and bottom surface. This factor must therefore be taken into account when a laminate formulation is designed as small pieces for aerial application. It is possible that a plastic film thicker than would be required for larger dispensers (destined for use in traps) might have to be used.

V. SUMMARY

Controlled release formulations have varied widely in design over the past 10 years. In general, they have progressed from the simple matrix containing the hand-applied solution of lure to the commercially produced formulations of microcapsules, plastic laminates, and hollow fibers.

During the development of these formulations, a wide variety of procedures have been utilized to evaluate the release rates of pheromone-type chemicals. These techniques have included static air collection, airflow plus trapping, radiochemical labeling, weight loss measurement, chromatographic analysis of residual lure content, meniscus measurement, and bioassay in field cages. Each of these methods can be used to obtain the comparisons needed to select the best material for a given application. However, none of the procedures produce an accurate description of the way the formulation will perform under the actual field conditions. Therefore, some compromises need to be made. In general, increased temperature and/or wind velocity increase the release rate. Particle size, polymer permeability, volatility extenders or suppressants, solvents, concentration of lure, polymer type, and thickeners all affect the release rate.

Table 2
APPARENT HALF-LIVES OF HERCON PLASTIC LAMINATE DISPENSERS

Lure	Test No.	Plastic film: Type	Plastic film: mil	Dispenser: Size (mm)	Dispenser: μg Lure/mm²	Dispenser: mg Lure/dispenser	$t_{1/2}$ (days)
Disparlure	1	Vinyl	2	3 × 3	29	0.3	29
Disparlure	1	Vinyl	2	13 × 13	31	5.2	>60
Disparlure	1	Vinyl	5	3 × 3	48	0.4	54
Gossyplure	1	Vinyl	4	13 × 13	13	2.2	22
Gossyplure	1	Vinyl	9	13 × 13	13	2.2	45
Gossyplure	1	Mylar/Vinyl[a]	0.5/2	13 × 13	13	2.2	43
Gossyplure	2	Vinyl	9	25 × 25	11	6.7	35
Gossyplure	2	Acrylic	3	25 × 25	12	7.2	66
Gossyplure	3	Vinyl	2	13 × 13	8	1.3	15
Gossyplure	3	Vinyl	8	13 × 13	9	1.6	23
1-Tetradecanol formate	1	Vinyl	9	3 × 3	104	0.9	28
1-Tetradecanol formate	1	Mylar/Vinyl[a]	0.5/2	3 × 3	100	0.9	22
1-Tetradecanol formate	2	Vinyl	9	25 × 25	109	68	37
1-Tetradecanol formate	2	Polymeric PVC	16	25 × 25	91	57	44
1-Tetradecanol formate	3	Vinyl	2	13 × 13	101	17	9
1-Tetradecanol formate	3	Vinyl	8	13 × 13	101	17	18
Virelure[b]	1	Vinyl	4	3 × 3	27	0.2	10
Virelure	1	Vinyl	4	13 × 13	19	3.2	17
Virelure	1	Rigid Vinyl	4	13 × 13	18	3.1	58
Virelure	1	Mylar/Vinyl[a]	0.5/2	13 × 13	21	3.6	31
Virelure	1	Acrylic	6	3 × 3	25	0.2	25
Virelure	1	Acrylic	6	13 × 13	17	2.9	>60
Virelure	3	Polymeric PVC	16	25 × 25	6	3.8	18
Virelure	4	Vinyl	2	4 × 20	10	0.8	12
Virelure	4	Acrylic	3	4 × 20	15	1.2	23
Virelure	4	Polymeric PVC	16	4 × 20	5	0.4	36

Note: Test 1: Aged in MD greenhouse, Jan-Feb 1979; 2: Aged in AZ field, Summer 1978; 3: Aged in MS field, Summer 1978; 4: Aged in MD lab, Nov-Dec 1978.

[a] Mylar (0.5 mil)-coated vinyl (2 mil).
[b] Data for C_{16} aldehyde only; 5% C_{14} aldehyde also present.

Variations in these factors can be used to modify the formulations to obtain the release profile that best fits the field specifications.

The technology of controlled release formulations has advanced with the need to use pheromones for survey, monitoring, and mating disruption of a variety of insects. In particular, the expense and difficulty of retreating large areas with pheromones or renewing the bait in isolated traps mean that controlled release formulations must be used. Thus the need exists to correlate field performance with reproducible formulation specifications and to develop evaluation techniques.

REFERENCES

1. **Marks, R. J.,** Field studies with the synthetic sex pheromone and inhibitor of the red bollworm *Diparopsis castanea* Hmps. (Lepidoptera, Noctuidae) in Malawi, *Bull. Entomol. Res.*, 66, 243, 1966.
2. **Bull, D. L., Coppedge, J. R., Ridgway, R. L., Hardee, D. D., and Graves, T. M.,** Formulations for controlling the release of synthetic pheromone (Grandlure) of the boll weevil. I. Analytical studies, *Environ. Entomol.*, 2, 829, 1973.
3. **Hardee, D. D., Graves, T. M., McKibben, G. H., Johnson, W. L., Gueldner, R. C., and Olsen, C. M.,** A slow release formulation of grandlure, the synthetic pheromone of the boll weevil, *J. Econ. Entomol.*, 67, 44, 1974.
4. **Campion, D. G., McVeigh, L. J., Murlis, J., Hall, D. R., Lester, R., Nesbitt, B. F., and Marrs, G. J.,** Communication description of adult Egyptian cotton leafworm *Spodoptera littoralis* (Boisd.) (lepidoptera Noctuidae) in Crete using synthetic pheromones applied by microencapsulation and dispenser techniques, *Bull. Entomol. Res.*, 66, 335, 1976.
5. **Campion, D. G.,** Sex pheromones for the control of lepidoterous pests using microencapsulation and dispenser techniques, *Pest. Sci.*, 7, 636, 1976.
6. **Mitchell, E. R., Jacobson, M., and Baumhover, A. H.,** *Heliothis* spp.: disruption of pheromonal communication with *(Z)*-9-tetradecen-1-ol formate, *Environ. Entomol.*, 4, 577, 1975.
7. **Fitzgerald, T. D., St.Clair, A. D., Daterman, G. E., and Smith, R. G.,** Slow release plastic formulation of the cabbage looper pheromone *cis*-7-dodecenyl acetate: release rate and biological activity, *Environ. Entomol.*, 2, 607, 1973.
8. **Daterman, G. E.,** Synthetic sex pheromone for detection survey of European pine shoot moth, *U.S. For. Serv. Res. Pap. PNW.*, 180, 1974.
9. **Sower, L. L. and Daterman, G. E.,** Evaluation of synthetic sex pheromone as a control agent for Douglas-fir tussock moths, *Environ. Entomol.*, 6, 889, 1977.
10. **Arn, H., Delley, B., Baggiolini, M., and Charmillot, P. J.,** Communication disruption with sex attractant for control of the plum fruit moth, *Grapholitha funebrana:* a two-year field study, *Entomol. Exp. Appl.*, 19, 139, 1976.
11. **Rauscher, S. and Arn, H.,** Mating suppression in tethered females of *Eupoecilia ambiguella* by evaporation of *(Z)*-9-dodecenyl acetate in the field, *Entomol. Exp. Appl.*, 25, 16, 1979.
12. **Beroza, M., Stevens, L. J., Bierl, B. A., Philips, F. M., and Tardif, J. G. R.,** Pre- and postseason field tests with disparlure, the sex pheromone of the gypsy moth, to prevent mating, *Environ. Entomol.*, 2, 1051, 1973.
13. **Schwalbe, C. P., Paszek, E. C., Webb, R. E., Bierl-Leonhardt, B. A., Plimmer, J. R., McComb, C. W., and Dull, C. W.,** Field evaluation of controlled release formulations of disparlure for gypsy moth mating disruption, *J. Econ. Entomol.*, 72, 322, 1979.
14. **Stevens, L. J. and Beroza, M.,** Mating-inhibition field tests using disparlure, the synthetic gypsy moth sex pheromone, *J. Econ. Entomol.*, 65, 1090, 1972.
15. **Glass, E. H., Roelofs, W. L., Arn, H., and Comeau, A. J.,** Sex pheromone trapping red-banded leaf roller moths and development of a long-lasting polyethylene wick, *J. Econ. Entomol.*, 63, 370, 1970.
16. **Roelofs, W. L. and Comeau, A. J.,** Sex pheromone perception: synergists and inhibitors for the red-banded leaf roller attractant, *J. Insect Physiol.*, 17, 435, 1971.
17. **Shorey, H. H., Kaae, R. S., and Gaston, L. K.,** Sex pheromones of Lepidoptera. Development of a method for pheromonal control of *Pectinophora gossypiella* in cotton, *J. Econ. Entomol.*, 67, 347, 1974.
18. **Fujiwara, H., Sato, Y., and Nishi, K.,** Physico-chemical studies on the sex pheromone of insects. II. Inclusion complexation of sex pheromone of *Spodoptera littura* F. with urea, *J. Takeda Res. Lab.*, 35, 217, 1976.
19. **Sato, Y., Nagano, M., Fujiwara, H., and Sakai, M.,** Formulation of the synthetic pheromone for attraction of the tobacco cutworm, *Spodoptera litura* F., *J. Takeda Res. Lab.*, 35, 227, 1976.
20. **Charmillot, P. J., Baggiolini, M., Murbach, R., and Arn, H.,** Comparison de differents pieges a attractif sexual synthetique pour le controle de vol du carpolapse *(Laspeyresia pomonella* L.), *Schweiz. Landw. Forsch.*, 14, 57, 1975.
21. **Kehat, M., Greenberg, S., and Tamaki, Y.,** Field evaluation of the synthetic sex pheromone, as an attractant for males of the cotton leafworm, *Spodoptera littoralis* (Boisd.), in Israel, *Appl. Entomol. Zool. (Tokyo)*, 11, 45, 1976.
22. **Gentry, C. R., Beroza, M., Blythe, J. L., and Bierl, B. A.,** Efficacy trials with the pheromone of the Oriental fruit moth and data on the lesser appleworm, *J. Econ. Entomol.*, 67, 607, 1974.
23. **Gentry, C. R., Beroza, M., Blythe, J. L., and Bierl, B. A.,** Captures of the Oriental fruit moth, the pecan bud moth, and the lesser appleworm in Georgia field trials with isomeric blends of 8-dodecenyl acetate and air-permeation trials with the Oriental fruit moth pheromone, *Environ. Entomol.*, 4, 822, 1975.

24. **Plimmer, J. R., Bierl, B. A., and Webb, R. E.,** Controlled release of pheromone in the gypsy moth program, *Controlled Release Pesticides,* ACS Symp. Ser. No. 53, Scher, H. B., Ed., American Chemical Society, Washington, D.C., 1977, 168.
25. **Mitchell, E. R.,** Recent advances in the use of sex pheromone for control of insect pests, in *Proc. 1977 Int. Controlled Release Symp.,* Goulding, R. L., Ed., Oregon State University, Corvallis, 1977, 41.
26. **Hendricks, D. E., Hartstack, A. W., and Shaver, T. N.,** Effect of formulations and dispensers on attractiveness of virelure to the tobacco budworm, *J. Chem. Ecol.,* 3, 497, 1977.
27. **Kydonieus, A. F. and Smith, I. K.,** Controlled release of pheromones through multilayered polymeric dispensers, in *Controlled Release Polymeric Formulations,* ACS Symp. Ser. No. 33, Paul, D. R. and Harris, F. W., Eds., American Chemical Society, Washington, D.C., 1976, 283.
28. **Kydonieus, A. F., Bierl-Leonhardt, B. A., Plimmer, J. R., Barry, M. W., and Quisumbing, A. R.,** Cotton Insects: Dispenser Development and Disruption of Mating Trials, presented at 1979 Controlled Release Pesticide Symposium, New Orleans, August 6 to 9, 1979.
29. **Tingle, F. C. and Mitchell, L. R.,** Controlled release plastic strips containing *(Z)*-9-dodecen-1-ol acetate for attracting *Spodoptera frugiperda, J. Chem. Ecol.,* 4, 41, 1978.
30. **Vick, K. W., Coffelt, J. A., and Sullivan, M. A.,** Disruption of pheromone communication in the Angoumois grain moth with synthetic sex pheromone, *Environ. Entomol.,* 7, 528, 1978.
31. **Ashare, E., Brooks, T. W., and Swenson, D. W.,** Controlled release polymeric formulations, in *Controlled Release Polymeric Formulations,* ACS Symp. Ser. No. 33, Paul, D. R. and Harris, F. W., Eds., American Chemical Society, Washington, D.C., 1976, 273.
32. **Brooks, T. W., Doane, C. C., and Staten, R. T.,** Experiences with the first commercial pheromone communication disruptive for suppression of an agricultural insect pest, in *Chemical Ecology: Odour Communication in Animals,* Ritter, F. J., Ed., Elsevier/North-Holland Biomedical Press, Amsterdam, 1979, 375.
33. **Gaston, L. K., Kaae, R. S., Shorey, H. H., and Sellers, D.,** Controlling the pink bollworm by disrupting sex pheromone communication between adult moths, *Science,* 196, 904, 1977.
34. **Iyer, P. P., Yates, W. E., Akesson, N. B., and Horgan, P. M.,** Cotton Pink Bollworm Control Using the Sex Pheromone, Gossyplure-I: Controlled Release of Gossyplure, presented at American Society of Agricultural Engineers Meeting, Chicago, Ill., December 18 to 20, 1978.
35. **Payne, T. L., Coster, J. E., and Johnson, P. C.,** Effect of slow-release formulation of synthetic endo- and exo-brevicomin on southern pine beetle flight and landing behavior, *J. Chem. Ecol.,* 3, 133, 1977.
36. **Nightingale, W. H.,** Improved means of obtaining sustained uniform emission of bioactive substances, in *Proc. 1979 British Crop Protection Conference-Pests and Diseases,* British Crop Protection Council, London, 1979, 401.
37. **Lewis, D. H. and Cowsar, D. R.,** Principles of controlled release pesticides, in *Controlled Release Pesticides,* ACS Symp. Ser. No. 53, Scher, H. B., Ed., American Chemical Society, Washington, D.C., 1977, 1.
38. **Lonsdale, H. K.,** Fundamentals of controlled release, in *Proc. 1975 Int. Controlled Release Pesticide Symp.,* Harris, F. W., Ed., Wright State University, Dayton, Ohio, 1975, 9.
39. **McDonough, L. M.,** Insect sex pheromones: importance and determination of half-life in evaluating formulations, *Agricultural Research Results,* APR-W-1, U.S. Department of Agriculture, Yakima, Wash., 1978.
40. **McKibben, G. H. and Johnson, W. L.,** The use of exponential and linear decay models in the evaluation of controlled-release pheromone formulations, *Bull. Entomol. Soc. Am.,* 22, 323, 1976.
41. **Kydonieus, A. F.,** The effect of some variables on the controlled release of chemicals from polymeric membranes, in *Controlled Release Pesticides,* ACS Symp. Ser. No. 53, Scher, H. B., Ed., American Chemical Society, Washington, D.C., 1977, 152.
42. **Caro, J. H., Bierl, B. A., Freeman, H. P., Glotfelty, D. E., and Turner, B. C.,** Disparlure: volatilization rates of two microencapsulated formulations from a grass field, *Environ. Entomol.,* 6, 877, 1977.
43. **Caro, J. H., Freeman, H. P., and Bierl-Leonhardt, B. A.,** Determination of *(Z)*-9-tetradecen-1-ol formate, a *Heliothis* spp. mating disruptant in air by electron-capture gas chromatography following photolytic cleanup, *J. Agric. Food Chem.,* 27, 1211, 1979.
44. **Look, M.,** Determining release rates of 3-methyl-2-cyclohexen-1-one anti aggregation pheromone of *Dendroctonus pseudotsugae* (Coleoptera: Scolytidae), *J. Chem. Ecol.,* 2, 481, 1976.
45. **Rothschild, G. H. I.,** A comparison of methods of dispensing synthetic sex pheromone for the control of Oriental fruit moth, *Cydia molesta* (Buck) (Lepidoptera:Tortricidae), *Bull. Entomol. Res.,* 69, 115, 1979.
46. **Weatherston, J., Huang, Y. Y., Golub, M. A., Benor, M. H., and Brooks, T. W.,** Methodology for Determining the Release Rates of Pheromones for Hollow-Fibers, presented at Gainesville Colloquium on Pest Management with Semiochemicals, Gainesville, Fla., March 24 to 28, 1980.

47. **Maitlen, J. C., McDonough, L. M., Moffitt, H. R., and George, D. A.**, Codling moth sex pheromone: baits for mass trapping and population survey, *Environ. Entomol.*, 5, 199, 1976.
48. **Butler, L. I. and McDonough, L. M.**, Insect sex pheromones: evaporation rates of acetates from natural rubber septa, *J. Chem. Ecol.*, 5, 825, 1979.
49. **Kuhr, R. J., Comeau, A., and Roelofs, W. L.**, Measuring release rates of pheromone analogues and synergists from polyethylene caps, *Environ. Entomol.*, 1, 625, 1972.
50. **Cross, J. H.**, A vapor collection and thermal desorption method to measure semiochemical release rates from controlled release formulations, *J. Chem. Ecol.*, 6, 781, 1980.
51. **Byrne, K. J., Gore, W. E., Pearce, G. T., and Silverstein, R. M.**, Porapak-Q collection of air-borne organic compounds serving as models for insect pheromones, *J. Chem. Ecol.*, 1, 1, 1975.
52. **Campion, D. G., Lester, R., and Nesbitt, B. F.**, Controlled release of pheromones, *Pest. Sci.*, 9, 434, 1978.
53. **Peacock, J. W., Cuthbert, R. A., Gore, W. E., Lanier, G. W., Pearce, G. T., and Silverstein, R. M.**, Collection on Porapak-Q of the aggregation pheromone of *Scolytus multistraitus* (Coleoptera: Scolytidae), *J. Chem. Ecol.*, 1, 149, 1975.
54. **Bierl-Leonhardt, B. A., DeVilbiss, E. D., and Plimmer, J. R.**, Rate of release of disparlure from laminated plastic dispensers, *J. Econ. Entomol.*, 72, 319, 1979.
55. **Beroza, M., Bierl, B. A., James, P., and DeVilbiss, E. D.**, Measuring emission rates of pheromones from their formulations, *J. Econ. Entomol.*, 68, 369, 1975.
56. **Beroza, M., Bowman, M. C., and Bierl, B. A.**, Improved ebullator for solution concentrator, *Anal. Chem.*, 44, 2411, 1972.
57. **Bierl, B. A., DeVilbiss, E. D., and Plimmer, J. R.**, The use of pheromones in insect control programs: slow release formulation, in *Controlled Release Polymeric Formulations*, ACS Symp. Ser. No. 33, Paul, D. R. and Harris, F. W., Eds., American Chemical Society, Washington, D.C., 1976, 265.
58. **Brooks, T. W. and Swenson, D. W.**, Design considerations in the formulation of controlled release product forms based on hollow fibers, in *Proc. 1976 Controlled Release Pesticide Symp.*, Cardarelli, N. F., Ed., University of Akron, Ohio, 1976, 45.
59. **Beroza, M., Bierl, B. A., Tardif, J. G. R., Cook, D. A., and Paszek, E. C.**, Activity and persistence of synthetic and natural sex attractants, *J. Econ. Entomol.*, 64, 1499, 1971.
60. **Plimmer, J. R., Bierl, B. A., DeVilbiss, E. D., and Smith, B. L.**, Evaluation of controlled-release formulations of insect pheromones for mating disruption, in *Proc. 1976 Controlled Release Pesticide Symp.*, Cardarelli, N. F., Ed., University of Akron, Ohio, 1976, 329.

Index

INDEX

A

B

C

D

E

F

G

H

I

J

K

L

M

N

O

P

Q

R

S

T

U

V

W

X

Y

Z